PAYLOAD AND MISSION DEFINITION IN SPACE SCIENCES

This book is intended for scientists and engineers involved in the definition and development of space science missions. The processes that such missions follow, from the proposal to a space agency, to a successful mission completion, are numerous. The rationale behind approval of a mission, its definition and the payload that it will include are topics that cannot be presented in undergraduate courses. This book contains contributions from experts who are involved in today's space missions at various levels. Chapters cover mission phases and implementation, launchers and cruise strategies, including gravity-assist manoeuvers and different thrust scenarios. The payload needed for remote sensing of the Universe at various wavelengths and for *in situ* measurements is described in detail, and particular attention is paid to the most recent planetary landers. Whilst the book concentrates on the ESA Program Cosmic Visions, its content is relevant to space science missions at all space agencies.

T0185415

PAYLOAD AND MISSION DEFINITION IN SPACE SCIENCES

V. MARTINEZ PILLET, A. APARICIO, F. SANCHEZ

Instituto de Astrofísica de Canarias
Tenerife, Canary Island, Spain

CAMBRIDGE
UNIVERSITY PRESS

CAMBRIDGE UNIVERSITY PRESS
Cambridge, New York, Melbourne, Madrid, Cape Town, Singapore,
São Paulo, Delhi, Dubai, Tokyo, Mexico City

Cambridge University Press
The Edinburgh Building, Cambridge CB2 8RU, UK

Published in the United States of America by Cambridge University Press, New York

www.cambridge.org
Information on this title: www.cambridge.org/9780521182454

First published 2005
First paperback edition 2010

A catalogue record for this publication is available from the British Library

ISBN 978-0-521-85802-1 Hardback
ISBN 978-0-521-18245-4 Paperback

Contents

Preface

The steps needed to define a successful space science mission are numerous. The science drivers, the unique advantages this mission provides over past missions or earth-based experiments, and the payload that it includes are the key factors to guarantee its success. Finding the required information on such topics is not so straightforward, especially as they are usually outside the scope of undergraduate courses. The 2003 Canary Islands Winter School of Astrophysics aimed at providing a focused framework that helps fill this need. Space agencies follow a necessarily complex path towards the selection of a specific mission, as required by the enormous costs that are associated with space activities. The steps towards its completion are elaborate and require careful assessment at every stage. The orbit that will be used and the requirements that are imposed have impacts on the science and the mission budget. Thus, knowing how to make the best use of propulsion technologies and gravity helps from solar system bodies plays a crucial role. The first two chapters of this book cover all these topics and illustrate the complexity of defining space missions as well as how and where look for help (i.e. other than the rarely receptive funding agencies).

The instruments on-board will in the end make the science that has driven the mission. How the science questions translate into specific requirements, and then, into the actual instruments are crucial aspects in the definition of the payload. Depending on the wavelength range that will be observed by the remote sensing instruments (from gamma rays to radio waves), the instruments must use broadly different technologies. In contrast to most other areas of astronomy, space science allows experimentation *in situ* with the objects under study, from interplanetary particles and fields to large planets of the solar system. These topics, with the different physical processes involved and corresponding technologies, are covered to various degrees in this book. Most importantly, the link between science questions and payload requirements is maintained throughout the book. The examples of specific missions in which specific payload was used clarify the process of mission and payload definition in space sciences.

The Winter School itself and the present book have focused heavily on the current and planned ESA Cosmic Vision program. This decision was taken not only on the basis of natural geographical considerations, but also because the European Space Agency (ESA) program offers a balanced set of missions that cover all aspects of space sciences. Thus, the resulting book is clearly based on the expertise gained by the european space community over the past few decades. However, the links with programs from other space agencies (mostly National Aeronautics and Space Administration NASA) make the contents of the book of general validity. The recurrent illustrative references to certain missions such as the ESA mission Solar Orbiter are related to the natural bias of the editor, but I trust the reader will forgive me for that.

January, 2005

Valentín Martínez Pillet
Instituto de Astrofísica de Canarias

Acknowledgements (and caveats)

I want to thank Nieves Villoslada (IAC), for her constant support to me during all phases of the organization of the Winter School (WS). This was the first (but not the last) of the schools that, although highly ranked by the European Union, obtained no funding (no thanks here). Nevertheless, it was her permanent enthusiasm that enabled us to make the WS a success.

As an aside, one result of this lack of funding was that my WS offered what I believe to be the worst meeting dinner ever. I apologize here to all participants and lecturers for such a nonsensical experience. One of the lecturers found a possible explanation for the outcome based on the supposed willingness of Spain to make food following UK standards (a British lecturer with a Scottish name). Although this may be true, I think the reasons were more doleful than that. One year after the event, I find it fantastic that, when meeting many of the participants and lecturers in different places, they continue to be my friends, and (caveat emptor!) have good memories of the WS. Thanks to all of them.

All the lecturers contributed to making the WS a success, but I would especially like to thank Prof. Balogh (Imperial College, UK), as he was able to stay with me throughout the whole event. I will always have good memories of my conversations with him. Special thanks as well to Prof. Alvaro Giménez (ESA), who had to take a total of six planes to participate in the WS. With his contribution to the event (explaining how to get involved in a space mission) during his lectures, he was able to create what I saw as an awe-inspiring atmosphere which must have convinced the participants to stay in this exciting field. Now we have to make the dreams come true!

Last, but not the least, I would like to thank Anna Fagan and Ramón Castro, for helping me with the edition of the different (and sometimes difficult) manuscripts and figures.

Thanks to my left ankle too!

1. The life cycle of an ESA science mission and how to get involved

By ALVARO GIMÉNEZ

Research and Scientific Support Department, ESA–ESTEC, The Netherlands

When I gave this talk in the Canary Islands Winter School of 2003, it was obvious that the interest of the audience was about how to make a successful proposal rather than finding out about the developing phases of a space mission. Unfortunately, I do not know how to make a 100% successful proposal. Success depends on a combination of bright ideas, creativity, timely response to the needs of a large scientific community, adequate system knowledge and, certainly, a bit of good luck. This presentation aims to make young scientists acquainted with the phases and challenges encountered in new space science missions. For that purpose these notes are organized in two sections. The first one establishes the phases of a mission, that is the process of carrying through a generic science project, while the second deals with the actual role of scientists in the whole process. Other talks in the Winter School focused in the science and the experiments that might be done, on how we can increase our knowledge of the Universe by means of space technologies. Here, we try to help making these, as well as other new ideas, real space science experiments.

1.1. The phases of a space mission

In this section, I want to bring to your attention the different phases of the process, starting with a bright scientific idea and finishing with the delivery of science data. This is, in other words, the life cycle of a European Space Agency (ESA) mission.

1.1.1. *The call for ideas*

The initial step of a space mission is normally a call for ideas periodically issued by the ESA science programme. These ideas are needed to allow for long-term planning in the agency and also to identify specific mission objectives and the technologies that may be required to actually carry them out. Through the call for ideas, ESA tries to identify possible flagship missions requiring concerted efforts as well as an advanced knowledge of smaller missions that may compete for selection following an eventual *Call for Proposals*. In this latter case, the response to the call for ideas allows ESA to evaluate the best rate and size of call for proposals and whether these will be completely open in scope or tailored to a specific budget envelope or even the use of a given spacecraft.

This process has been put in place by ESA several times, basically every decade. The one that took place in 1984–5 crystallized in the Horizon 2000 long-term programme that brought cornerstone missions (as flagship projects), and medium and small missions to be selected through dedicated calls for proposals and open competition. That plan consolidated a European-level approach to space science and led to a steady enhancement of the programme funding. Around 1994–5, a second call for ideas was issued that helped the definition of the Horizon 2000 programme extension by introducing additional cornerstone missions as well as new opportunities for calls for proposals with the scope of the then called flexible and Small Missions for Advanced Research and Technology (SMART) missions. The projects that are being launched now, within the period of time between 2004 and 2014, were actually identified after the latter *call for ideas*. The selection of the order in which the approved cornerstone missions would be implemented as well as the identification of the next flexi missions took place in 2000. A full revision was made in 2002, following the continuing degradation of the *level of resources* of the programme, and an imposed de-scope in 2003 was needed due to unforeseen financial problems. Nevertheless, in all discussions the programming of launches was kept open within a 10-year cycle, that is with launches up to 2014, and the long-term programme, already well beyond 2000, was renamed Cosmic Vision. In 2004, the period of 10 years finalizes again, just before an ESA Council meeting at Ministerial level in 2005. Consequently, the process of defining the missions leading to launches in the period between 2015 and 2025 has to be started, and the new programme receives the name Cosmic Vision 2020.

The call for ideas process starts again, but this time ESA has decided to begin with a *call for themes*. This was a result of a preliminary discussion with the scientific community that took place under the structure of a cross-disciplinary working group (XPG) and the difference is obvious: rather than looking for specific missions immediately, with a name attached to them, ESA is looking for priorities in scientific research within open areas of space sciences. It is, of course, difficult to forecast the best science to come, but after a number of years of proposing and even developing space science missions, it is worrying that the community could lose its sense of prioritization in scientific object-ives and merely focus on trying to recover old ideas or even reuse more or less available payloads. The analysis of the responses to the call for themes will allow ESA, through the advisory structure of the scientific programme, to identify the possible new flagship missions as well as the need for a call for proposals restricted to smaller projects. In this process, the analysis of possible mission scenarios to answer the questions put forward within the call for themes, as well as the identification of technology developments, will be performed. Eventually, the missions to be launched after 2014 will be selected in steps; thus completing the Cosmic Vision 2020 science programme. This call for themes has already been issued with a deadline to receive the inputs by the 1 June 2004. The advisory structure of ESA is involved in prioritizing the themes and the community at large will be consulted through specific workshops.

Once the call for ideas or themes is issued and the proposals from the scientific community are received, the following steps are taken. With internal resources, ESA evaluates the broad technical feasibility of the proposed ideas, rejects those technically unfeasible or clearly unrealistic, and brings the rest to a survey committee or the advisory structure for:

(*a*) evaluation of the scientific merits,
(*b*) assessment of timeliness and scope of mission,
(*c*) identification of scale of the mission and overall feasibility,
(*d*) identification of possible international cooperation.

In the case of call for ideas the proponents already draw a mission scenario to ESA. In the case of the call for themes, several mission scenarios are provided with internal resources, to analyse the feasibility of properly addressing the questions raised, without compromising the final selection of a specific mission. In other words, the thematic approach provides the necessary flexibility to optimize the scientific output of the programme and the use of international cooperation efficiently and science driven.

1.1.2. *The advisory structure of the science programme*

The importance of involving the scientific community in these early phases of the life cycle of an ESA science mission has been mentioned. This can be done through an *"ad hoc"* survey committee that scrutinizes the proposals received from the scientific community and then makes recommendation to the advisory structure of the programme, or directly by the latter with the necessary inputs from the corresponding ESA groups (Figure 1.1).

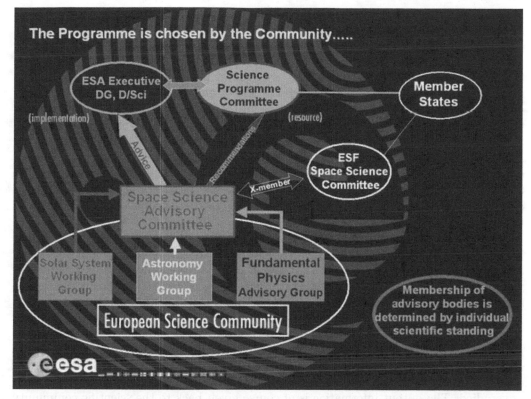

The Programme is chosen by the Community.....

ESA Executive DG, D/Sci

Science Programme Committee

Member States

(implementation)

(resource)

ESF Space Science Committee

Space Science Advisory Committee

X-member

Advice

Recommendations

Solar System Working Group

Astronomy Working Group

Fundamental Physics Advisory Group

European Science Community

Membership of advisory bodies is determined by individual scientific standing

esa

FIGURE 1.1. The components of the advisory structure of ESA science programme.

The advisory structure is composed of two Working Groups, devoted to Astronomy (AWG) and Solar System (SSWG) missions, an Advisory Group focused on Fundamental Physics missions (FPAG) and a senior body, the Space Science Advisory Committee (SSAC), that embraces the inputs from the three previous groups, and makes final recommendations to the science programme and its governing body, the Science Programme Committee (SPC). The AWG and the SSWG are formed by European scientists actively working in space sciences and selected partly by cooptation and partly by the Executive for a period of 3 years. The working groups each include 15 members and a chairman while the FPAG is composed of nine members and the chairman. Each of these groups convenes to make recommendations in their expert fields. These recommendations are then passed to the SSAC, which is composed of six experts from all three fields appointed by the ESA Director General and the Chairmen of the SSWG, AWG and FPAG. In addition there are four *ex officio* members: the Chairman of the SPC, the Chairman of the European Science Foundation – European Space Science Committee (ESF–ESSC) and two representatives of advisory bodies of other ESA programmes. Finally, the SPC is composed of delegates from each Member State of ESA. The SPC has the power to decide on the activities of the programme and recommends the yearly budget to the Council, which approves it.

One of the results of the process initiated by the call for themes is the selection of important scientific topics, or long-term goals, requiring the definition of flagship missions. The following step for these potential types of missions is a detailed assessment study (see Section 1.1.5). The other output is to identify the scope, scientifically and programmatically, to set up a call for proposals in the science programme, out of which the missions to be assessed will be selected (see Section 1.1.4).

1.1.3. *Other options*

If a good scientific idea is available, ESA is of course not the only option. The possible mission scenario may not fit the way ESA works or the idea possibly cannot afford to wait for the call to be issued. Scientists normally proceed following all possible routes. One of them is to propose a national mission, within the scope of a given Member State programme or a bilateral cooperation with other, or even non-European, agencies. Nowadays this does not seem to be a very promising route, given the budgets of national agencies, but the support of ESA can be requested once the mission has been secured as a nationally led proposal to the science programme with certain conditions: among others, that ESA will always only play a junior role in these proposals, as it will not be the driving force.

Ideas can also be presented for evaluation as unsolicited proposals. They may not get into the science programme of ESA, but a scientific and technical evaluation is performed allowing the proposal to be presented to other agencies, or other directorates within ESA, with an adequate independent assessment of feasibility. For example, this is the case of some experiments proposed to fly on board the International Space Station.

Internally, the ESA science programme also works on what are now called Technology Reference Studies to ensure innovative ideas within the community, so that scientists are not self-censored by what they think is available technology, rather they think openly about new possibilities for missions. In this case, without any previous call or competition, "wild" or "blue-sky" ideas proposed by scientists who are either inside or outside ESA are studied. The output information is of course given back to the scientific community for their evaluation of usefulness in future calls for proposals.

A final possibility is to present scientific ideas to ESA or national agencies involving a contribution to a mission driven by other agencies (generally the National Administration for Space and Aeronautics (NASA), but also Russia, Japan, China and India). In all cases a scientific evaluation is possible, an assessment of technical feasibility may be done and, eventually, ESA may get involved.

1.1.4. *The call for proposals*

What used to be called the pre-phase A of a mission starts with the formal call for proposals to the wide scientific community. These are issued asking the community for new mission proposals within a given budgetary envelope and, sometimes, with specific scientific objectives or technical requirements.

Letters of intent (LoI) may be requested to scope the expected response before the actual deadline. This is important so that ESA can set up the evaluation panels with no conflict of interest as well as identify the levels of competition and needs for technical support. As a result of the peer-review process, a number of proposals (typically four per slot) are selected and the SSAC recommends them for the assessment study phase. Proposals not selected for assessment may be proposed again in modified form to a later call, and they can also seek other opportunities within ESA (outside the science programme) or elsewhere as pointed out in Section 1.1.3.

1.1.5. *The assessment phase*

The objective of this important phase of a mission life cycle is to look for a definition of the project to a level showing scientific value, technical feasibility as well as to be programmatically realistic.

Assessment studies normally require some limited industrial support and usually make use of the Concurrent Design Facility (CDF) at ESTEC. From the point of view of the science directorate organization, assessment studies are led by Science Payload and Advance Concepts Office (SCI-A), with the coordination of Science Coordination Office (SCI-C). A study manager is appointed in SCI-A and a study science team is formed with external scientists, including the mission proposers, to monitor all the activities. The Research and Scientific Support Department (RSSD) provides a study scientist to chair the study science team and to make sure that the ultimate scientific goals of the proposal are well respected and supported by the scientific community.

The competitive assessment phase generally takes less than 9 months. Larger, long-term goal missions (cornerstones) may stay in this phase for significantly longer time since specific technology developments are generally needed before going on and no competitive phase is required. During this phase, only a model or "strawman" payload is considered.

For each proposed mission, ESA produces a final assessment report. This includes confirmed science objectives and top-level science performance together with a model payload as defined in a Payload Definition Document (PDD). Of course, an overall mission profile (launch vehicle, orbit, bus, model payload, technology map and preliminary operations profile) is also provided as well as a technical feasibility assessment and an implementation scenario, including the evaluation of possible international cooperation. Another point to be addressed is the potential reuse of technology/hardware from previous missions. As a final result of all these considerations, a schedule outline is produced and preliminary cost estimates to ESA and to Member States (payload) are drawn.

Out of this competitive phase, the advisory structure, and eventually the SPC, recommend one, or may be two missions (one being kept for further assessment and as a backup), to go into definition phase. Technology readiness and programmatics drives the handover of cornerstone missions from assessment into definition phase after the SPC decision (Figure 1.2).

1.1.6. *The definition phase*

This includes the old phase A, pre-phase B and phase B1. It is characterized by the startup of real industrial studies. In fact, two parallel industrial studies are initiated through an Invitation To Tender (ITT) for mission definition. Deviations from the parallel competitive baseline can only be done in special cases (e.g. spacecraft reuses or specific international cooperation). Industrial studies are first done with the model payload and with the real one only after selection. On the other hand, industry also incorporates technology developments into the system design as a result of studies underway and planned within long-term technology programmes of ESA. Towards the end of this phase, once the payload is selected, industry comes to an agreement about payload interfaces with the principal investigators (PIs) and makes binding proposals for the implementation phase in response to a (restricted) ITT.

The selection of the payload to achieve the science goals of the mission is the other main objective of the definition phase. The first step is the preparation of a Science Management Plan (SMP), which has to be approved by the SPC, framing the responsibilities of the respondents to the announcement of opportunity (AO) for the provision of the payload, including the Science Requirements Document (SRD). The Science Directorate of ESA then issues the AO and the different components of the payload, submitted as proposals, are peer reviewed. Selection is based on three criteria: the science case, technical feasibility and maturity, and management and financial plans. The results of the selection process are then to be endorsed by the advisory structure and confirmed by the corresponding funding agencies. Eventually, the SPC makes the final payload decision with specific PIs. Soon after, the PI teams come to an agreement with ESA and industry on an instrument delivery programme. The understanding of the interfaces is generally done through a very important document for PIs, the experiment interfaces document (EID), containing the mission characteristics and constraints (e.g. in terms of spacecraft power, observing restrictions or radiation environment), the instruments design and requirements (e.g. telemetry, thermal control or data handling), the schedule for deliveries (including model philosophy) and the management structures (including reporting procedures and decisions on changes). Since changes are in fact unavoidable during the different phases of the mission, the EID is a living document serving as the reference to ensure that the spacecraft and the payload are suitable to achieve the approved scientific goals. Therefore, it is very important to keep the EID updated to reflect reality rather than design goals.

In order to make sure that the science of the mission covers the required range of objectives, and is not just driven by the specific scientific interests of the selected PIs, ESA also issues a parallel AO to involve mission scientists, or multidisciplinary scientists, in the project. These are scientists from the community who are not providing hardware but rather their scientific expertise. In this way, they contribute significantly to the excellence of the mission performance by ensuring that the science not covered by the PIs is taken into account and by providing cross-disciplinary thinking.

During definition phase, a study science team is kept to monitor the initially approved science goals and, after payload selection, a Science Working Team (SWT) is formed including the PIs and mission scientists. A study scientist provided by the RSSD, who evolves into a project scientist after payload selection, again chairs these teams. Within the science directorate, the overall responsibility shifts from SCI-A (assessment) to the Science Projects Department, SCI-P (definition). This phase lasts some 2–3 years.

When definition phase is nearing completion, the mission reaches a critical point. The SPC should now decide on starting the implementation phase or not, having the payload selected and committed, and a cost at completion (CaC) properly evaluated. At this point a decision about issuing an ITT for the spacecraft main contractor has to be made. Possible problems shown by the definition phase are that science might not be properly scoped, the payload not properly funded or the required CaC may be larger than initially foreseen. Any of these may introduce delays in the decision to go full speed into implementation since, once started, this is a clear budget-consuming phase and delays have dramatic effects (Figure 1.3).

1.1.7. *Comments on payload procurement*

Before going on with the different phases of missions, I considered it important to make some comments on payload procurement due to the relevance of these aspects in the scientific performance of the project as a whole and the interfaces with the scientific community in particular. Things are generally not black and white in this area. The continuous erosion of national funding for payloads as well as the increasing complexity of space instruments are leading to large consortia being formed to procure payloads that exclude real science competition. In practice, we are finding that some payloads have to be directly procured by ESA with industry and others may require ESA involvement at a later phase due to technical or financial problems not properly addressed by the PI. As a partial solution, a non-competitive consortium may have to be formed to ensure the proper payload delivery. Nevertheless, in these cases peer review is still needed to verify that payload packages meet science requirements and are going to be technically and managerially feasible.

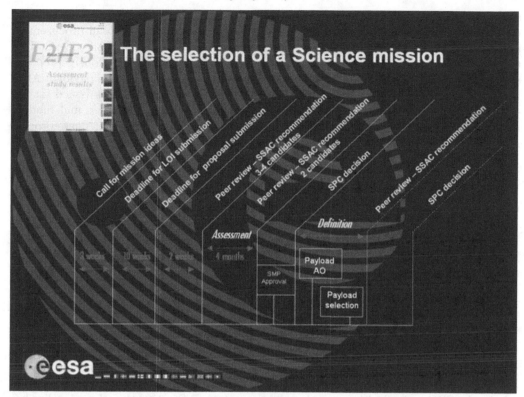

FIGURE 1.2. The selection process of a science mission.

The traditional approach to payload procurement through open AO and national funding has many advantages. Among them, the full control of payload development by the scientific community and keeping a "user-driven" programme may be cited. Moreover, this approach compensates for the lack of expertise in industry to provide control of the scientific performances while ensuring close cooperation between spacecraft developers and the instrument building community. Finally, mission cost sharing between ESA and Member States (typically 30%) is always welcome.

However, this traditional approach also presents clear disadvantages. Some of these are that it becomes easy to arrive at the payload AO with low maturity levels, the spacecraft and mission design may be based on a poorly defined payload, or even that the availability of funding, rather than scientific and technical merits, may drive payload consortia to be formed. Moreover, shortfalls in Member States are picked up by ESA during implementation under cost and schedule pressure, and such ESA funding lacks visibility and effective management control during critical payload development phases. Within this context, smaller instruments may be welcomed, but in fact, isolated instrument development may well lead to unbalanced maturity and duplication of efforts, resulting in problems during implementation and further complications during final assembly, integration and verification (AIV) activities. The lack of technology funding in some Member States adds to the potential difficulties, while ESA's control, or oversight, on payload activities is very limited due to lack of binding agreements.

During the past few years new approaches have been discussed to circumvent the problems mentioned. The basic idea has been to promote iterations between ESA, science teams and funding agencies, leading to a high-level formal agreement. This agreement is signed by ESA and the funding agencies responsible for the payload delivery. The process starts with the approval by the SPC of a baseline mission and payload procurement approach. Then a peer-review team is put in place with the support of the SSAC. This team reviews the adequacy of a reference payload (based on the previously prepared science requirements and PDD) leading to the specification of a definitive payload. After the approval by the SPC of the SMP, a Request For Proposals (RFP) is issued for the payload procurement. Then, instrument consortia involving PIs, institutes and funding agencies are formed, and the proposals reviewed by the advisory structure. These proposals should in principle reflect the initially defined payload complement but new ideas can be accommodated if found scientifically interesting and technically/managerially feasible. Finally, the SPC is asked to endorse the instrument consortia and the level of ESA involvement through ESA-provided elements. This is done through the signature of multilateral formal agreements for each consortium.

Examples can now be found for payloads developed following all kinds of approaches. For instance, ESA procured some astronomy payloads, as for HST (FOC), EXOSAT, Hipparcos, GAIA or JWST (NIRSPEC), whereas some others were supported at different levels by ESA, as in the case of Integral, Mars Express and Rosetta. Finally, certain payloads have been selected for different reasons via non-competitive processes as in the cases of the JWST (MIRI), LISAPF and Venus Express.

1.1.8. *Implementation phase*

This phase includes the old phases B2 (design), C/D (development, integration and verification) and E1 (launch and in-orbit spacecraft commissioning). At the end of the definition phase, the confirmation of the CaC envelope leads to a ratification of the mission by SPC and a decision to go into implementation. Starting this phase implies "no return" since cancellation may involve higher costs to the programme than the delivery of the mission. That is why commitments on schedule and delivery of all components of the project (payload, spacecraft, launcher and ground segment) have to be well evaluated and ensured before starting.

At the beginning of implementation phase the prime contractor is selected among those involved during definition. Following the selection of the prime contractor and its core team, industry executes all the spacecraft development activities, integration, test and delivery of the flight spacecraft, launch preparation, launch and in-orbit commissioning. Within ESA, the overall responsibility remains in SCI-P where a project manager is appointed together with a project team. The duration of this phase is around 4–5 years. Meanwhile, RSSD maintains its support to the project through a project scientist and starts preparation activities for the exploitation phase. In fact, science operations are defined through a Science Implementation Requirements Document (SIRD) and the subsequent Science Implementation Plan (SIP). For interfaces with the scientific community, the SWT continues with its activities, chaired by the project scientist, and including the payload PIs and mission scientists. During this phase, the PIs are generally extremely busy developing their contribution to the payload (and worried about how real the assumed national funding was), but it is important for them to keep the scientific research active to ensure the optimal exploitation of the final data. Again, I would like to emphasize the importance of keeping an updated EID alive.

1.1.9. *Exploitation phase*

This phase includes the old phase E2 (operations and archiving). After the successful integration and verification, the launch, and the commissioning in orbit of the payload and spacecraft, ensuring that all performances are within the scientific requirements, the real

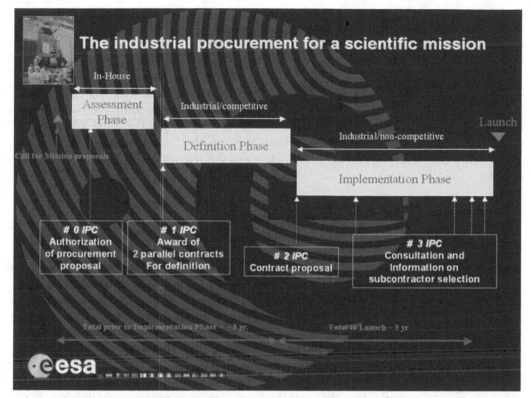

FIGURE 1.3. The development of a project and its industrial milestones.

science starts. As a matter of fact, this is the time for which scientists have been working all the previous years. Data are being delivered and have to be properly processed and calibrated to be useful. In this phase it is very important, for the benefit of the scientific community at large, that the full set of science data is put into useful format, with physical units, and recorded in an archiving system that can be accessed by all those interested in carrying out scientific research with them. In the case of observatory-type missions, calls are issued regularly to select the best-observing proposals through competition and peer review. In all kinds of missions, a detailed science operations planning is needed to ensure the best possible science-driven use of the available resources in orbit.

Within ESA, after commissioning, the overall responsibility of the mission falls on the RSSD. A mission manager is appointed in charge of the mission performance and the CaC while the project scientist remains responsible for the science delivery (e.g. users group, science working group or the issuing of AO for observing time allocation).

1.2. How *you* may be part of it

Since the reader of these notes is most probably a young scientist keen to get involved in space sciences, I found it pertinent to add some comments about science and engineering. A science mission can obviously not be carried out without the scientists, but it is important to understand that it can also not be implemented without the involvement of space engineers. We have to work together, and of course talk to each other, which is something the scientist aiming to get involved in a mission has to be prepared to do. One of the problems in this working relationship is the different ways of setting priorities, and another problem is the use of quite different languages. The latter is generally solved through extensive documentation; the former requires an effort to quantify the scientific objectives through a set of scientific requirements and priorities. This is not a one-off task and involves the continuous monitoring of the scientific performances throughout the life cycle of the mission. Engineers often see lack of communication as poor knowledge of the scientists about what they really want to achieve. Scientists, on the other hand, accuse the engineers of incompetence: they believe the engineers jump too easily to the conclusion that they have to reduce the scientific performance of the mission in order to solve technical problems. It is very important to break the "us versus them" syndrome.

Let us see how to face the most critical phases of the scientific definition of a potential mission.

1.2.1. *The first questions*

Before you even propose an idea, it is necessary to be sure about what you want to do. What is the *science*? What are the questions to be answered? For example, do you want to know what is the radial velocity of a given quasar? Do you possibly want to analyse the chemical profile of the atmosphere of a given planet? Perhaps, the objective is to have a final answer to the origin of the Universe? Then, you have to evaluate whether answering the identified primary questions can really be considered first-rate science or merely useful to know. The other basic point is whether or not it could be done from the ground and if use of space is fully justified.

Once all these questions are correctly answered, you have to evaluate your particular level of involvement and how far you are willing to go. There are many ways of using space sciences to answer interesting questions and they may require very different levels of involvement. You have to assess the investment you are ready to make for the sake of the science return. First, you may consider making a proposal for observing time with a space observatory (e.g. Newton or Integral). In this case, though you should have a good understanding of the performance of the instrument you intend to use, no hardware development is needed. Of course, using a mission in orbit with an existing payload puts constraints on the scientific wishes you may have but the preparation of a response to an AO for observing time could satisfy your needs and the required effort is not very different from that of requesting observing time for ground-based facilities.

Another option you may like to consider is to apply for a mission or multidisciplinary scientist position on an approved mission. If successful, you will be able to influence the development of the mission by making sure that it can achieve the best possible scientific results. In this case, you will have to deal with PIs for guaranteed time and gain a better knowledge of the performance of the instrumentation before launch, thus placing yourself in an optimum situation to apply for observing time in response to AOs. In addition, you gain exclusive access to science data from the mission in orbit.

A third level of involvement in a mission is to make a proposal for a contribution to the payload. In this case, you have to make sure that the proposed science is within the objectives of an approved mission and, therefore, understand well the core science goals of the mission and reference payload. Then, you have to respond to the AO (or RFP) writing a LoI and presenting a *proposal*. Perhaps, this is the highest level of involvement possible from the point of view of engineering, financial and managerial responsibilities, though the level can be somewhat tuned to your capabilities by playing either the role of PI or the co-investigator (Co-I). What you gain by submitting a proposal and being selected is access to influence the science of the mission, large amounts of science data and possibilities of developing state-of-the-art hardware.

A final possibility is to consider the eventuality that nobody has thought of your scientific research problems before and present a proposal for a completely new mission. Responding to a call for ideas and/or proposals does this. Depending on the nature of the mission proposed the involvement might be very different. Is it going to be a PI or an observatory-type mission? What you gain is the possibility to make your dreams come true, very good chances to become PI or mission scientist depending on the interest in the hardware and, above all, the possibility to achieve unique science.

What you are supposed to do next, after the scientific idea is clear and your level of effort assumed, is to evaluate the required support you are likely to procure. Who is the scientific community that you expect will support the idea? Would they kill other proposals to support yours? And, from the point of view of the feasibility of the proposal, do you have enough technical support? Do you expect to obtain the necessary funds?

In summary, to embark on a challenging space project, you need to have clear ideas about the final goal, strong determination and leadership, access to adequate funding, management skills, and high levels of patience and persistence.

1.2.2. *Looking for opportunities*

As mentioned in the first part of these notes, the initial step is to respond to a call for ideas or a call for themes. These calls allow ESA to carry out long-term planning and the identification of reference missions. For you, they may end up in getting involved in subsequent calls for proposals or in the definition of cornerstone type of missions. The call for proposals may be for missions within a given budget envelope or for payload contributions to approved missions.

In addition, the possibility of using the road of unsolicited proposals could be appealing. These are not responding to actual opportunities in the current planning but an evaluation might be needed for other purposes. In fact, it could be a good opportunity to start the process within national agencies or to trigger bilateral cooperation. In practice, the use of nationally led missions also opens the opportunity to get ESA involved "*a posteriori*".

But let us return to the standard call for proposals. When issued, they normally require you to send an LoI first. Responding to the call for proposals is not really necessary, but it helps in the preparation of the peer-review process. You can also obtain additional information from the agency to prepare the proposal and it helps the process of putting together the team of scientists involved. Sending an LoI is therefore highly recommended, clearly indicating in one page the scientific objective, an identification of the type of the mission, and the support of other scientists (forming an embryo consortium). Later, while preparing the proposal, you will have to discuss all the scientific constraints within the group and consolidate the consortium, get advice on mission analysis and feasibility, and get support from national funding agencies.

1.2.3. *The proposal*

During the time devoted to the preparation of the actual proposal several options are possible. You may decide to forget about further work on it if it is not found to be scientifically relevant enough, if the support of the scientific community and/or funding agencies is not found to be realistic, or if in the end the scope of the mission does not actually fit the call. Another possibility is that you may effectively focus the proposal and improve it by joining forces with other scientists not previously considered or getting advised about better ways to address the scientific issues and the possible interfaces within the group. In any case, it is important to make proposals, even if they are not

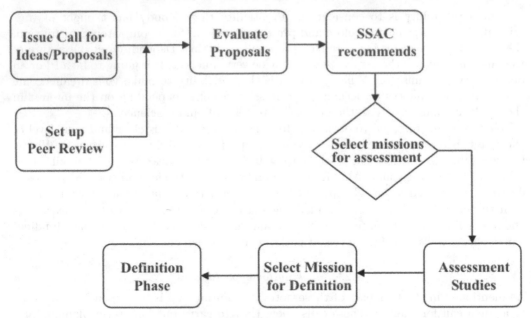

FIGURE 1.4. The assessment phase.

successful, to ensure that future calls take into account your ideas and that you are considered for participation in reference studies.

The proposal should contain first a mission statement, that is a sentence summarizing the objective of the mission or payload instrument. This may obviously repeat what was used for the LoI. Then, the scientific justification should be further detailed by means of an explanation of the scientific objectives, including the primary objectives, the secondary objectives or parallel science, and the comparison with other current and planned missions. This discussion leads to the definition of the mission requirements that include observational requirements, target/subject selection as well as a good justification for the need to go to space. Then, the mission performance should be evaluated by discussing the required duty cycle, proofs of feasibility as well as tradeoffs and performance comparison. Finally, a preliminary mission design should be included. This part should explicitly address the payload, system design, system engineering and environment. As a complement to all the above inputs, it is also convenient to add a preliminary mission analysis and mission budget discussion, a first-cut approach to the scientific operations and the planned management structure.

1.2.4. *Assessment studies*

After the proposals are received and a selection is made about which of them merit an assessment study, the next phase starts with the definition of the mission objectives (Figure 1.4). At this point, you are supposed to define broad objectives and constraints, and to estimate quantitative mission needs and requirements. A more detailed definition of the primary and secondary objectives of the mission is expected together with a clear understanding of its nature (observatory/experiment type).

The primary objective now is to transform objectives into requirements. This implies the definition of functional requirements, operational requirements and constraints. Functional requirements are essentially linked to an understanding of how well the system has to perform to meet its objectives. This implies a discussion of performance (by science payload, orbit, pointing), coverage (by orbit, satellites, scheduling) and responsiveness (by communications, processing, operations). The operational requirements refer to how the system operates and the users interact with it; its duration (by nature, orbit, redundancy), availability (by redundancy, access to data), survivability (by orbit, hardening, electronics), data distribution (by communication architecture) as well as data content, form and format (by level and place of processing, payload).

The constraints are generally linked to the limitations imposed by cost, schedule and implementation techniques. Here we refer to cost (affecting number of spacecraft, complexity, size), schedule (affecting technology readiness), political (customers, funding agencies), interfaces (users and existing infrastructure) and development (like programmatics or spacecraft availability).

But, how do we establish the mission requirements? In order to achieve a requirements baseline, we have to identify properly the users (scientific community) as well as their objectives and priorities for the mission. Then, you should define the internal and external constraints, and translate user needs into functional attributes and system characteristics. Then, we can establish functional requirements for the system and provide inputs for decomposition to system elements. In order to establish quantifiable requirements, different mission concepts should be reviewed and expected data deliveries evaluated. From the point of view of the scientist, the primary requirements are those related to how to collect as much good scientific data as possible and the definition of spectral range for observations as well as spectral and time resolution. From these primary requirements, specific characteristics should be derived for the choice of optics, focal length and detectors, collecting area or telescopes, pointing, orbit and data rates, and volume.

In this evaluation, it is important to address the issues of how mission data are collected, distributed and used, the benefits of space versus ground, the need for central versus distributed processing and the required level of autonomy in the spacecraft. Another question to think about is the communications architecture. How do the various components talk to each other? Discussion about data rates and bandwidths is needed here as well as the acceptable time delays for communications. In addition, some thinking will be required on tasking, scheduling and control; that is, how the system decides what to do. Again the level of autonomy requirements will be impacted. Finally, the mission timeline should be presented with an overall schedule, from planning to end-of-life.

In order to play during the assessment phase with different mission concepts, you should identify the common system drivers, that is those parameters that will be repeated for all of them. The first is the size (limited by shroud and weight) and the second is the on-orbit weight (limited by launch vehicle and orbit). Then, power (limited by size and weight) and data rate (limited by storage and processing capabilities and antenna) as well as communications (limited by station availability) have to be considered. Finally, pointing requirements (limited by cost, weight), the number of spacecraft (limited by cost) and operations (limited by cost, communications or team size) should be evaluated as possible drivers.

Once the system drivers are known, we can proceed with a number of design cases (including alternative platforms). For all of these cases, a mission characterization will be needed. After the definition of the preliminary mission concept, the subject characteristics are defined. These are either controllable subjects (what we change, we can experiment) or passive subjects (what we do not change, we only measure). The determination of possible subjects should meet the scientific objectives and is linked to the determination of ways to detect/interact with them. A result of this discussion will also settle whether multiple subjects and payloads should be used. You should carefully define and document the initial subject selection.

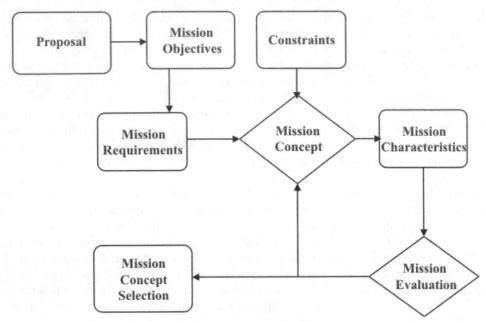

FIGURE 1.5. The selection process.

The following step is to determine the orbit characteristics: altitude, inclination, eccentricity, delta V budget for orbit transfer and maintenance, number and spacing of spacecraft, observation/experiment time, duration and duty cycle, minimum solar aspect angle (angle between line Sun/satellite and line of sight) and roll angle, around the line of sight. Given the main spacecraft and orbit characteristics, the definition of a strawman payload has to be made based on the subject selection (what we want to measure). This part includes size and performance through physical parameters (size, weight), viewing (field of view, wavelength range, telescope design), pointing (primary direction) and sensitivity (noise levels, frequency resolution). Then, the requirements for the performance in terms of electrical power (voltage, average and peak power), telemetry and commands (data rate, memory size), thermal control (active/passive, temperature limits) should be addressed and defined for each payload element.

Finally, the selected mission operations approach has to be outlined including the communications architecture, ground stations (LEOP and cruise, routine and backup), communications link budget, ground system (mission operation centre, MOC, and science operations centre, SOC) and operations (resources, data distribution).

Once the mission concept is clear in our mind, with all elements and drivers well understood, we can address the problems related to actual deployment of resources, logistics and possible strategies and, finally, provide costing estimates.

1.2.5. *Mission evaluation*

The evaluation of all possible mission concepts is actually an analysis of how well the different options satisfy the fundamental mission objectives (Figure 1.5). Critical requirements dominate the mission's overall design through the coverage or response time (payload FOV, communications, scheduling, staffing), resolution (instruments size, attitude control), sensitivity (payload size and complexity, processing, thermal control) and in-orbit lifetime (redundancy, weight, power, propulsion budgets).

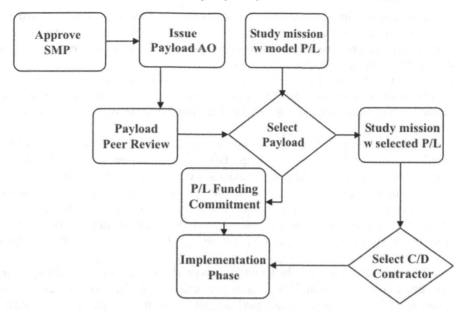

FIGURE 1.6. The definition phase.

A mission utility analysis then quantifies mission performance as a function of design, cost, risk and schedule. Performance parameters (how well the system works), measures of effectiveness (how well the system can be operated, e.g. searching for GRB) and performance versus requirements (simulations) are now fully evaluated. Finally, cost assumptions and overall assessment are required. The mission CaC is estimated at rough order of magnitude (ROM) level and a risk assessment is carried out together with an overall schedule assessment.

The result of the evaluation of all possible mission concepts is to select a baseline mission. This is done with all the inputs of the characterization exercise and the comparison of all mission concepts. As a summary, affirmative answers to the following questions will determine a successful assessment phase of your proposal:

- Does the proposed system meet the overall mission objectives?
- Is it technically feasible?
- Is the level of risk acceptable?
- Are the schedule and budget within the established constraints?
- Do preliminary results show this option to be better than other solutions?
- Does the proposed system meet more general programmatic objectives?

If there is a concept which fits these criteria, this is obviously the one to be carried forward to the definition phase (Figure 1.6). If not, a process of iterative corrections in one or more of the concepts should be initiated. It is again very important to document all options and decisions properly. For example, this iterative procedure is applied at the ESA Concurrent Design Facility (ESTEC) in real time.

1.2.6. *Further into the mission*

During definition phase the selected mission concept is further refined through parallel industrial studies and the payload goes through a similar process of proposal (after an AO), assessment and selection. If you are not involved in the development of a payload element, then your role now becomes a bit more relaxed unless the previous process was not properly documented. Objectives and requirements are clearly stated, and the scientists just monitor and ensure that the engineering work does not jeopardize the science. The main task of the scientists in the study team is to ensure that industrial studies satisfy the scientific requirements and the payload selection process ends up with the best possible instrument complement.

If you are part of the payload team, the situation does not improve, indeed it gets worse. Continuous interaction with the project team is needed and a full procurement management of your contribution has to be put in place containing the right engineering and financial items.

When entering implementation phase, the situation of the mission is rather safe; it is very difficult to cancel missions at this stage. Even so, things may get complicated. This is a sweat and tears business! Critical situations are likely to come up due to technical or financial problems, generally external to the project, which could not be foreseen. Moreover, if involved in hardware (PI) this is a very active phase where the pressure of the spacecraft schedule tries to drive your development planning. If not involved in hardware, you should prepare the coming phase by developing tools, support ground observations, etc.

Eventually the mission is launched and we enter the commissioning phase, when a reality check is made and the mission science performances evaluated. After a successful commissioning of both spacecraft and payload, the exploitation phase starts with the science operations and recording phases. This is the time when you get your science done, but you must be sure not to overlook the importance of records. Only properly calibrated and accessible data ensure the full scientific exploitation of the mission as well as the collaboration of the community at large and not just those directly involved.

Before closing these notes, I would like to refer to a final item. If you propose a mission or an instrument, think of a name or acronym. You can choose an acronym without any additional meaning, like X-ray multi-mirror mission (XMM), or with some additional meaning, like GAIA or Hipparcos. Another option is to forget about acronyms and choose the name of a relevant scientist, like Newton, Herschel or Planck. Think about it, you will have to use the chosen name for many years and everybody will refer to the mission by this name or acronym. Presentations in scientific meetings or project documentation require an easy, catchy name for the mission, and it is also part of the team-building process for the project.

2. Design issues for space science missions

By YVES LANGEVIN

Institut d'Astrophysique Spatiale, 91405 Orsay, France

This set of lectures starts with a general view of space science missions, from their rationale to the sequence in which they have been defined in the last decades. A crucial aspect in the definition of the mission is its launch and cruise strategies. In the case of solar system bodies, also orbital insertion becomes a major issue in the mission planning. The different strategies based on gravity assists maneuvers, chemical and electric propulsion are detailed. As case examples the Rosetta, Bepi-Colombo and Solar Orbiter missions are studied in their different scenarios.

2.1. The evolution of space science missions: past, present and future

2.1.1. Introduction

Astronomy, defined as the study of the universe beyond Earth's atmosphere, is the oldest established science discipline. When asked about the need to build large telescopes, an astronomer explained half jokingly that the race for the best possible observing tools started thousands of years ago with Stonehenge. During the second half of the 20th century, the exploration of the solar system by space probes has emerged as one of the science enterprises with the most direct appeal to the general public. However, the costs are very high, and there is a high risk of failure, as demonstrated by missions to Mars (with a meagre 50% full success rate) and the mixed record of "faster cheaper better" missions of National Aeronautics and Space Administration (NASA). It is therefore interesting to investigate the scientific rationale of space missions for astronomy and solar system exploration.

Let us first restrict the discussion to remote sensing, which consists of collecting information using photons coming from the object to be studied. Ground-based observations have significant limitations in terms of wavelength coverage. At sea level, the atmosphere represents a mass equivalent to 10 m of water, which blocks all high-energy photons. Minor constituents, such as water vapour and carbon dioxide, have extensive absorption bands.

The transmission of the atmosphere is very low in many regions of the wavelength range, from the Ultra Violet (UV) to the Infrared (IR) (Fig. 2.1), in particular in the UV and in major water bands (1.35, 1.85 and 2.7 μm). As for the transmission windows, the window from 0.4 to 0.7 μm corresponds to the highest flux of solar energy. It is therefore not surprising that evolution has selected it for optical sensors of living organisms, including our own eyes, which is why it became the visible range. The thermal IR range (10–100 μm) and the sub-mm range (100 μm–1 mm) are also absorbed by the atmosphere or contaminated by its thermal emissions. Further on, the ionosphere blocks a major part of the radio signals, and radioastronomers wage a desperate war against mobile phone and satellite TV companies competing for increasingly crowded wavelength ranges. At ultra low frequencies, anthropogenic noise drowns out any possible signal from the sky. Atmospheric transmission can be improved by setting observatories at high altitudes. ALMA (Atacama Large Millimeter Array), the international interferometer at millimetric wavelength, is being built on a plateau in South America at an altitude of more than 5000 m, with considerable difficulties as workers cannot stay there for long.

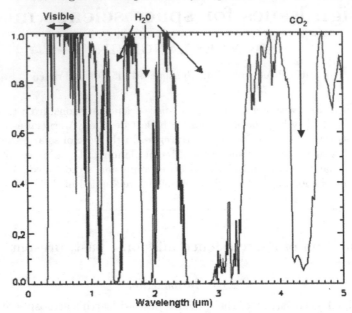

FIGURE 2.1. Transmission of the atmosphere at sea level from 0 to $5\,\mu$m.

The first asset of space observations is therefore to open up whole regions of the electromagnetic spectrum for astronomic observations, and the farther from the Earth, the better. This is the main argument for space astronomy, which focuses on objects beyond our solar system. The improved spatial resolution due to the removal of atmospheric turbulence (scintillation) constitutes a second asset. However, the advances in adaptive optics have much improved the situation for ground-based observatories, which can now provide resolutions close to the diffraction limit, and these techniques are constantly improving (multi-conjugate methods and laser star). One of the design parameters for space observatories is the operating orbit. If the orbit extends far enough from the Earth, such an instrument can provide full sky access and continuity of observations, while a given ground-based site has restrictions in declination (hence the interest of combining northern and southern sites) and a visibility period of at most 10 h every day for a given object.

The capability of space missions to get close to solar system objects drastically improves the observation capabilities in terms of spatial resolution. Let us consider a telescope in the 10 m class operating at $1\,\mu$m. Its maximum resolution is 10^{-7} radians assuming that the diffraction limit is reached using adaptive optics. The best observing conditions are at quadrature for Mercury and Venus (maximum angular distance from the Sun) and at opposition for outer bodies. With a space mission, one can easily implement a 10 cm aperture instrument, with a diffraction limited resolution of 10^{-5} radians. The closest distance for a fly-by or an orbiter is typically 300 km for most planets and asteroids, 30,000 km for Jupiter and 80,000 km for Saturn (outer edge of the A ring). In Table 2.1, we can list the corresponding spatial resolution for major classes of solar system bodies.

It is immediately apparent that a visit to any solar system body provides improvements by a factor of more than 1000 in terms of spatial resolution with a similar photon flux, hence signal to noise ratio. Indeed, any telescope working at its diffraction limit will provide the same signal to noise for an extended source: the angular resolution is λ/D, where λ is the wavelength and D is the diameter of the telescope, hence the solid angle of the resolved surface element is $(\lambda/D)^2$, the collecting area is proportional to D^2 and the photon flux is the product of these two terms.

TABLE 2.1.

Target	Mercury	Venus	Mars	Vesta	Jupiter	Saturn
Minimum distance (AU)	0.92	0.69	0.37	1.38	4.10	8.50
Earth-based resolution (km)	13.7	10.3	5.5	20.6	61	127
Orbiter resolution (m)	3	3	3	3	300	800

Observing the Sun is a very specific challenge: photon fluxes are extremely high for direct observations, hence one does not need large collecting areas, but it is much more difficult to improve the spatial resolution by getting close to the Sun than for any other solar system body, due to the thermal environment. As an example, the solar orbiter mission, which will be discussed later in this chapter, has a minimum heliocentric distance of 0.2 AU (30 solar radii). This provides an improvement by a factor of 5 in spatial resolution when compared to a solar observatory at 1 AU (e.g. Solar and Heliospheric Observatory, SOHO), but the thermal flux reaches $35 \, \text{kW/m}^2$. The solar probe studied by NASA (minimum distance of 3 million km, 4 solar radii) is a formidable challenge with a thermal flux of $3 \, \text{MW/m}^2$.

The solar system provides a unique opportunity of performing *in situ* investigations, which by definition requires a space mission. *In situ* investigations constitute the main source of information for space plasma physics, which requires measuring the local electromagnetic field as well as the composition, energy and direction of ionized and neutral constituents so as to characterize a planetary environment (in particular the Earth magnetosphere) or the interplanetary medium. Some constraints can however be obtained on the density of ionized particles by monitoring the propagation from a spacecraft antenna to a ground station ("radio science"). *In situ* investigations of planetary magnetospheres are combined with planetary science goals when designing planetary missions. Major results on the interplanetary medium have also been obtained during the cruise phases of planetary missions, but specific goals can require dedicated missions. The study of the Earth magnetosphere requires dedicated missions.

For planetary science, the implementation of *in situ* methods implies the deployment of surface modules (landers and rovers). If an atmosphere is present, airborne modules (balloons and airplanes) can be implemented. It is also possible to lower temporarily the orbit of a satellite so as to study the upper atmosphere with *in situ* methods such as mass spectrometry or aerosol collection and analysis. The capabilities of highly miniaturized *in situ* techniques is constantly improving. The ultimate *in situ* mission is the sample return, as techniques available in the laboratory will always maintain an edge over what can be achieved *in situ* due to limited mass and energy resources.

2.1.2. *The logical sequences of space missions*

In the fields of space astronomy, solar-terrestrial physics and planetary exploration, space missions can be classified in natural sequences, from the missions which are easiest to implement, to the most complex missions requiring larger budgets and capability build-up through technological developments.

2.1.2.1. *Space astronomy*

A space observatory can observe many different types of objects. For space astronomy, the main driver is to cover the full range of the electromagnetic spectrum with missions of increasing capabilities. As shown in Fig. 2.2, each wavelength range provides specific information on the observed object. There is a link between the most useful wavelength range and the temperature of the observed object. High-energy ranges are most interesting for active objects, such as novae, supernovae, black holes or active galactic nuclei. "Cold" objects, such as molecular clouds and protostars are best observed in the IR. Furthermore, each wavelength range requires specific techniques, such as grazing incidence mirrors or coded masks for high-energy astronomy. Therefore, until a few years ago, space astronomy developed autonomously in each wavelength range.

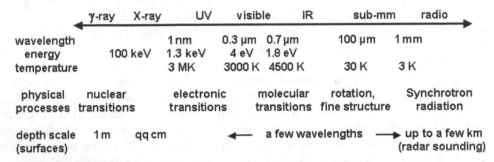

FIGURE 2.2. Limits of wavelength ranges with the equivalent photon energy and black body temperature (peak of the emission curve). The physical processes associated with each wavelength range and the probing depth for remote sensing of surfaces are also indicated.

Similarly to ground-based astronomy, advances in space astronomy relies on the improvement of focal plane instruments (quantum efficiency and resolution) and the increase in collecting areas.

One can consider three levels of missions in space astronomy:

• The pioneer missions which open up a new wavelength range, often with a whole sky survey at selected wavelengths.

• The second generation missions, which provide pointing and imaging capabilities.

• The observatory class missions, which implement high-efficiency detectors to provide extensive capabilities for both imaging and spectroscopy.

From Fig. 2.3, it is clear that by 2010 the observatory class level would have been reached in all the major wavelength ranges except for radioastronomy (an orbiting antenna for VLBI has been studied for many years, but no project has as yet been fully funded). In sharp contrast to the situation in the early 1980s, bidimensional detectors are now available from the lowest energies to gamma rays with good quantum efficiencies.

Space astronomy missions beyond 2010 need to be justified by a very significant improvement in terms of spatial resolution, collecting area or both. With the exception of the James Webb Space Telescope (2012), the next step beyond the highly successful Hubble Space Telescope, and dedicated missions such as GAIA (μarcsec astrometry, also programmed in 2012), most of the presently studied concepts for missions beyond 2010 require constellations of satellites either for implementing interferometry techniques or for aggregating large collecting areas without incurring the cost and complexity of a single large satellite (e.g. for a new generation X-ray telescope beyond Newton and Chandra). Non-photonic space astronomy is now considered very seriously, with a major observatory for gravity waves, Laser Interferometer Space Antenna (LISA), being planned for 2012–13. It requires three satellites maintained in a precisely controlled configuration at distances of several million kilometers. A precursor technology mission (Smart Missions for Advanced Research and Technology, SMART-2) is programmed for 2007. For all long wavelength missions, avoiding the contamination by the thermal radiation from the Earth

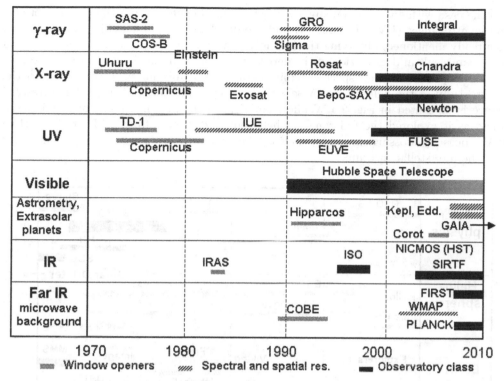

FIGURE 2.3. Major missions in space astronomy.

will strongly favor implementing the missions at the Sun–Earth L2 point (1.5 million km away from the Sun on the Sun–Earth line). The L2 point has already been selected for WMAP (NASA), Planck and Herschel (ESA). It is the leading candidate for ambitious interferometry missions such as Terrestrial Planet Finder (TPF) Darwin, dedicated to the direct imaging of extrasolar planetary systems.

2.1.2.2. *Solar astronomy and Solar-terrestrial physics*

The Earth magnetosphere and the solar atmosphere expanding into the interplanetary medium (the solar wind) constitute remarkable natural laboratories for space plasma physics, with an enormous range of densities and temperatures. The magnetic activity of the Sun dominates events all the way to the boundaries of the heliosphere. Solar activity has its roots deep inside the Sun, hence all aspects of solar physics are now considered in combination with solar-terrestrial physics. Similarly to the Earth magnetosphere, the solar wind has a strong influence on the characteristics of the ionized environments of other planets. However, as their investigations require planetary missions, these goals are considered in combination with the other scientific objectives of planetary science.

Solar-terrestrial physics developed early on due to the easy access to the Earth magnetosphere and the solar observation capabilities of low-Earth orbits. The early single satellite missions such as those of the Explorer program were followed by increasingly sophisticated multi-satellite missions. The International Solar-Terrestrial Physics (ISTP) program which coordinated the efforts of NASA (Wind and Polar), Institute of Space and Astronautical Science (ISAS) (Geotail) and ESA suffered a setback when Cluster, a mission combining observations from four satellites at variable distances, was lost on the launch pad in 1996. A recovery Cluster mission was successfully launched in 2001. As was also the case with space astronomy, solar astronomy benefited from the opening of new wavelength ranges (UV, EUV and X) and the development of high-efficiency detectors. As already mentioned, observing the Sun is a very specific challenge as the high-photon fluxes can result in degradation of the detectors or optics, in particular in the UV range. The development of deep space capabilities made possible missions dedicated to the *in situ* study of the heliosphere. A selection of important missions for solar-terrestrial science is presented in Fig. 2.4. While primarily a planetary mission, Voyager is mentioned as it explored the full range of heliocentric distances from 1 AU to more than 50 AU, possibly crossing these very days the heliopause which separates the solar cavity from the interstellar medium.

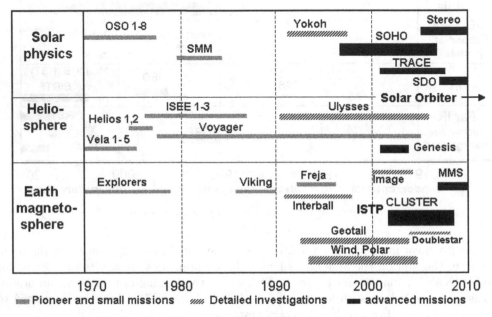

FIGURE 2.4. Major missions in solar-terrestrial physics.

The present generation of solar-terrestrial missions is dominated by SOHO (for solar physics) and Cluster (for the study of the Earth magnetosphere). A new four-satellite mission, MMS, is planned for 2007. Apart from the scientific objectives, there are operational interests for detailed monitoring and forecast capabilities in the Earth magnetosphere ("space weather"). It is therefore likely that the 2010–20 period will focus on large constellations of micro-sats monitoring the local conditions. The remarkable success of SOHO has demonstrated the interest of the Sun–Earth L1 lagrangian point for solar observation (1.5 million kilometers towards the Sun on the Sun–Earth line), as it makes possible continuous observations of the Sun in stable conditions. The L1 point will be a strong candidate for the location of a new generation solar observatory as broad-based as SOHO if it emerges in the time frame 2010–20. The major prospect for solar and heliospheric studies is Solar Orbiter, a mission scheduled in 2013 in the recently revised program of ESA. This mission requires extremely large capabilities in terms of velocity changes (delta V), as it plans to explore the inner regions of the heliosphere at inclinations of more than 30° so as to provide observation capabilities of the polar regions of the Sun at close range. Solar Orbiter will be examined in further detail when discussing ambitious missions in terms of delta V. An even more challenging mission, the solar probe, has been studied extensively by NASA, but the technical problems are such that it is unlikely to be implemented before Solar Orbiter.

2.1.2.3. *Planetary exploration*

As discussed in Section 2.1.1, planetary science strongly benefits from space missions: for remote sensing, not only are new wavelength ranges being made available, spatial resolution can also be improved by several orders of magnitude by sending a spacecraft to a planet. Furthermore, space missions provide the unique capability of studying a planetary body with surface modules (if there is a surface), entry probes or airborne modules (if there is an atmosphere). Finally, one can consider bringing samples back from the surface or the atmosphere, so as to benefit from the most sophisticated laboratory techniques.

The logical sequence for planetary exploration therefore involves four steps:

(1) *First exploration mission*: one or several fly-bys make it possible to observe the major characteristics of a planet and its environment (satellites, rings, particles and fields). The very first such mission, Luna 3, was launched less than 2 years after Sputnik 1, and it revealed for the first time the farside of the Moon, dominated by highlands in sharp contrast with the nearside. The paramount exploration mission has been Voyager 1 and 2, launched in 1976, which was able to visit in turn the systems of Jupiter, Saturn, Uranus and Neptune. These exploration missions completely changed our vision of the solar system.

(2) *Comprehensive study mission*: a spacecraft in orbit (preferably on an inclined orbit and with a low eccentricity) covers the whole planet with increasingly sophisticated remote sensing investigations and studies its environment. The Cassini mission, which will reach Saturn in July 2004 is one of the most complete missions of this type.

(3) *In situ study mission*: one or several *in situ* modules study the interior, surface and/or atmosphere of a planetary body. Many such missions are carried to their target by a mission of the previous type (e.g. the Beagle lander by Mars Express, or the Huygens entry probe into Titan's atmosphere, by Cassini), using the orbiter as a relay to transmit their data back to the Earth. Once satellites are in orbit, one can consider direct entry missions such as Pathfinder or the Mars Exploration Rovers which will reach Mars in the first days of 2004.

(4) *Sample return mission*: samples are brought back from the surface and atmosphere to the Earth. Such missions are particularly complex as the return trip requires large delta V capabilities.

Planetary science covers an enormous range of disciplines, from Earth sciences (internal geophysics, geochemistry, geology, ...) to atmospheric sciences and space plasma physics (ionized environment and interaction with the solar wind). This means that more than one comprehensive study mission is usually needed (e.g. an aeronomy mission, focussing on the ionosphere and environment, a mission focussing on the atmosphere and a surface-oriented mission).

As shown in Fig. 2.5, this sequence with four stages has been the general trend for all solar system objects apart from the Moon: the Apollo program had mainly political and strategic goals (such as closing the "missile gap" opened by the first Sputnik) and was only marginally (at least as far as the first missions were concerned) dedicated to science. This resulted in a 8-year race which brought back several 100 kg of samples from the Moon by 1972, leaving major gaps in the global coverage of the satellite (e.g. 7% coverage for the elemental composition by X-ray spectroscopy).

The first fly-bys of asteroids by Galileo in 1991–1993 completed the "first exploration" phase, with the exception of Pluto and Charon. It is interesting to note that since then Pluto has changed status: it is no longer a planet, but the largest of a new family of solar

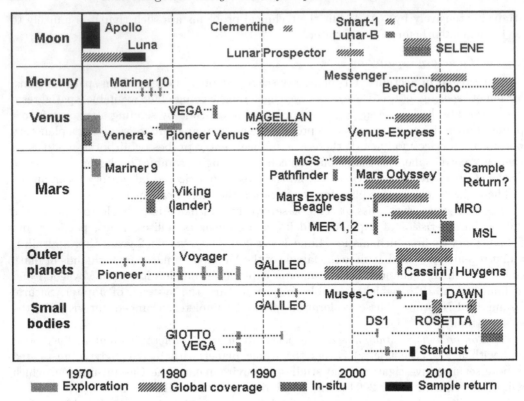

FIGURE 2.5. Major missions in solar system exploration.

system bodies, the Kuiper Belt objects, dynamically related to short-period comets. This means that the first survey is complete as far as systems of planets are concerned. The planetary missions of the last 15 years and the missions already programmed (Dawn for main belt asteroids, Rosetta for comets, Messenger and BepiColombo for Mercury, which will yield results in the 2009–2015 time frame) address most of the goals of the comprehensive study phase. The major missing elements will be the detailed investigation of the systems of Uranus and Neptune, which require very long-travel time with classical techniques. A strong argument can also be made for a new mission of this type to the system of Jupiter, with the new interest generated by Galileo discoveries, such as the buried ocean of Europe, the silicate nature of Io's volcanism or the dipolar field of Ganymede. Furthermore, remote sensing investigations on Galileo were hobbled by the extremely low-data rate (a few 10 bits/s) due to the failed deployment of the high-gain antenna.

The 2000–2010 time frame is a transition period during which an increasing emphasis is given to *in situ* modules. Such missions are technically more challenging than orbiters. Europe has been at the forefront of this evolution, providing the Huygens entry probe to the joint exploration mission to the Saturn System (Cassini/Huygens) and implementing landers on Mars Express as well as Rosetta. The recent cancellation of the Mercury Surface Element of the BepiColombo cornerstone mission is a setback which results both from the specific technical problems of landing on Mercury and the budget crisis of the mandatory program. It is clear that new planetary exploration missions designed for the post 2010 time frame will most likely focus on *in situ* investigations and sample return missions. One of the most promising themes is the characterization of the internal structure of planetary bodies, which requires a network of seismometers. There is a very active debate for Mars exploration on whether the development of *in situ* analysis techniques will alleviate the need for a Mars sample return. I have strong doubts that radiochronology, which is the only solution for determining the absolute ages of surface units, or isotopic anomalies, which are critical for our understanding of planetary formation processes, can be measured *in situ* with accuracies similar to what can be done in the laboratory.

2.1.3. *Programmatic context and design strategy for future missions*

Figures 2.3 to 2.5 demonstrate that Europe is today a major player in space science, with quite a respectable program with respect to NASA, in spite of the ratio of 1–4 in terms of budgets: 5 of the 10 recent or planned observatory class missions (ISO, Newton, Integral, Herschel and Planck) are ESA missions, the 2 most important recent missions for solar-terrestrial physics (SOHO and CLUSTER) result from European initiatives (SOHO is however a 50–50 collaboration with NASA). Additionally, ESA has built up a very significant planetary exploration program, with leadership in cometary science (GIOTTO and Rosetta), missions to all four inner planetary bodies (Mars Express, SMART-1, Venus Express and BepiColombo) and major contribution to Cassini (Huygens probe, 30% of the science teams on the orbiter). This favourable situation can be credited to the significant increases made in space science budgets in Europe from 1980 to 1995, in particular for the ESA science program.

Maintaining such an ambitious program beyond 2010 is becoming more difficult due to several factors: the gradual decline of resources in ESA since 1995, the more rapid decrease of resources in major member states and the increasing sophistication of cutting edge missions as demonstrated in the previous section. The science program of ESA faces a severe crisis, with missions such as Eddington being jeopardized and cornerstone missions (GAIA and BepiColombo) being delayed to 2012. These difficulties can be linked to a transition period from the multi-national approach which has proven very successful in the 1980–2000 time frame and the painstaking process towards a more unified European strategy in science and technology. The recent cancellations of DIVA (astrometry) in Germany and of the Mars program led by France demonstrate that it will be more and more difficult to justify major undertakings in space science at the national level. National programs should therefore focus on smaller scale endeavours (mini- or micro-missions). Thus, despite its current problems, there is little doubt that ESA will provide the mainstay of science mission opportunities in Europe. There are strong political signs that budget increases could be sought once ESA assumes its inevitable transformation into the space agency of the European Union in the post 2012 time frame, which is relevant for new projects. The trend towards a more important role for the European Union in science and technology was indeed emphasized in the proposed constitution, with a shared responsibility for space activities.

The ESA selection system gives a critical role to science in the selection process. It would be naive to ignore the role of programmatic and industrial policy arguments, but they only come into the picture among projects which have "passed the cut" in terms of their science relevance at the level of the scientific advisory structure of ESA. The clearly expressed support of the relevant community is of course a prerequisite. However, the most important step is to convince the committee members belonging to other communities, of the scientific and technical timeliness of the proposed mission: given the broad range of represented disciplines, a mission does not need (and cannot get) unanimous support, but it needs nearly unanimous acceptability both within the subcommittees (Astronomy, Solar System, Fundamental Physics) and at the Space Science Advisory Committee level. The best strategy to achieve this is to be consistent with the logical development of each discipline as outlined in the previous section. A time-tried recipe for failure is to propose the next step when the previous mission in the same area has not yet provided the results on which to build the new science case. Of course, as was the case for the BepiColombo Cornerstone mission to Mercury, nothing can prevent NASA to try and steal some limelights from an already approved ESA mission by rushing through a less-ambitious project ahead of it, but ESA does not have the resources to overhaul its program at every twist and turn of NASA's programmatics. This becomes even more evident when considering that the next twist and turn may be just around the corner, as witnessed by the multiple stop and go episodes for Pluto Express, Europa orbiter or FAME (astrometry).

Complexity and its impact on cost as well as technological readiness is a major stumbling block for a space mission. The study team must therefore walk a narrow path between the Charybdis of underdesign, with science objectives which fail to secure the required level of acceptability by the other communities, and the Scylla of overdesign, which results in a very appealing mission at an unaffordable cost level or which has to wait for new technological developments.

The major technical elements to be defined when designing a new space science mission are summarized in Fig. 2.6. In an ideal world, the science team would set the science goals and define the model payload, then this would be translated by the technical team into constraints for spacecraft design, mission analysis and propulsion and the ground segment. In real life, the arrows go both ways: as an example, the first item to be questioned when a mass crisis arises is the payload mass, which has a direct impact on science. Therefore, the science team and the technical team must work closely together, being ready to redefine the science requirements (without compromising on the essentials) if they turn out to have an unacceptable impact on the budget.

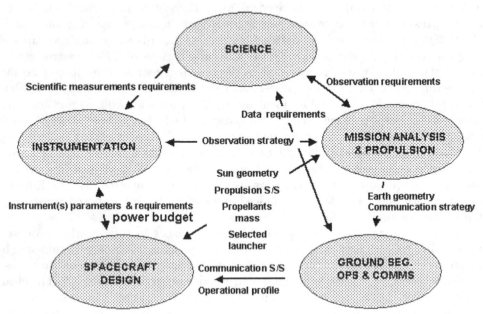

FIGURE 2.6. Major elements of mission design. The two dimensional arrows stress the required trade-offs between science requirements and their impact in terms of complexity and cost when designing a viable project.

Such a relationship is much easier to build if the members of the science team have some understanding of the technical issues which are involved in mission design, and if the members of the project team have some grasp of the critical science issues. The major goal of this chapter is to discuss this relationship between requirements and resources, in particular mass and cruise time, the most important parameters of mission design.

2.2. Basic concepts of mission analysis

2.2.1. *Launch requirements for science missions*

2.2.1.1. *Introduction*

Most future science missions (except for Earth observation) require launching on escape orbits or weakly bound orbits (astronomy). The most effective way to leave the Earth gravity field is to increase the velocity as close to the Earth as possible. One has to stay clear of the Earth atmosphere, hence the perigee distance R_{per} on the escape orbit is typically 1.05–1.1R_t (altitude: 317–630 km for a mean radius R_t of 6370 km for the Earth). All velocities on orbits close to the Earth scale with the circular velocity at altitude 0, $V_c = 7.904$ km/s. Taking R_t as the unit of distances, the velocity at perigee for bound orbits is:

$$V_{per} = V_c \left(\frac{2}{R_{per}} - \frac{1}{a} \right)^{0.5}$$

where a is the semi-major axis of the orbit.

The circular orbit is of particular interest, as it has to be reached by the launcher in a single thrust arc, otherwise the spacecraft will collide with the Earth somewhere on its orbit. Once on a low-Earth orbit, it is possible to consider strategies in which thrust is applied close to the Earth on successive perigees ("delayed thrust" strategies). Other bound orbits of interests are the Geostationary Transfer Orbit (semi-major axis of 3.8R_t), which is the target orbit for the Ariane family of launchers and the starting orbit for any mission launched as a companion to a commercial satellite, and orbits encountering the Moon at velocities of 1 km/s or larger ("Moon crossing orbits") with a semi-major axis of about 50R_t, from which one can escape the Earth gravity field using a lunar gravity assist (Section 2.3).

The upper limit of bound orbits correspond to $a = 0$ (parabolic orbit), with a velocity at pericentre $V_{esc} = 11.178/(R_{per}^{0.5})$ km/s. For higher velocities, the conservation of energy requires that the velocity far from the Earth (V_{inf}) is linked to the velocity at perigee:

$$V_{inf}^2 = V_{per}^2 - V_{esc}^2 = << C3 >>$$

The $C3$ parameter, expressed in (km/s)^2, is used in tables defining the mass which can be launched to a given orbit by a specific launcher. Negative $C3$ correspond to bound orbits:

$$V_{per} = \left(\frac{11.178^2}{R_{per}} + V_{inf}^2 \right)^{0.5}$$

For $R_{per} = 1.05$ (altitude: 318 km):

V_{circ}	7.71 km/s	escape at 3 km/s	11.31 km/s
GTO	10.14 km/s	escape at 5 km/s	12.00 km/s
Moon crossing orbit	10.85 km/s	escape at 7 km/s	12.96 km/s
Weakly bound orbits	10.91 km/s	escape at 9 km/s	14.14 km/s

It can already be noted that the pericentre velocities can become very large for large velocities at infinity. If we want to leave the Earth at 9 km/s, the velocity at pericentre is nearly twice that of a circular orbit. The next steps in the analysis are therefore to assess what velocity at infinity is needed to reach a given target, then to examine what is the best launch strategy for each class of velocity.

2.2.1.2. *The coplanar, circular approximation*

Solar system bodies have in general relatively low eccentricities and inclinations, with the exception of comets. For estimating the difficulty of reaching a given target, it is useful to first consider an approximation in which its orbit lies in the ecliptic (orbital plane of the Earth) and is considered circular with a radius equal to the semi-major axis. The reference units for this problem are the astronomical unit (149.5 million kilometers) for distances, the year for durations and the mean orbital velocity of the Earth ($V_{orb} = 29.8$ km/s) for velocities.

For such a configuration, the optimum transfer between the Earth and a planetary body with a radius R_2 in astronomical units is an ellipse tangent to the two circles (Hohman transfer). From Kepler's equation, it is easy to derive the relative velocities at departure and arrival:

$$\text{Departure velocity: } 29.8 \left(\left(2 - \frac{2}{1 + R_2} \right)^{0.5} - 1 \right)$$

$$\text{Arrival velocity: } 29.8 \left(\frac{1}{R_2^{0.5}} - \left(\frac{2}{R_2} - \frac{2}{1 + R_2} \right)^{0.5} \right)$$

From the previous section, one can then define the additional velocity which has to be provided at perigee compared to the escape velocity ("delta V at departure" in Table 2.2). Similarly, one can evaluate the reduction of velocity which is required at pericentre when reaching the target body so as to capture in a weakly bound orbit ("delta V at arrival" in Table 2.2). This delta V at arrival is particularly critical for the mass budget as it has to be provided by the spacecraft propulsion system or by aerocapture if there is an atmosphere. Adding these two values provides the total delta V for a marginal capture, a measure of the difficulty of getting into orbit around a given target.

The Hohman transfer can only be performed at regular intervals, when the two bodies are in the proper relationship in terms of timing (phasing). The duration of a Hohman transfer for the coplanar, circular approximation is half a period of the transfer orbit, $0.5((1 + R_2)/2)^{1.5}$ from the 3rd law of Kepler. As an example, for a transfer to Mars, the favourable period (planetary window) occurs before each opposition. The repeat rate of this favourable configuration is called the synodic period P_{syn} of the planetary body, which is linked to its period P_2:

$$P_{syn} = \frac{P_2}{(1 - P_2)} \text{ for inner bodies } (P_2 < 1 \text{ year})$$

$$P_{syn} = \frac{P_2}{(P_2 - 1)} \text{ for outer bodies } (P_2 > 1 \text{ year})$$

It is interesting to note that the synodic period is largest for bodies with periods closest to that of the Earth, Venus (1.6 years) and Mars (2.1 years).

For Jupiter family comets (aphelion at 5.5 AU), which are highly elliptical, the optimum transfer is tangent to the orbit of the comet at aphelion and to the orbit of the Earth. For a "good" comet, with its perihelion close to 1 AU (Wirtanen: $a = 3.15$ AU, $p = 1.04$ AU),

TABLE 2.2. Delta V requirements in the coplanar, circular approximation (coplanar only for P/Wirtanen).

Target	Repeat rate (years)	Departure (km/s)	Delta V over escape (departure)	Arrival (km/s)	Delta V over escape (arrival)	Total delta V
Moon	NA	NA	−0.10	0.80	0.14	0.04
L2 point	NA	NA	0.00	0.15	0.15	0.15
Venus	1.60	−2.50	0.28	2.71	0.34	0.62
Mars	2.14	2.94	0.39	−2.65	0.68	1.07
Wirtanen	5.45	8.81	3.11	−0.15	0.15	3.26
Jupiter	1.09	8.80	3.10	−5.64	0.45	3.55
Saturn	1.04	9.60	3.63	−5.44	0.61	4.24
Vesta	1.38	5.51	1.31	−4.43	4.43	5.74
Mercury	0.32	−7.53	2.35	9.61	6.35	8.70

one has:

$$\text{Departure velocity: } 8.81 \text{ km/s} = 29.8 \left(\left(2 - \frac{1}{3.15} \right)^{0.5} - 1 \right)$$

$$\text{Arrival velocity: } 0.15 \text{ km/s}.$$

The comet moves very slowly at aphelion, hence a transfer relatively close to the Hohman transfer can be obtained with a launch 3 years before each aphelion, therefore with a repeat rate equal to the period of the target comet.

From Table 2.2, the Moon, near-Earth space (L2 point), Venus and Mars appear relatively easy to reach (although Mars windows are few and far between). In terms of delta V requirements, it is interesting to note that Venus and Mars are much closer to the Moon than to Jupiter or even more so to Mercury. Short-period comets with perihelia close to 1 AU are relatively cheap in this approximation, but their large inclination will much change the picture. Jupiter and Saturn are difficult to reach ($\sim$9 km/s at departure), but their very large gravity field makes it relatively easy to stop. Mercury is easier to reach than outer planets, but it would represent the most difficult challenge for a direct rendezvous mission, with a very large on-board delta V required (6.35 km/s).

2.2.1.3. *Inclined planetary orbits and the quality of windows*

The coplanar, circular approximation provides a good first estimate of the requirements for each target. Apart from comets, for which non-circular orbits have already been considered, the eccentricity of orbits is small, except for Mars (0.09) and even more so Mercury (0.205), for which departure and arrival velocities differ markedly from window to window. As an example, the coplanar velocity at infinity required to reach Mars is 3.5 km/s when it is reached at aphelion (1.665 AU) and only 2.3 km/s when it is reached at perihelion (1.38 AU). This corresponds to a range of only 300 m/s when translated into delta V at departure. A much more important problem results from the small but non-zero inclination of the orbit of planetary bodies with respect to the orbit plane of the Earth (the ecliptic).

The target body is in the ecliptic only at two points 180° apart on its orbit (the nodes). The transfer orbit is in the plane containing the Sun, the Earth at departure and the target body at arrival. This means that unless one is very lucky, and the target body is at one of the nodes, the Hohmann transfer is forbidden as it corresponds to an angle of 180° between the Earth and the target: the orbit plane is then perpendicular to the ecliptic, with huge relative velocity to the Earth (>42 km/s !). The out-of-ecliptic velocity component is useless in terms of energy as it is only required to rotate the orbit plane. Therefore, one has to reduce it by moving away from the 180° angle between departure and arrival. The arrival velocity is no longer tangent to the orbit, with a radial velocity component in the ecliptic at arrival which increases as one moves further away from 180°. There is a trade-off which leads to angles around 150° (short transfers, or "type I") and around 210° (long transfers, or "type II"). Transfers with an additional full orbit are also possible (type III = type I + 360°, type IV = type II + 360°). They can provide additional possible departure dates, but they take longer, hence can only be considered if the corresponding geometry is particularly favourable (target close to the nodes).

Each window is quite specific with respect to the inclination. In particular, in a given year, types I and II windows are not equivalent. So as to understand this, it is useful to consider a graphical representation of the orbit planes: to each plane is associated a representative point with polar coordinates $(r, \theta) = (i, \Omega)$ where i is the inclination (in degree) and Ω is the ecliptic longitude of the ascending node, where the target crosses the ecliptic towards the North. The representative point of the Earth is obviously at the origin. The origin of ecliptic longitudes, γ, corresponds to the Sun–Earth direction at the fall equinox. The representative points of major targets in the solar system are presented in Fig. 2.7. If one draws a line from the Earth to a given target, this gives the ecliptic longitude of the two arrival positions for which the out of ecliptic velocity would be 0 (target at the node).

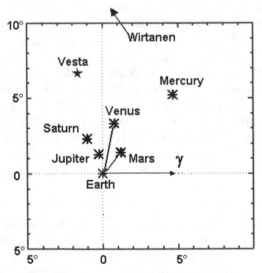

FIGURE 2.7. Representative points of the orbit planes of the four inner planets, Jupiter and Saturn, a large asteroid (Vesta) and the initial target comet of Rosetta (Wirtanen). Targets with high inclinations (Mercury, Vesta) require large launch velocities if the arrival is not close to one of the nodes (45° and 225° for Mercury).

All inclinations in Fig. 2.7 are lower than 10°. This means that $\sin(i)$ is close to i (in radians) and $\cos(i)$ close to 1. In this approximation, the representative point of the transfer plane is at the intersection of the two lines drawn from the Earth in the direction of its ecliptic longitude at departure and from the target in the direction of its ecliptic longitude at arrival. The distance between the representative point of the body

(Earth or target) and that of the transfer plane gives an indication of the out-of-plane component for the encounter. Two examples are given in Fig. 2.8 for transfers to Venus. For a good window, the transfer plane is very close to the Earth. The out-of-ecliptic component of the launch velocity is very small, which means that the escape velocity is close to the evaluation in the coplanar approximation. If the representative point also lies between the Earth and Venus, this reduces the out-of-plane component of the arrival velocity.

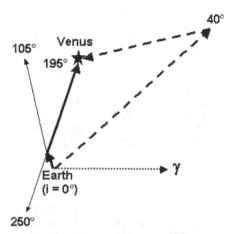

FIGURE 2.8. Examples of a good window to Venus (solid arrows: 01/2009, type I, launch at L_s 105°, arrival at L_s 250°) and a bad window to Venus (dashed arrows: 10/2005, type I, launch at L_s 40°, arrival at L_s 195°). The escape velocity at launch is much smaller for a good window (2.8 km/s) than for a bad window (4.6 km/s), as the out of ecliptic component is very small.

An important parameter for launcher performances is the declination, which is the angle of the escape velocity to the equatorial plane of the Earth: the rotational velocity of the Earth is far from negligible (m/s at the equator) and it adds to the velocity if one can launch due East. This angle is a non-trivial combination of the season, the angle of the escape velocity with respect to the direction of motion of the Earth, and the angle of the escape velocity to the ecliptic. As an example, for a very good window such as the Venus window in January 2009, the escape velocity is very close to the ecliptic and it points nearly 180° away from the direction of motion of the Earth. As we are close to the winter solstice, this direction is not far from the equator, and the resulting declination is only 3.6° North (a very favourable declination for a launch from Kourou). Conversely, for a good window with a launch in March or September, the escape velocity will also be close to the ecliptic, but the declination will now be close to 23.5° North or South.

2.2.1.4. *Actual launch requirements for science missions*

One has to consider the actual launch windows for defining launch requirements. As can be seen from Table 2.3, the total delta V's as derived from the coplanar circular approximation are close to that of the best windows for Mercury, Venus, Mars, Jupiter and Saturn. For small bodies, which have large inclinations, the actual windows are significantly worse, with 0.8–1.8 km/s larger delta V requirements.

The range of total delta V requirements is very significant: for Mars and Venus, they can differ by upto 800 m/s from window to window. In the next section on the rocket equation, we will see that the impact on the final mass is very significant, in the range of 30% of more.

TABLE 2.3. Characteristics of actual windows to solar system bodies.

Launch	V departure km/s	Launch declination	Delta $V/$ V escape	Arrival	V arrival km/s	Delta $V/$ V escape	Delta V total
Venus							
Coplanar approximation	2.500		0.280		2.710	0.340	0.620
2009/01/02 (I)	2.680	3.6°N	0.324	2009/05/10	3.855	0.690	1.014
2008/12/07 (II)	3.370	30.2°N	0.508	2009/05/19	2.700	0.343	0.851
2010/08/18 (I)	3.035	5.7°N	0.413	2010/12/08	4.710	1.012	1.425
2010/07/07 (II)	3.865	51.1°S	0.663	2011/01/06	3.950	0.722	1.385
2012/02/21 (II)	3.615	54.8°N	0.583	2012/08/23	4.665	0.994	1.577
2013/10/30 (II)	2.795	26.6°S	0.352	2014/04/06	4.730	1.020	1.372
2005/11/05 (II)	2.795	21.2°S	0.352	2006/04/11	4.650	0.987	1.339
2015/06/03 (I)	3.440	13.8°S	0.528	2015/10/06	3.015	0.426	0.954
2015/05/21 (II)	2.480	9.3°N	0.278	2015/10/27	3.780	0.662	0.940
Mars							
Coplanar approximation	2.940		0.390		2.650	0.680	1.070
2009/10/14 (II)	3.200	20.6°N	0.459	2010/09/03	2.465	0.594	1.053
2011/11/11 (II)	3.000	32.3°N	0.404	2012/09/17	2.720	0.714	1.118
2013/12/06 (II)	3.075	24.5°N	0.424	2014/09/27	3.180	0.955	1.379
2016/03/07 (I)	3.355	46.1°S	0.503	2016/09/24	3.855	1.353	1.856
2018/05/13 (I)	2.800	35.2°S	0.353	2018/12/05	2.945	0.829	1.182
Sun (Solar probe, 4 solar radii)							
Once per day	24.07	<23.5°	15.5	Every day	NA	NA	15.500
Jupiter							
Coplanar approximation	8.80		3.100		5.640	0.450	3.550
2016/12/25 (II)	8.715	1.6°S	3.050	2020/02/14	5.770	0.472	3.522
Saturn							
Coplanar approximation	9.610		3.630		5.440	0.610	4.240
2013/01/10 (II)	10.275	8.7°S	4.037	2020/03/09	5.223	0.723	4.760
Wirtanen							
Coplanar approximation	8.810		3.110		0.150	0.150	3.260
2007/12/05 (II)	10.198	36°N	4.020	2014/12/26	0.830	0.830	4.850
Vesta							
Coplanar approximation	5.510		1.312		4.430	4.430	5.740
2013/12/28 (II)	5.192	3°S	1.168	2014/12/26	5.428	5.428	6.596
Mercury							
Coplanar approximation	7.530		2.350		9.610	6.350	8.700
2010/05/12 (4)	9.496	0.3°S	3.550	2011/03/09	7.980	4.875	8.420

We can now associate ranges of escape velocities to each type of science missions, hence define the required velocities close to the Earth which have to be provided by launchers:

Bound orbit	10.85 km/s for the Moon, the outer magnetosphere, ...
Escape at 0 km/s	10.9 km/s for near-Earth space (L1, L2, drift orbits)
Escape at 3 km/s	11.3 km/s for Venus, Mars
Escape at 6 km/s	12.8 km/s for Vesta
Escape at 9+ km/s	>14.0 km/s for direct launches to Jupiter, Saturn, Mercury, comets and Kuiper Belt Objects

2.2.2. *The rocket equation*

When defining a science mission, the science team should always remain aware of the potential for runaway effects which is built in the rocket equation, in particular for planetary and heliospheric missions. Spacecraft being propelled by the reaction effect, for a small increment of speed we have:

$$dV = \frac{dm V_{ej}}{\text{mass}} \qquad (2.1)$$

where V_{ej} is the ejection velocity and dm the fuel consumption.

Integrating (2.1) over all velocity increments after separation from the launcher leads to the rocket equation:

$$\frac{\text{Mass\_i}}{\text{Mass\_f}} = \exp\left(\frac{dV_{tot}}{V_{ej}}\right) \qquad (2.2)$$

where Mass_i is the initial mass (or "wet mass"), Mass_f the final mass (or "dry mass") and dV_{tot} the total velocity change.

If restricted for the time being to classical propulsion, the best ejection velocities are in the range of 3.1 km/s (400 N engine most commonly used for European spacecraft). This means that the mass after launch (Mass_i) needed is 2.7 times larger than the dry mass if a total velocity change of 3.1 km/s is required. Unfortunately, this is not the full story: in such an hypothesis, more than 60% of the initial mass is constituted by propellant. Tanks are needed for this propellant, additional pipes and valves, as well as a stronger structure to support the propellant mass at launch. The impact on dry mass of each additional kilogram of fuel is called the "tankage factor". It ranges between 10% and 14%. If we take the most conservative value, and if we call Mass_p the mass of propellant and Mass_u the "useful mass" of the spacecraft (with the engine, but without propellant-related items such as tanks), we have:

$$\text{Mass\_f} = \text{Mass\_u} + 0.14\,\text{Mass\_p}$$

$$\text{Mass\_i} = \text{Mass\_u} + 1.14\,\text{Mass\_p}$$

$$\text{Mass\_p} = \text{Mass\_u}\,\frac{\exp(dV_{tot}/V_{ej}) - 1}{1.14 - 0.14\exp(dV_{tot}/V_{ej})}$$

$$\text{Mass\_u} = \text{Mass\_i}\frac{1.14 - 0.14\exp(dV_{tot}/V_{ej})}{\exp(dV_{tot}/V_{ej})} \qquad (2.3)$$

From (2.3), it appears clearly that the propulsion system is just able to push itself and its propellant to a velocity of 2.1 times the ejection velocity (6.5 km/s for an ejection velocity of 3.1 km/s) as the useful mass drops to 0. When considering that the velocity needed to leave the Earth is more than 11.2 km/s, it is no surprise that launchers rely on a multi-stage strategy, in which the empty tanks (and a complete propulsion system) are jettisoned before igniting the next stage. Fortunately, as we shall see, modern launchers have higher ejection velocities and lower tankage factors than the bipropellant systems used on spacecraft. For a spacecraft, the practical limit for the delta V is around 1.4 times the ejection velocity (4.4 km/s) as the useful mass is already down to one-seventh of the initial mass. This capability allows us to discriminate between easy targets such as the Moon, Mars or Venus (total delta V requirements $<<$4.4 km/s), marginal targets such as Jupiter or Saturn (total delta V requirements $\sim$4.4 km/s) and targets such as comets, asteroids or Mercury (delta V requirements $>$4.4 km/s) requiring either a propulsion system with a higher ejection velocity or a strategy using gravity assists.

2.2.3. *Direct launch capabilities*

2.2.3.1. *Specific characteristics of the Ariane 5 launcher for science missions*

The Ariane 5 launcher characteristics provide a vivid illustration of the runaway mass factors built in the rocket equation. This launcher is designed to inject large masses (8 tons) into the geostationary orbit (10.14 km/s at perigee). It is launched from Kourou, close to the equator (5°N), so that the rotation of the Earth provides 470 m/s for free, but during the early launch phases most of the thrust is wasted in fighting the Earth gravity and atmospheric drag. The delta V requirement to get into circular orbit is therefore over 8.5 km/s, with 2.3 km/s remaining to be provided to reach the GTO orbit.

The composite on the launch pad has a very large mass: 675 tons. Two large boosters represent nearly two-thirds of this mass. They constitute a de facto first stage with a thrust of 1000 tons (one has to be able to lift the composite) and an ejection velocity of 2.8 km/s, which is jettisoned after 135 s. The second stage, the Vulcain, has a wet mass of 170 tons, a dry mass of 12.2 tons (27 tons including the third stage and the payload), a thrust of 116 tons and a very effective ejection velocity of 4.25 km/s. It burns from the start (adding 10% to the thrust of the two boosters so as to provide much needed useful acceleration once the 675 tons of gravity drag are subtracted) until 600 s after launch. The third stage (Aestus) is a bipropellant stage with a wet mass of 11.2 tons, a dry mass of 1500 kg to which the mass of the payload must be added. The exhaust velocity is 3.18 km/s and the thrust is of 2.96 tons (29 kN), so that the thrust arc lasts 1000 s. It can be stopped and restarted, which much improves performances for escape, but this has not yet been tested in flight.

As can be seen in Fig. 2.9, even with staging and a very effective cryogenic stage, the compounded factors from rocket equation lead to a ratio of 90 between the payload set into GTO (8 tons) and the mass at launch. The situation becomes rapidly worse for escape orbits, in particular if the third stage thrust is continued until the desired escape energy is reached. An example for an escape at 3 km/s is given in Fig. 2.10.

The effectiveness of the third stage decreases as one goes away from the Earth: for each velocity increment dV, the gain in energy is $V \cdot dV$, hence delta V is optimum close to perigee, where V is largest. At $1.5R_t$, the escape velocity is 9.13 km/s instead of 10.91 km/s at $1.05R_t$, hence a 20% penalty in terms of delta V efficiency. This gravity loss issue will be even more important when using the much lower thrust of the spacecraft itself (Section 2.2.3.3).

Kourou is at a latitude of 5°. The nearly 180° angle between Kourou and the departure asymptote with a direct injection strategy means that reaching large declinations (angle between the departure asymptote and the equator) is very costly. The optimum strategy consists of launching due East. From Kourou, one cannot launch towards the South (the drop zones of boosters would be over Brazil), but even if one considers North declinations,

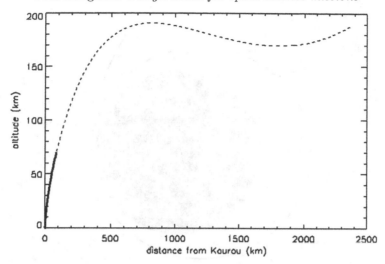

FIGURE 2.9. Schematics of the launch strategy with Ariane 5. The initial phase with the boosters is the thick line immediately after launch. The launcher altitude decreases before starting to increase again once orbital velocity has been achieved. It is optimum to launch due East to get the most benefit from the 470 m/s rotational velocity of the Earth near the equator.

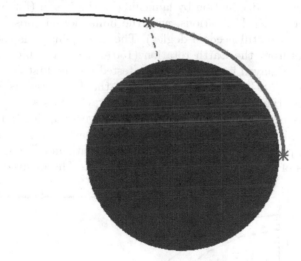

FIGURE 2.10. Direct injection by Ariane 5 to an escape trajectory with a Vinf of 3 km/s (transfer to Mars). It can be seen that part of the thrust of the third stage is delivered at large distances from the Earth (>3000 km). Furthermore, the asymptote of departure is nearly 180° away from Kourou.

launching to the North does not help as the escape asymptote is only slightly changed while performances decrease. One has to use the launcher itself in the late phases to rotate the orbit plane ("dogleg"), which is very ineffective for large rotation angles, so that declinations above 20° can only be reached with much reduced masses.

A delayed ignition strategy, in which the third stage is only ignited after a full orbit, and even better a restart strategy (in which it is ignited twice) provides much better performances for escape trajectories. The third stage is ignited (or reignited) on the way down to the second pericentre. As can be seen from Fig. 2.11, the gravity loss in now marginal, as the maximum altitude when thrusting is now less than 1200 km. Furthermore, as this strategy opens the angle between Kourou and the escape asymptote, one

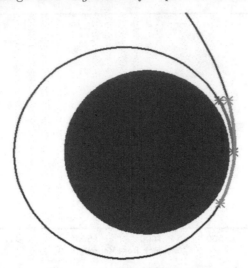

FIGURE 2.11. Injection by Ariane 5 with delayed ignition to an escape trajectory. All the thrust is delivered very close to the Earth (dark grey and light grey thrust arcs). Furthermore, the asymptote of departure is now only 130° away from Kourou.

can effectively change the declination by launching to the North (for North declination) or to the South (for South declinations, with the limitations imposed by booster drop zones; beyond them, one still needs a dogleg). The mass penalty is then simply due to the reduced benefits from the Earth rotation (150 m/s for a launch 45° to the North instead of due East). Delayed ignition has been tested for the first time with the launch of Rosetta in February 2004. In this particular case, the declination is low, and the main purpose is to improve the mass budget by reducing gravity losses.

The performances of the Ariane 5 launcher (ESV version, 2003) for interplanetary missions as presented in Fig. 2.12 demonstrate the drastic impact of even moderate declinations on the performances for direct injection, with more than a factor of 2 for departure velocities of 2–4 km/s between 5°S and 20°N. The delayed ignition strategy

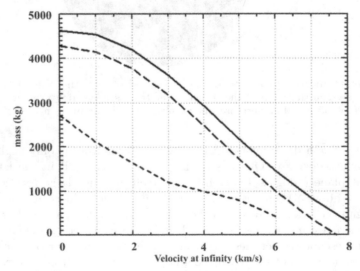

FIGURE 2.12. Performances of Ariane 5 ESV; solid line: delayed ignition, optimum declination (5°S); long dashes: direct injection, 5°S; short dashes: direct injection, 20°N.

provides significantly improved performances for the optimum declination, in particular for large departure velocities. The declination penalties are much lower with delayed ignition, in particular for North declinations (only 8% for a departure velocity of 3 km/s with a declination of 20°N), which makes delayed ignition mandatory for most planetary windows.

An immediate conclusion of Fig. 2.12 is that Ariane 5 can deliver very large masses to the Moon, near-Earth space (departure velocity close to 0), Mars or Venus (departure velocities from 2.7 to 4 km/s). However, a direct launch to outer planets, comets, Mercury or the close vicinity of the Sun cannot be achieved. This conclusion applies to any launcher using a bipropellant upper stage, as it is a direct consequence of the rocket equation.

2.2.3.2. *The Soyuz launcher: launching from high latitudes*

The Soyuz launcher was developed in the 1960s from the Vostok launcher (the one used on Gagarin's launch). It uses the Baikonur launch pad, at a latitude of 47.5°N. With direct injection, the asymptote would be nearly 180° away from Baikonur, hence only high south declination could be reached. As a consequence, a restartable third stage had to be developed for the early soviet missions to Mars and Venus, two planets which require a very wide range of declinations (see Table 2.3). The present version of the Soyuz upper stage (Fregat) has a dry mass of 1100 kg (+payload), a wet mass of 6550 kg, a thrust of 20,000 N and an ejection velocity of 3.28 km/s, which is high for bipropellant systems. This launcher has already been used (successfully) for two major ESA missions (Cluster 2 and Mars Express). It will also be used for Venus Express in 2006, and it is the present baseline for BepiColombo and Solar Orbiter.

The strategy for high-latitude launches (Fig. 2.13) is significantly different from that required for launches from Kourou: one launches due East so as to fully benefit from the reduced rotational velocity of the Earth at the latitude of Baikonur (310 m/s), using the Soyuz, then a small burn from the Fregat is implemented to reach a nearly circular orbit. By design, this orbit has a declination of 47.5° as its maximum latitude which is that of Baikonur. By selecting the start of the second burn of the Fregat, one can

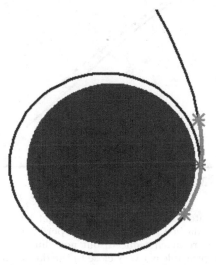

FIGURE 2.13. Launch strategy with a Soyuz–Fregat upper stage. The first orbit is nearly circular, which make it possible to restart the Fregat stage anywhere on this orbit without penalties in terms of gravity loss. The departure asymptote can then be rotated freely within the orbit plane.

rotate the departing asymptote to any direction of this orbit plane, hence from 47.5°S to 47.5°N. Within this range, there is therefore no penalty for escape at large declinations. With this strategy (Fig. 2.13), thrust is always applied close to the Earth, which reduces gravity losses. As demonstrated by Fig. 2.14, the performances are typically one-third of that of Ariane 5 ESV with delayed ignition at its optimum declination, but the ratio is actually smaller for most targets due to the flat performance with respect to declination from 47.5°S to 47.5°N. As with Ariane 5, departure velocities of 9 km/s or more which are required to reach outer planets, comets or Mercury cannot be achieved.

There are perspectives for launching Soyuz from Kourou starting in 2008 2009. If this possibility materializes, the benefits will be most substantial for low declination: the additional 150 m/s provided by the Earth rotation would improve by about 5% (rocket equation) the mass available for the third stage and its payload on a circular orbit. By using the same strategy (nearly circular first orbit, then timing the re-ignition so as to reach the required asymptote), the declination is directly defined by the angle between the launch direction and the East. The performance for a declination of 47.5° is exactly the same as that from Baikonur (same reduction in terms of rotational velocity), and can actually be worse from Kourou for large South declinations if the same constraints are set on booster drop zones. With the same Fregat version as for Mars Express, the launch capability from Kourou for a low declination escape at 3 km/s should reach 1400 kg (Mars Express launch: 1200 kg from Baikonur). The benefits would be similar in proportion for the upgrades of the Soyuz–Fregat which are presently considered.

The mass of the dry Fregat stage (1100 kg) exceeds that of the departing spacecraft for departure velocities larger than 3.5 km/s (Fig. 2.14). This large overhead to which the

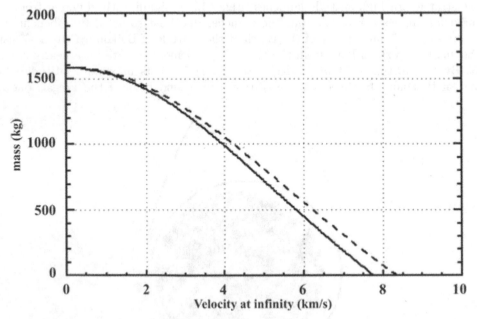

FIGURE 2.14. Solid line indicates the available mass budget with Soyuz–Fregat as a function of the velocity at infinity. The dashed line indicates the performances which would be achieved without gravity losses, which are much reduced, but not eliminated by the implementation of two thrust arcs. One could in principle get very close to the dashed line by splitting the thrust in three segments, but this would require restarting the Fregat after first 90 min, then for example, 20 days (the third thrust would be after a long-bound orbit). The additional 50 kg at 4 km/s is not worth the added complexity and risk.

rocket equation applies as well as to the spacecraft mass severely limits the maximum departure velocity. In the next section, we will consider launch strategies which get rid of the Fregat so as to improve direct launch capabilities.

2.2.3.3. *Launch strategies using a bound orbit*

Most spacecraft dedicated to solar system exploration require an on-board propulsion system so as to perform an insertion manoeuvre into a planetary orbit or (as will be discussed in Section 2.2.4) to perform the deep space manoeuvres required for a delta-V gravity assist strategy. An idea worth considering is therefore to use this on-board engine as a de facto fourth stage providing part of the escape energy. Typical thrust levels for on-board engines are much smaller than that provided by the three stages. As an example, the Mars Express main engine provides 400 N, to be compared with 20,000 N for the Fregat. The ejection velocity is also a bit smaller (3.1 km/s instead of 3.28 km/s). As limiting gravity loss is desirable, it is optimum to use the Fregat to reach a long period-bound orbit (e.g. a Moon crossing orbit with a semi-major axis of 320,000 km). This orbit adds only 20 days to the mission and it is only 60 m/s slower than escape in terms of perigee velocity. At the next perigee, it is therefore easy to provide escape energies by thrusting with the main engine. Such a thrust arc close to a planet is called a "powered swing-by", in this case of the Earth. The launch capability of the Mars Express version of the Soyuz launcher is about 1610 kg for such an orbit. This must now include the bipropellant mass needed for reaching the required departure velocity. Reaching 3 km/s would require an additional delta V of 510 m/s if it could be provided instantaneously at perigee, hence a final mass of 1365 kg could be available, a clear improvement with respect to the 1200 kg launch capability of the Fregat at 3 km/s.

As can be seen from Fig. 2.15, the mean mass during thrust is close to 1500 kg, as a result close to 1900 s of thrust are needed, and the mean distance of the spacecraft to the

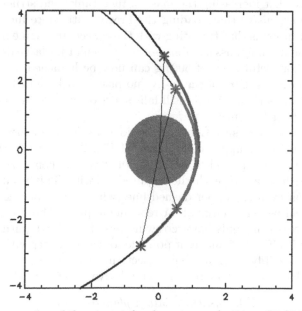

FIGURE 2.15. Two examples of thrust arcs for departure velocities of 3 (light grey) and 5 km/s (dark grey). The maximum distance from the Earth reaches $1.8R_t$ for 3 km/s, nearly $3R_t$ for 5 km/s. Gravity losses are therefore very significant, in particular for departure velocities of 5 km/s or higher, for which they decrease the final mass by more than 15%.

Earth is significantly larger than the pericentre distance: gravity losses are very significant (Fig. 2.15) in particular for large departure velocities. A fair comparison with the direct launch with Fregat requires taking into account the tankage factor (Section 2.2.2) as this reduces the useful mass. The results are given in Fig. 2.16.

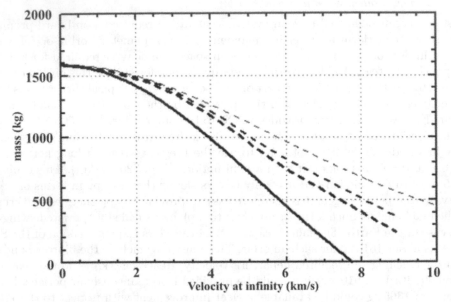

FIGURE 2.16. Launch capability with Soyuz–Fregat using a bound orbit strategy. Upper dashed line: available dry mass without taking into consideration gravity losses; middle dashed line: dry mass (with gravity losses); lower dashed line: useful mass (taking into account the tankage factor).

It should be noted that increasing the mass of fuel, tanks and structure on the spacecraft is much more expensive than loading the Fregat with more fuel. Even with these limitations, the conclusion is that thrusting with the spacecraft engine from a large bound orbit does provide additional mass capabilities, in particular for large departure velocities as a spacecraft with a useful mass of 500 kg can now be launched at more than 7 km/s instead of less than 6 km/s. Even for a fly-by (no post launch delta-V requirements and their corresponding mass impact), this still falls short of the requirements of missions to outer planets, comets or Mercury.

For further improving the situation, one could consider reducing the gravity losses (hence getting closer to the upper dashed line of Fig. 2.16) by launching at a smaller velocity on an orbit with a period of exactly 1 or 2 years, then performing a powered swing-by of the Earth so as to reach the required velocity. This would indeed improve the mass budget, but even the upper dashed line (which does not take into account the tankage factor) would be very marginal for reaching Jupiter. The conclusion is therefore that there is no strategy using only powered swing-bys of the Earth (either by a third stage or by the spacecraft itself) which makes it possible to reach Jupiter with a large spacecraft mass. Fortunately, as will be seen in the next section, there are much more effective ways of using post-launch delta V for increasing the relative velocity to the Earth.

2.2.4. *Insertion into a planetary orbit*

The problem of capturing a spacecraft with a given approach velocity into a large bound orbit around a planet has close similarities with the transition from a large bound Earth orbit to a hyperbolic orbit (providing a significant departure velocity) using the main engine of the spacecraft, which was discussed in Section 2.2.3.3. The only difference is that the mass is smaller on the bound orbit, while it was larger for the launch strategy mentioned in Section 2.2.3.3. Similarly to the Earth, relatively low-gravity losses are obtained when the thrust arc does not extend farther than two planetary radii from the center. This manoeuvre is critical for most planetary missions. An example is given in Fig. 2.17 with the insertion of Mars Express.

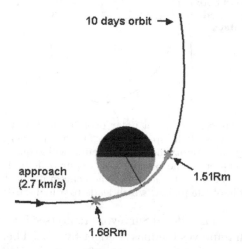

FIGURE 2.17. Insertion of Mars Express (25/12/2004); the mass at approach was 1130 kg (the Beagle lander had already been jettisoned), 2200 s of thrust with the 400 N engine were required to brake down to an orbit with a period of 10 days. The insertion delta V was 800 m/s, a gravity loss of only 40 m/s as the whole thrust arc was relatively close to Mars. The 10-day orbit had its pericentre 31° away from the subsolar point and it was nearly equatorial. After half an orbit, a plane rotation was performed at apocentre to obtain a near-polar orbit (I = 87°), at a cost of 120 m/s.

Again, as in the case of an optimum launch strategy, it pays to split the reduction of the orbital period into small thrust arcs. The only drawback is a delay in reaching the operational orbit (30 days for Mars Express). For this mission, the sequence of orbital period was 10 days, then 2 sol (1 sol = 24 h 37 min, the Mars day), then 1 sol, then a series of periods down to 4/13 of a sol, the operational orbit. The cost in delta V of apocentre reduction was 460 m/s, with only a 1 m/s gravity loss. Depending on the final orbit, a Mars orbiter mission requires from 1.4 km/s (Mars Express, low arrival velocity and 7.6 h elliptical orbit) up to 2.5 km/s (circular orbit, such as Mars Odyssey). Such a large on-board delta V reduces the useful mass by a factor of 1.7–3 unless propellant-free orbit-reduction approaches are considered.

The insertion of Messenger (Fig. 2.18), a NASA Discovery mission to Mercury, gives an example of the limitations resulting from large approach velocities. When the mass budget is marginal, as in the case of Messenger, one has to insert directly into a polar orbit: a plane turn on the first bound orbit, similar to that implemented for Mars Express, would have cost more than 150 m/s. The initial pericentre is then close to the North pole (or the South pole). The high-pericentre latitude (70° for Messenger) restricts the high-resolution coverage capabilities on an elliptical orbit. A similar orbit is achieved by Venus Express (launched in 2005), which overcomes the coverage problem by implementing several experiments with different spatial resolutions in the same spectral range.

When there is an atmosphere, a very interesting strategy in terms of mass budget consists of progressively reducing the orbital velocity by dipping into the upper atmosphere, after capturing into a long-period-bound orbit with the main engine. This strategy, called

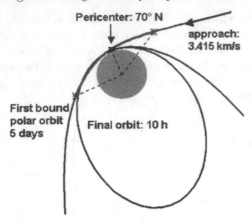

FIGURE 2.18. Insertion of Messenger into a polar orbit around Mercury, scheduled for 2009. Even with the very low dry mass of the spacecraft (<500 kg), the very large approach velocity (3.415 km/s) and relatively low-gravity field of Mercury will require a thrust arc extending beyond 5000 km from the center of the planet. The required delta V is 1.47 km/s, including 170 m/s of gravity loss. For the orbit to be polar, the spacecraft must pass directly above the North pole or the South pole of the planet, which leads to a high-latitude pericenter.

"aerobraking", was used by Mars Global Surveyor (1996) (see Fig. 2.19). It saves 1.4 km/s for a circular orbit, which improves the mass budget by 70%. There are some drawbacks: a symmetrical design is required for the spacecraft, in particular its solar arrays, there is an element of risk if the solar panels bend (with a narrow escape for MGS), optical experiments may require covers so as to avoid contamination, and, most importantly, science operations are delayed by up to 1 year, as was the case for MGS.

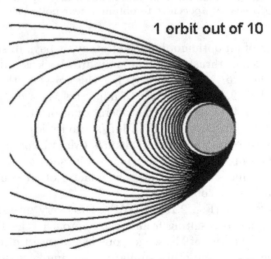

FIGURE 2.19. Aerobraking strategy for reaching a low circular orbit around a planet with an atmosphere (Mars and Venus). The stresses on solar panels limit the equivalent thrust to a few Newtons. A period of 6 months – 1 year can be needed to reach the operational orbit.

Aerocapture, which consists of using atmospheric drag for the capture manoeuvre itself, seems attractive as it can save at least 800 m/s (this corresponds to the insertion manoeuvre for Mars Express, which benefited from a nearly optimal window). Aerocapture represents a much more challenging problem than aerobraking: by definition, it has to be performed in a single pass, while aerobraking can recover from underperformance on one pass at the next orbit. When considering the required energy reduction and the time during which the spacecraft can stay in relatively dense regions of the atmosphere, the power to be dissipated corresponds to several MW/m^2. Furthermore, a very precise control of the trajectory within the atmosphere is required so as to secure capture without risking an impact on the planet. It is actually easier to design an entry module for a lander. Therefore, while aerocapture has been considered for several Mars missions, it has not yet been implemented. Even if the direct entry of surface modules has in principle a wider margin of error, the failures of several Mars missions shows that this manoeuvre remains risky.

2.2.5. *Conclusions*

The constraints imposed by celestial mechanics and the rapid decrease of the useful mass which results from the rocket equation have potentially devastating effects on mass budgets depending on launch velocity requirements. Getting out of Earth orbit or in planetary orbit are nearly symmetrical problems for chemical propulsion. From the preceding chapters, it clearly appears that science missions can be classified into three categories in terms of mission analysis:

(1) *Missions to near-Earth space including the Moon*: launch velocity requirements are close to that of a marginal escape. Mass budgets are large, and the major mission analysis problems deal with reaching specific locations (e.g. the lagrangian points) as cheaply as possible and maintaining reliable operational orbits for lengthy intervals. These problems will be briefly discussed in Section 2.3.

(2) *Missions to Venus and Mars*: Soyuz–Fregat and even more so Ariane 5 can deliver very significant masses to these targets, which can be improved by using elaborate launch strategies. The main issues deal with the specifics of each launch window (every 2.1 years for Mars and every 1.6 years for Venus), in particular the launch declination (for Ariane 5), and the on-board delta V requirements when reaching the target planet. For orbital missions the delta V requirements can be reduced by using atmospheric drag at a cost in terms of mission duration and complexity. *In situ* missions rely on direct entry techniques, which require a thermal shield but no on-board delta V. These specific problems will be presented in Section 2.6.

(3) *Missions to giant planets, small bodies, Mercury and the vicinity of the Sun*: Even the most elaborate launch strategies fail to directly deliver significant useful masses to these targets, even before considering the large on-board delta V requirements (asteroids, Mercury). New mission strategies relying on gravity assists had to be developed to reach these targets, which will be discussed in Section 2.4. Large on-board delta V requirements cannot be met by chemical propulsion. Propulsion methods with high ejection velocities will be discussed in Section 2.5.

2.3. Missions to near-Earth space

As shown in Section 2.2, for a given launcher, the mass budgets available for missions to the Moon, the outer magnetosphere of the Earth and regions close to 1 AU are very similar. There are two specific regions of interest in near-Earth space, the Lagrangian points L1 and L2 on the Sun–Earth line, where a spacecraft can stay at nearly fixed location with respect to the Earth. The location of these points depends on the mass of the Sun, M_s, that of the Earth, M_e, the distance between the Sun and the Earth, R, and the distance between the Earth and the spacecraft along the Sun–Earth line, r, which is much smaller than R. The balance of forces for the Earth in the rotating frame is:

$$F = M_s w^2 R - \left(\frac{GM_s}{R^2}\right) = 0$$

If we consider a point on the Earth–Sun line close to the Earth, we have

$$dF = r\left(M_s w^2 + \frac{2GM_s}{R^3}\right) = r\left(\frac{3GM_s}{R^3}\right)$$

This component, which points outwards farther from the Sun and inwards closer to the Sun, can be balanced by the Earth attraction (pulling out closer to the Sun, pulling in farther from the Sun):

$$\frac{GM_e}{r^2} = r\left(\frac{3GM_s}{R^3}\right)$$

$$r = R\left(\frac{M_e}{3M_s}\right)^{\frac{1}{3}} \tag{2.4}$$

The mass ratio between the Sun and the Earth is close to 300,000, hence the two lagrangian points L1 and L2 are 1.5 million kilometers inwards and outwards of the Sun, respectively. The outer Lagrangian point, L2, is a privileged location for astronomy missions, as observations are nearly free from perturbations by the Earth or the Moon, with an excellent communication configuration. The L1 point is very well suited for solar observations or observations upstream in the solar wind.

Missions at L1	SOHO (Solar observatory)	1995
	Smart-2 (Technology for LISA)	2007
Missions at L2	W-map (Cosmology)	2001
	Planck (Cosmology)	2007
	Herschel (Sub-mm observatory)	2007
	GAIA (μarcsec astrometry)	2012
	NGST (Advanced space telescope)	2012

Another group of missions with similar launch requirements drifts slowly away from the Earth at 1 AU. The increasing angle with respect to the Sun is of obvious interest for the Stereo mission, which observes the Sun simultaneously with two spacecrafts.

SIRTF (IR observatory)	2003
STEREO 2 (two solar observatories)	2005
LISA (gravity wave observatory)	2012

For all these missions, the required perigee velocity is only 50 m/s beyond that required for a large Moon crossing orbit. Even a relatively small launcher such as Soyuz–Fregat can launch a mass close to 1600 kg for these conditions (Fig. 2.16). It is obvious that three body effects must be accurately taken into account for transfer trajectories to Lagrangian points, as they are responsible for these unstable equilibrium regions. This is critical for assessing on-board delta V requirements.

The behaviour of trajectories in "weak stability" regions dominated by three body effects can be extremely complex. A comprehensive discussion of these issues is presented in Hechler and Cobos (2001). Figure 2.20 gives an example of transfer trajectories for Herschel, launched in 2007, showing major evolution as a function of the launch hour, which defines the relationship with respect to the Sun.

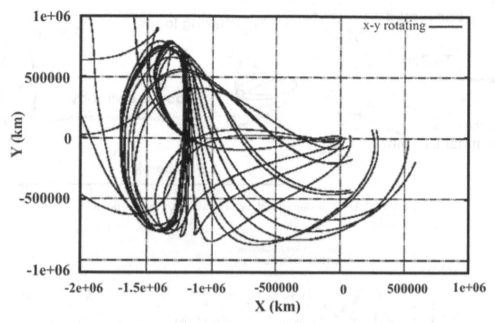

FIGURE 2.20. Trajectories obtained at different launch hours with the same launch velocity, when drawn in the rotating frame of the Earth with respect to the Sun. The results range from bound orbits to escape from the Earth, and a range of launch conditions lead to free injection into a large (and unstable) "halo orbit" around the L2 point (to the left). From Hechler and Cobos (2001).

Some delta V is required for orbit maintenance, as the "orbits" around L1 or L2 are not stable. However, these requirements are relatively small ($<20\,\text{m/s}$ in most cases). The overall delta V budget is therefore relatively small, and depends only weakly on the launch window. As an example, for GAIA, $150–210\,\text{m/s}$ of on-board delta V are required (Hechler and Cobos, 2001), with a relatively minor impact on the mass budget ($<10\%$, including overheads). This made it possible to consider a Soyuz–Fregat launch for GAIA in 2012, as the new versions are expected to deliver close to 2 tons to near escape orbits in this time frame.

The Moon, 35% of the way out to the Lagrangian points, can be used effectively to reduce both the launch requirements and the on-board delta V requirements. This strategy was used by WMAP, launched in 2001 (Fig. 2.23). The Earth–Moon system has also Lagrangian points, situated quite far from the Moon (60,000 km, from Equation 3.1) due to the small mass ratio (81:1) between the Earth and the Moon. This creates extensive weak stability regions, from which the Moon can pull out the spacecraft and eject it into near-Earth space (Farquhar, 1991).

An example is given in Fig. 2.21 with WMAP, a cosmology background mission launched in 2001. After a series of phasing orbits, which could be adjusted to provide a large launch window without mass penalties, WMAP was pulled out of the vicinity of the inner Lagrangian point L1 by the attraction of the Moon, which transferred it to a near escape trajectory. Similarly to the trajectories in Fig. 2.22, three body effects in the Sun–Earth system provided a cheap insertion into a halo orbit around L2. With such strategies, round trip missions such as Genesis can be devised with very little expenditure of delta V (Fig. 2.22).

The examples of Herschel, GAIA, WMAP and Genesis demonstrate that it is now possible to design very complex missions to near-Earth space with nearly the same

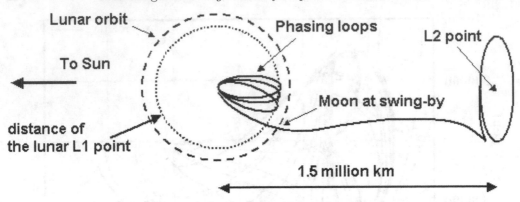

FIGURE 2.21. Transfer trajectory of WMAP to the L2 point.

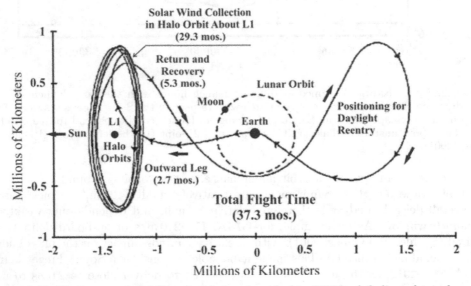

FIGURE 2.22. Trajectory of the Genesis mission, launched in 2000 and dedicated to solar wind collection, which performs a complicated round trip to a L1 halo orbit, then back to the Earth using three body effects.

launch requirements as for a large bound orbit and with very small on-board delta V requirements. The mission design for such missions is now very mature, with few further improvements to be expected.

These combinations of lunar swing-bys and weak boundary strategies can provide escape velocities of up to 1.8 km/s from the Earth–Moon system (Belbruno and Amata, 1996). The best opportunities are highly constrained in terms of the Moon, the Sun and the departure velocity. We will see in the next section that without relying on solar perturbations, a single lunar swing-by can provide an escape at ~1.5 km/s with any departure velocity vector close to the lunar orbit plane, hence to the ecliptic (the inclination of the lunar orbit to the ecliptic is only 5°). This does not look very useful as it falls short of the launch requirements for even the best Mars or Venus windows. However, it will prove very interesting when combined with Delta V Gravity Assists (DVGA) (Section 2.4) using low-thrust propulsion (Section 2.5).

2.4. Gravity assist strategies

2.4.1. *Basic principles*

In Section 2.2, it was shown that missions beyond Mars and Venus were either marginal or impossible with even the most effective direct launch strategies. The solution to this energy problem has been provided since 1974 (Mariner 10) by planetary gravity assist methods. A gravity assist consists of passing close to a planet to change the heliocentric energy of the spacecraft. If the spacecraft passes behind the planet, there is a forward pull, and the orbital energy increases. If the spacecraft passes in front of the planet, the orbital energy decreases. One needs to consider the heliocentric reference frame before and after the encounter, and the planetocentric reference frame during the encounter. In this reference frame, the controlling parameters are the relative velocity and the impact parameter, which is the distance at which the spacecraft would intersect the plane perpendicular to the arrival velocity going through the centre of mass if there was no gravity. When the impact parameter is small, the swing-by will be tight.

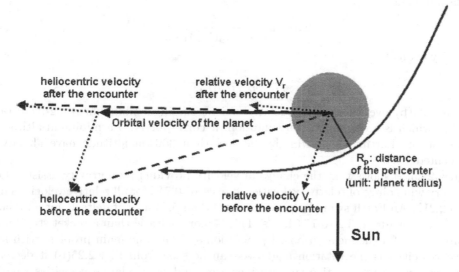

FIGURE 2.23. Example of a gravity assist which increases the orbital energy (e.g. to go to Jupiter). The spacecraft passes behind the planet, and the relative velocity is rotated so as to be nearly aligned with the orbital velocity.

Figure 2.23 presents a typical gravity assist used to increase orbital energy. It is important to understand that a planet cannot modify the relative velocity, as the energy must be conserved in the planetocentric reference frame. It can however rotate it by a large angle, and it is when the orbital velocity of the planet is added back, that the net effect of the forward pull of the planet becomes apparent.

The trajectory of the spacecraft in the planetocentric reference frame is very close to a hyperbola, as solar perturbations are very small. The heliocentric trajectories are usually ellipses. This approximation is therefore called the "patched conics" approach, which much speeds up the first stages of mission analysis when compared to a full 3-D integration taking into account all massive objects in the solar system.

The most important parameter to assess the gravity potential of a planet is the circular velocity V_c, which is the velocity on a circular orbit grazing the surface of the planet. For the Earth, V_c is close to 7.9 km/s. A swing-by at such a low altitude would not be a good idea, in particular if there is an atmosphere (Earth, Venus, Mars, ...) or rings (Saturn).

Therefore, the minimum approach distance R_p expressed in planetary radii must keep a margin which is smaller for atmosphereless bodies. The minimum approach distance R_p and the ratio between the relative velocity V_r and the circular velocity V_c fully define the encounter hyperbola. The dependence of the rotation angle and the resulting velocity change is given in Equations 4.5 and 4.6 (Fig. 2.24).

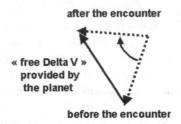

after the encounter

« free Delta V » provided by the planet

before the encounter

FIGURE 2.24.

Rotation angle:

$$2\arcsin \frac{1}{(1 + R_p(V_r/V_c)^2)} \tag{2.5}$$

Velocity change:

$$\frac{2V_r}{(1 + R_p(V_r/V_c)^2)} \tag{2.6}$$

Figure 2.25(b) gives the results in relative units assuming a minimum swing-by distance of $1.1R_p$, which is a conservative assumption: it corresponds to a minimum altitude of 635 km for the Earth, while swing-bys at less than 300 km altitude have already be implemented.

Figure 2.25(a) shows that the change of velocity provided by a gravity assist is large over a very wide range of relative velocities: it exceeds $0.7V_c$ for all relative velocities from 0.42 to $2.2V_c$. An Earth swing-by provides more than 5.5 km/s of delta V for all relative velocities ranging from 3.2 to 17.5 km/s. This is more delta V than a rocket stage or the full capability of a chemical on-board propulsion system. The main problem with large relative velocities is the rotation angle. As can be seen from Fig. 2.25(b) it decreases very rapidly when the relative velocity increases, and only relative velocities $<0.62R_c$

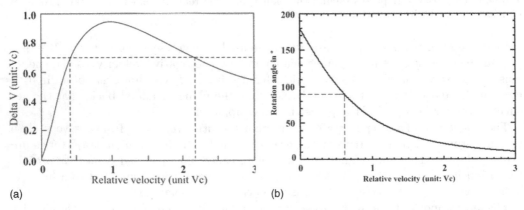

(a)   (b)

FIGURE 2.25. (a) Change of velocity provided for free by a planetary encounter as a function of the ratio between the relative velocity and the circular velocity. (b) Maximum rotation angle for a swing-by at a distance of at least 1.1 planetary radii as a function of the ratio between the relative velocity and the circular velocity.

TABLE 2.4. Gravity assist potential of planetary bodies.

	V_c (km/s)	Lowest altitude (km)	Maximum delta V (km/s)	0.2 x orbital velocity (km/s)
Mercury	3.1	200	2.9	10
Venus	7.6	300	7.4	7
Earth	7.9	250	7.7	5.8
Mars	3.55	200	3.4	4.8
Jupiter	45	30,000	37	2.6
Saturn	27	80,000	17	1.8
Moon	1.7	200	1.6	0.2

(Earth: 4.8 km/s) can be rotated by more than 90°. This is an important constraint for the specific case of 180° singular transfers, which require rotation angles close to 90° to go out of the orbital plane and then back.

From this discussion, the gravity potential of solar system bodies can easily be evaluated and compared to the velocity changes which are required to modify the orbit. Table 2.4 gives the circular velocity, the minimum fly-by altitude, the maximum delta V provided by the gravity assist, taking into the minimum fly-by altitude and 20% of the orbital velocity, which gives an evaluation of delta V requirements for large orbital changes in each region of the solar system. As an example, if 20% is added to the velocity of the Earth (29.8 km/s), an orbit with a period of more than 2.5 years can be reached.

From Table 2.4, three groups of bodies can be identified:

(1) The Moon, Saturn and even more so Jupiter have a gravity assist potential which exceeds the orbital velocity, even when minimum distance limitations due to the rings are considered. Once they are reached, one can go anywhere in the solar system (for Jupiter) or the earth-moon system (for the Moon). It is even possible to reach hyperbolic orbits. Saturn is much more difficult to reach than Jupiter (see Table 2.3), hence using this planet to go somewhere else is always less effective than using Jupiter. The tremendous orbital change capability of Jupiter was already used for a grand tour of outer planets (Voyager) and a rotation of the orbit plane perpendicular to the ecliptic (Ulysses). A Jupiter swing-by is also the simplest solution to "fall" into the Sun (Solar Probe).

(2) Venus, the Earth and Mars (to a lesser extent) have gravity assist capabilities which are similar to that required for large orbital changes. Single swing-bys can therefore be very effective.

(3) Mercury has a very significant gravity assist capability (2.9 km/s, which is equivalent to a large fuel load for a chemical propulsion system), but the orbital velocity is much larger. Multiple gravity assists are often required, which is made possible by the short orbital period of Mercury (89 days).

2.4.2. *Lunar swing-bys*

As noted from Table 2.4, the Moon has a large gravity assist potential (1.6 km/s) with respect to its orbital velocity (1 km/s). Indeed, as is the case of Jupiter, the Moon can eject spacecraft on a bound orbit from the Earth–Moon system, hence could help in improving the mass budget for solar system exploration missions.

Launching on a bound orbit with a period of about a month provides large mass capabilities, as noted in Section 2.2 (more than 1600 kg with the Mars Express version

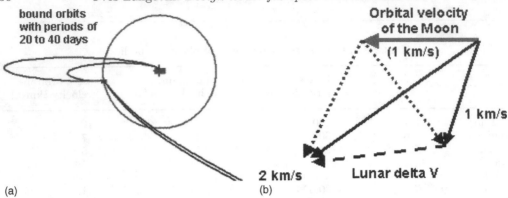

FIGURE 2.26. (a) Geometry of an inbound lunar swing-by which ejects a spacecraft from the Earth-Moon system. The same interplanetary trajectory can be obtained for a wide range of orbital periods (e.g. from 20 to 40 days). As in a direct launch, there is only one passage in the radiation belts of the Earth. (b) Typical gravity assist geometry for a lunar assisted launch from an elliptical orbit to a low heliocentric velocity. The relative velocity (dots) is rotated by more than 60°, and the lunar swing by increases the velocity of the spacecraft in the Earth reference frame from 1 to 2 km/s. This corresponds to 1.4 km/s after escape.

of Soyuz–Fregat). Encountering the Moon inbound after nearly a full orbit increases the cruise time, but it makes it easy to open a launch window of 20 days: for each launch date, the period of the orbit is selected so as to encounter the Moon at the same location after 20–40 days (see Fig. 2.26(a)). This provides nearly identical heliocentric escape conditions. This range of orbits is similar to the semi-major axis of the Moon. Therefore, the velocity of the spacecraft is close to that of the Moon (1 km/s). The relative velocity is smaller than 1.6 km/s, which means that the rotation angle can be larger than 60° (Fig. 2.26(b)). The new velocity in the Earth referential is close to 2 km/s. This provides a departure velocity of 1.4 km/s once the 2 $(km/s)^2$ corresponding to the square of the escape velocity at the Moon (1.4 km/s) have been subtracted. This looks like a big advantage, but it should be noted that a direct transfer from the bound orbit to the departure orbit could also have been achieved with a delta V at pericentre of 150 m/s, a significant but not overwhelming improvement of the mass budget (∼5% if the on-board engine is used, ∼10% if the launcher itself is used). Slightly higher departure velocities (up to 1.8 km/s) can be reached from a bound orbit by implementing two or three lunar swing-bys, but this much restricts launch flexibility as the configuration is strongly dependent on the Earth–Sun–Moon relationship. Encountering the Moon on the outbound leg of a direct hyperbolic escape trajectory can also be considered. However, this totally constrains the launch date (an unacceptable feature), and the delta V advantage provided by the lunar swing-by decreases with the departure velocity (100 m/s for 3 km/s).

With a departure velocity of 1.4 km/s, not even our closest planetary neighbours, Mars and Venus, can be reached. It would therefore seem that there is very little interest in using the Moon. This is not fully correct: as we shall see in the next paragraph, a 1.4 km/s velocity is ideal to initiate a 1 year delta V gravity assist manoeuvre, and this approach will prove very effective when we go on to consider low-thrust propulsion missions in Section 2.5.

2.4.3. *Velocity build-up strategies using only the Earth: the delta V gravity assist*

In this chapter, we will discuss orbit strategies which involve only the Earth (and exceptionally the Moon), ignoring for the time being the gravity assist potential of other planetary bodies. A planetary swing-by cannot change the relative velocity to the planet, so as not to violate energy conservation in the local referential. As we have seen in Section 2.4.1, such an encounter can only reorient the relative velocity vector towards a new direction. This means that a swing-by of the Earth during the cruise is not helpful *per se* as the relative velocity to the Earth will not be modified, hence cruise time would be saved and a similar mass budget would be obtained by launching directly at the required velocity at the time of the swing-by.

An Earth swing-by can however eliminate the large declination penalty incurred with a non-restartable launcher: the launch can be made 1 year earlier than the required departure date at the most favourable declination (5°S for Ariane 5) on an orbit with a period of exactly 1 year, then an Earth swing-by can be used to orient the velocity to the required declination. This strategy enormously increases the launch capability of Ariane 5 with direct injection at non-optimum declinations: the mass limit for a departure velocity of 3 km/s is 3200 kg at 5°S and only 1200 kg for a declination of 20°N (which is far from extreme for Mars or Venus windows). Such a strategy has provided one of the options for the Messenger NASA mission to Mercury. The launch velocity cannot exceed ∼5.5 km/s, as the rotation angle is close to 90° and the swing-by would become too tight.

An Earth swing-by 1 year after launch is also potentially interesting when considering the launch strategy from a bound orbit to large departure velocities. As demonstrated by Fig. 2.16, gravity losses are small (30 m/s) for velocities up to 3 km/s, but they reach more than 250 m/s for a departure velocity of 5 km/s, reducing the effective mass by nearly 10%. One could therefore consider a launch on a bound orbit, then a first Earth-powered swing-by (465 m/s ideally, 490 m/s with gravity losses) to escape on an orbit with a 1-year period and a departure velocity of 3 km/s, then a second Earth-powered swing-by after 1 year to increase the relative velocity to 5 km/s or more, reducing gravity losses and improving the mass budget. There are however hidden traps: for this strategy to be effective, the second powered swing-by must be at a low altitude, while the low arriving velocity is likely to result in a much wider swing-by unless it is itself highly inclined (with a loss in launch capability from Kourou, but not from Baikonur). In ideal conditions, 150 m/s can be saved for a departure at 5 km/s (1.27 km/s instead of 1.42 km/s from the bound orbit). The improvement in terms of effective mass, while not negligible (from 950 to 1000 kg for a departure velocity of 5 km/s) is not worth the additional cruise time.

Apart from these very specific problems, it is therefore clear that to be effective a gravity assist by the Earth requires increasing the relative velocity when going back to the Earth, which means that an orbital change must be performed before this encounter. If we do not consider encounters with Mars or Venus on the way, such an orbital change requires a deep space manoeuvre. This sequence of a deep space manoeuvre for increasing the relative velocity followed by an Earth swing-by to set this larger velocity in the desired direction is called DVGA.

So as to understand the basic physical principles which make DVGA so effective, let us consider the 2-year DVGA, the most appealing "Earth only" strategy for reaching Jupiter. It consists of launching the spacecraft on an orbit with a period of nearly 2 years, which requires a departure velocity of ∼5.1 km/s. At aphelion, 1 year after launch, one performs a 550 m/s braking manoeuvre. The balance of angular momentum and the balance of energy determine the consequence of this manoeuvre for the Earth encounter after 2 years.

Orbit after launch:

V_{dep} = 5.1 km/s, aligned with the orbital velocity of the Earth (29.8 km/s)

V_t = 29.8 km/s + V_{dep} = 34.9 km/s (velocity at departure in the heliocentric system)

a = 1.59 AU (semi-major axis for a period slightly larger than 2 years)

e = 0.37 (eccentricity)

At aphelion:

$$R_a = \frac{1+e}{1-e} = 2.17 \text{ AU (heliocentric distance)}$$

$$V_a = \frac{1-e}{1+e}V_t = 0.46 V_t \text{ (conservation of angular momentum)}$$

If we perform a braking manoeuvre dV:

Change in orbital energy: $dV\,V_a = 0.46\,dV\,V_t$
Change in angular momentum: $dV\,R_a = 2.17\,dV$

For a 550 m/s braking manoeuvre:

Change in orbital energy: $-0.25 V_t$
Change in orbital momentum: -1.2 (km/s AU)

Back to 1 AU:

Balance of momentum: $dV_t = -1.2$ km/s
Balance of energy: $-0.25 V_t = -1.2 V_t + 0.5 V_r^2$ (where V_r is the radial velocity)
$$V_r = (1.9 V_t)^{0.5} = 8.1 \text{ km/s}$$

The radial velocity component must therefore reach 8.1 km/s to balance the energy. The forward relative velocity is 3.9 km/s (5.1–1.2 km/s) and the relative velocity is now 9 km/s. This is not much larger than the circular velocity of the Earth, hence it is possible to rotate it by 50° at a very safe altitude of 320 km. The overall picture after the manoeuvre and the gravity assist is then the following:

Relative velocity	Forward	Radial	Total
At launch	5.1	0	5.1
At return	3.9	8.1	9.0
After GA	8.7	2.2	9.0

In terms of orbital energy, the forward component is what counts most as it adds to the orbital velocity of the Earth (29.8 km/s). In this example, it is very close (at 8.7 km/s) to the full relative velocity (9 km/s). With a 2-year DVGA strategy, a 550 m/s manoeuvre increases the relative velocity by 3.9 km/s, a "leverage factor" of 7.1 which makes it possible to reach Jupiter in favourable windows (see Table 2.3) with more than 80% of the effective mass launched at 5.1 km/s, hence 760 kg instead of the paltry 250 kg which would be available with a very extended powered swing-by after a bound orbit (Fig. 2.16). By spending an additional 2 years of cruise time, a mission to Jupiter which was not achievable with a direct launch has become possible. This is the golden rule of gravity assist strategies: mass can always be traded with cruise time, and very little is not possible if one accepts extremely long missions.

The 2-year DVGA strategy has two options as shown in Fig. 2.27(a), which provide two opportunities 90° apart for the same launch date. It is the most frequently used strategy for exploration missions to the outer solar system, as it combines a large leverage factor with a departure velocity of 5.1 km/s (within reach of a direct launch) and a (relatively) short increase of 2 years in terms of cruise time. The same approach can be used for any resonant orbit which goes back to the Earth after a given number N of years. In the following, we will designate by ($N{:}M$) a resonant orbit which goes back to the Earth

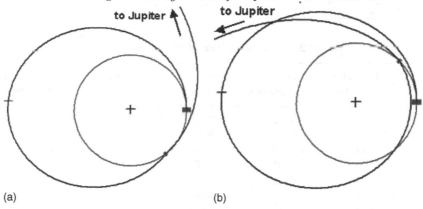

(a)   (b)

FIGURE 2.27. (a) "2−" gravity assist strategy. The Earth is encountered with a large inbound radial component after the gravity assist manoeuvre at aphelion, slightly less than 2 years after the launch. (b) "2+" gravity assist strategy. The Earth is encountered on the outbound leg, slightly more than 2 years after the launch. It requires a larger launch velocity (5.3 km/s instead of 5.1 km/s), but the maximum departure velocity is also larger.

after N years and M round trips of the spacecraft around the Sun. The strategy we have just discussed is therefore based on the (2:1) resonance.

The whole idea behind a DVGA manoeuvre is to significantly decrease the orbital momentum while only slightly decreasing the orbital energy. Whatever the magnitude of the deep space manoeuvre at aphelion, we have:

$$\frac{\Delta\text{energy}}{\Delta\text{momentum}} = \frac{(1-e)^2}{(1+e)^2} V_t$$

Therefore, the leverage factor of a DVGA manoeuvre in terms of velocity gain versus delta V increases with the eccentricity, hence with the period of the orbit. At this stage, it should be noted that a DVGA works just as well with a resonant orbit which has a period smaller than that of the Earth: instead of braking at aphelion, one has to accelerate at perihelion. The efficiency of a DVGA using the 1:2 resonance (0.5 years, 1 year additional cruise time) is nearly the same as that of the 2:1 resonance we have just considered. However, the required launch velocity is much higher (10.7 km/s instead of 5.1 km/s) hence DVGA with periods shorter than one orbital period are only interesting in very specific cases where one wants to go very close to the Sun (it will provide interesting strategies for reaching Mercury after several encounters of Venus) or for the 1-year DVGA (which will be discussed separately). The leverage factor decreases with the resulting relative velocity as the increase of the radial velocity is most effective when it remains small when compared with the orbital velocity.

From Table 2.5, we can define the cheapest "Earth-only" strategy for a mission to Jupiter in a moderately favourable window:

Launch at 2.55 km/s	365 m/s from a $50R_t$ bound orbit
4:3 DVGA to 3.44 km/s	130 m/s
3:2 DVGA to 5.10 km/s	245 m/s
2:1 DVGA to 6.93 km/s	220 m/s
3:1 DVGA to 9.20 km/s	220 m/s
Total	1180 m/s

TABLE 2.5. Characteristics of DVGA strategies.

Resonance	Launch velocity	Delta V at aphelion	Relative velocity	Leverage factor	Additional cruise time (years)
4:3	2.55	−0.143	3.52	6.8	4
	2.68	−0.490	5.10	5.3	4
3:2	3.44	−0.245	5.10	6.8	3
	3.50	−0.400	6.00	6.2	3
2:1	5.11	−0.235	7.03	8.2	2
	5.14	−0.585	9.20	7.0	2
3:1	6.93	−0.220	9.20	10.3	3

This very reasonable delta V (one-third of the ejection velocity) provides a significant effective mass approaching Jupiter even with small launchers: 1050 kg with the Mars Express Soyuz from Baikonur, 1270 kg with a Soyuz T2 from Kourou (both values assuming a 10% tankage factor). However, the cruise time is 11 years (for the 4 DVGAs) + 3 years (Earth to Jupiter), hence 14 years!

As demonstrated by our initial example, one can go much faster by reaching directly the 2:1 DVGA orbit. A Mars Express Soyuz can launch 1610 kg in a bound orbit; 1480 m/s of on-board delta V are needed to escape at 5.14 km/s. A 2-year DVGA manoeuvre of 580 m/s is needed to reach the same relative velocity of 9.2 km/s as the previous strategy. The total delta V is now 2.08 km/s and the effective mass of the spacecraft approaching Jupiter is ∼750 kg. The cruise time has been reduced to 5 years (2 years for the DVGA, 3 years to go from the Earth to Jupiter), a much more reasonable value. The previous strategy provides an increase of 40% of the available mass at a staggering cost of 9 years of travel time, a definitely unappealing trade-off.

An surprisingly effective alternate solution to reach the 2:1 resonance is to use the 1:1 resonance, starting with a period which is smaller than that of the Earth. The eccentricity is now low, but the trade-off between orbital momentum and orbital energy still results in a significant leverage factor of 3 as shown by the actual example of Fig. 2.28. Furthermore, as the required departure velocity is 1.4 km/s, a lunar swing-by strategy is now quite appropriate, as it can provide this departure velocity from a 30 days bound orbit without any delta V expenditure.

For the (1:1) DVGA, only the "1+" strategy is available. The cruise time cost is 1.22 years, as the Earth is encountered 80° further on its orbit (the cruise time to the Moon is similar to that spent on a full bound orbit). It is important to note that the angle between departure and return can vary from 65° to 95° (1 month) with a penalty of less than 30 m/s in delta V, so that for any swing-by date, a departure date can always be selected with the Moon in the proper relationship with the Earth. The total delta V cost of a (1:1) DVGA + (2:1) DVGA sequence is now 1.83 km/s instead of 2.08 km/s for a direct departure from a bound orbit, hence the effective mass is 830 kg instead of 750 kg, but the cruise time is now 6.30 years instead of 5.08 years. The (1:1) DVGA when coupled with a lunar launch is indeed more effective than a thrust arc on a bound orbit for departure velocities ranging from 4 to 5.6 km/s. Beyond that, the rotation capability of the Earth swing-by becomes too small and the 1:1 DVGA strategy becomes less effective.

The advantage of a (1:1) DVGA for this range of departure velocities over a bound orbit launch strategy is marginal with a chemical propulsion system, and it is not worth the

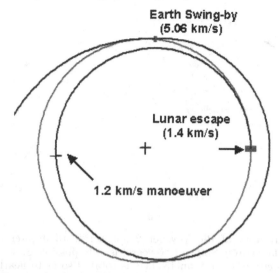

FIGURE 2.28. Most effective (1:1) DVGA strategy for reaching a 2:1 resonant orbit. It provides an encounter velocity of more than 5.1 km/s with a departure velocity of 1.4 km/s (which can be provided for free by the Moon) and a manoeuvre at perihelion of 1.22 km/s which increases the semi-major axis. The optimum leverage factor is 3.04, and it is only slightly lower if departure is 15 days earlier or later if the position of the Moon on its orbit is not adequate.

increased cruise time (1.22 additional years) and complexity which result from a lunar swing-by. However, this advantage becomes overwhelming with low-thrust propulsion systems (see Section 2.5), which have typical capabilities of 20 m/s per day. It is totally ineffective for power swing-by strategies. However, the delta V for deep space maneuvers such as that required for a (1:1) DVGA can be built up over a period of months with minimal penalties, and this strategy is by far the most effective for any departure velocity up to 5.6 km/s.

2.4.3.1. *Jupiter and Venus as stepping stones to the outer and inner solar system*

The most straightforward way of taking advantage of planetary swing-bys is to use them to go even further away from the Earth. Mars provides good opportunities for the main asteroidal belt, but its gravity assist potential is not large enough for reaching the outer solar system. Furthermore, its synodic period of 2.1 years is the largest of all planets, with corresponding limitations on launch opportunities. Jupiter is a much more appealing stepping stone: as demonstrated in the previous section, it can be reached in a reasonable time with substantial mass, and its gravity potential is so large that any relative velocity can be turned to whatever direction is required. Furthermore, the synodic period is now only 1.1 years. The best possible example is provided by the Voyager mission to the outer solar system: two spacecraft were launched in 1976, and they used swing-bys of Jupiter to perform fly-bys of Saturn, Uranus and Neptune (for Voyager 1) and Saturn (for Voyager 2). The Voyager 1 sequence required an exceptional staggered configuration of all four outer planets. A Voyager 2 type mission to Saturn via Jupiter can be performed every 19 years, which corresponds to repeat period of the favourable Jupiter/Saturn configuration of Fig. 2.29(a). It is therefore no surprise that the next mission to Saturn, Cassini, was launched in 1997, 21 years later than Voyager.

Jupiter has such a large gravitational potential that it is possible to obtain an orbital plane which is nearly perpendicular to the ecliptic. Such a strategy was implemented for the Ulysses mission (Fig. 2.29(b)), launched in 1989, which is now completing its second

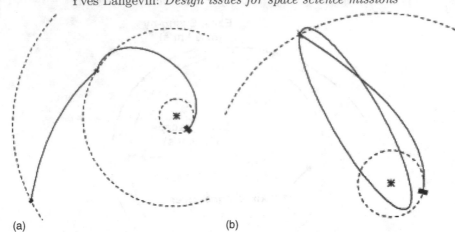

(a)   (b)

FIGURE 2.29. (a) Trajectory of the Voyager 2 spacecraft to Jupiter and Saturn. Such a remarkably cheap and fast trip to Saturn (<6 years) is only possible every 20 years. (b) Ulysses mission. A relative velocity of more than 15 km/s is rotated so as to nearly kill the orbital velocity of Jupiter. The resulting orbit was highly inclined with respect to the ecliptic, making it possible to investigate high-latitude regions of the heliosphere.

full orbit around the Sun, each with a passage over the North and South poles of the Sun at a distance of nearly 2 AU. A similar strategy makes it possible to fall into the Sun.

For the inner solar system, Venus is the obvious candidate as it has a large gravity assist potential while being easy to reach from the Earth. Actually, the Mariner 10 mission to Mercury was the first example of a mission based on gravity assists: after a launch at 3.8 km/s in the autumn of 1973, a swing-by of Venus in early 1974, at a very safe altitude of 3500 km, reduced the semi-major axis from 0.8 to 0.6 AU, and the first close range observations of Mercury were performed after less than a year of cruise time (Fig. 2.30).

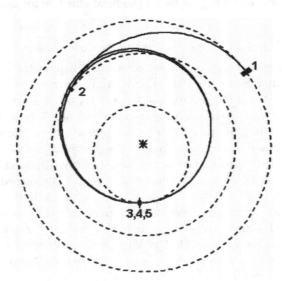

FIGURE 2.30. Mariner 10, the first exploration mission to Mercury. A swing-by of Venus (2) made possible a first swing-by of Mercury (3) less than 1 year after the launch (1). This first swing-by transferred the spacecraft to a (2:1) resonant orbit, so that two additional Mercury swing-bys (4,5) were performed 6 months and 1 year later. Mercury was encountered close to apocentre, which is the cheapest configuration for a fly-by mission, hence the relative velocity was large (10.3 km/s).

An Italian scientist, Bepi Colombo, pointed out that it was possible to transfer to a (2:1) resonant orbit (period: 0.48 years) after this first swing-by. This made it possible to perform three successive swing-bys of Mercury at 6-month intervals without any delta V expenditure.

The Mariner 10 strategy, in which Mercury is encountered close to aphelion, is the cheapest possible strategy for a fly-by. However, it results in very large relative velocities (10.3 km/s for Mariner 10), therefore it is not at all appropriate for a rendezvous mission, which requires an encounter velocity as small as possible, hence an encounter not too far from the perihelion of Mercury ($L_s = 77.5°$).

For such a mission, it is interesting to note that Venus lies half-way between the Earth and Mercury not only in terms of semi-major axis, but also in terms of orbital planes. When considering the diagram of inclinations (Fig. 2.31), it can be seen that the inclination of Mercury with respect to Venus (4.3°) is much smaller than that with respect to the Earth (7°). The intersection of the orbit planes of Mercury and Venus is at L_s 28° or 208°. The latter position is much farther away from the pericentre of Mercury (130° instead of 50°). From these considerations, Yen (1974) found out that it was possible to reach Mercury at relatively low velocities (5.8 km/s) by leaving Venus around L_s 208° with a relative velocity of ~8 km/s pointing mainly backwards, ~1.7 km/s to the South and ~2 km/s outwards. As Mercury is not necessarily at the right place at the right time, the spacecraft may have to go around the Sun more than once before reaching Mercury (phasing orbits). Each phasing orbit increases the cruise time by 0.4 years.

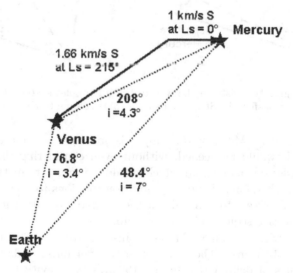

FIGURE 2.31. Orbit plane matching strategy for a rendezvous mission to Mercury. By leaving Venus close to the intersection of the orbit planes (in this case L_s 215°) with a moderate South pointing velocity component (1.66 km/s), Mercury can be reached close to L_s 0° with a low out of plane velocity component (1 km/s).

Reaching Venus with a 8 km/s velocity is not very demanding: it can be done with a departure velocity of 3.7–4 km/s (easily within reach of direct launches), and a type II transfer will result in a relative velocity at Venus which is pointing mainly outwards, 2 km/s to the North and ~1 km/s forward, as the semi-major axis (0.82 AU) is larger than that of Venus (0.723 AU). There are two major problems: the first one is that this encounter is not at the right location for most windows, hence there is a need to move around the orbit of Venus. The second problem is that the maximum rotation angle provided by Venus is 54° for a relative velocity of 8 km/s, when more than 90° separate the required departure and arrival velocities. For these two reasons, more than one swing-by of Venus must be included in the mission strategy.

The fastest strategy consists of inserting a 1:1 resonant orbit, which encounters Venus again after 0.6 years. The semi-major axis of this orbit (0.723 AU) is close to that at arrival (0.82 AU), and a large plane change can be implemented (Fig. 2.32(a)). Unfortunately, the second swing-by cannot perform the full rotation needed to reach Mercury, and typically 700 m/s of on-board delta V are needed. An alternate strategy consists of transferring to a 3:4 resonant orbit (period $= 0.45$ years, semi-major axis $= 0.6$ AU). The second Venus encounter is 1.8 years later, but the second swing-by can now easily set the spacecraft on its way to Mercury (Fig. 2.32(b)). With these two strategies, mass can be traded with time, in this case 25% more useful mass (reduction by 700 m/s of the on-board delta V) against a cruise time which is 1.2 years longer.

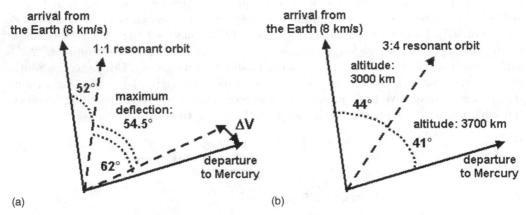

FIGURE 2.32. Venus gravity assists on the way to the inner solar system. (a) Route with a 1:1 transfer. (b) Strategy using a (3:4) resonant orbit.

The arrival velocity at Mercury (5.8 km/s) is far too high for a direct insertion: 3.45 km/s of delta V would be needed, without even considering the staggering gravity losses for an insertion manoeuvre which would be larger than the 3 km/s circular velocity of Mercury. Therefore, chemical propulsion missions to Mercury must use a sequence of DVGA maneuvers to bring down the relative velocity with a large leverage factor. The most effective sequence is the following.

Messenger, the NASA mission to Mercury, inserts after the 4:3 resonance with an approach velocity of 3.45 km/s. The insertion maneuver into a 5-day long elliptical orbit requires 1.45 km/s of delta V (including 150 m/s for gravity losses), hence a total on-board delta V for this phase of 1.75 km/s. The cruise time after the initial Mercury encounter is 1.7 years.

The two prime opportunities for the Messenger launch in 2004 are presented in Fig. 2.33. They demonstrate the elaborate strategies which are needed to finally encounter Venus close to L_s 210° (encounter 3 in Fig. 2.33(a), encounter 4 in Fig. 2.33(b)) with the required outbound and backwards relative velocity of ~ 8 km/s. It is interesting to note how two completely different strategies can achieve a very similar result. Such a flexibility is not customary with planetary gravity assist strategies.

A more ambitious mission to Mercury, BepiColombo, is programmed by ESA for a launch in 2012, using ion propulsion as a baseline (Section 2.5). The BepiColombo chemical back-up strategy implements the full sequence of 3 DVGA's, which take nearly 3.2

TABLE 2.6. Chemical propulsion insertion to Mercury.

Initial encounter velocity (km/s)	Resonance	Delta V (m/s)	Final encounter velocity (km/s)	Leverage factor	Duration (years)
5.80	(3:2)	73	5.20	8.5	0.72
5.20	(4:3)	230	3.45	7.5	0.96
3.45	(6:5)	182	2.27	6.5	1.44

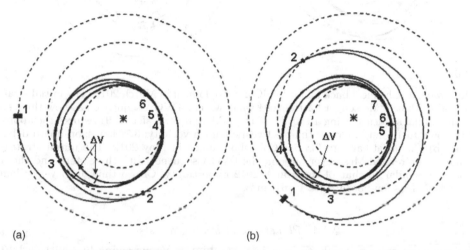

(a)   (b)

FIGURE 2.33. (a) Messenger mission to Mercury, launch in March 2004. Two swing-bys of Venus separated by a 4:3 non-resonant transfer (2–3) are followed by a transfer to Mercury with three phasing orbits (3–4), then a two-step delta V Mercury GA sequence (4–6). Cruise time: 5 years; launch velocity: 3.83 km/s; on-board delta V: 330 m/s. (b) Messenger mission to Mercury, launch in May 2004. A series of three swing-bys of Venus separated by a 1:1 non-resonant transfer (2–3), then a 2:3 non-resonant transfer (3–4) are followed by a transfer to Mercury with two phasing orbits (4–5), then a two-step delta-V Mercury GA sequence (5–7). Cruise time: 5.15 years; launch velocity: 4.08 km/s; on-board delta V: 380 m/s.

years after the initial Mercury encounters. A total of 485 m/s is needed to bring down the relative velocity to 2.27 km/s. An interesting end of mission strategy is then implemented: the encounter at 2.27 km/s is very close to perihelion. It is possible to transfer to an inclined (1:1) resonant orbit, then a last swing-by rotates the relative velocity perpendicular to the orbit plane, so as to set the spacecraft on an orbit identical to that of Mercury, but rotated around the Sun–Mercury line at encounter. Mercury can then be encountered after a half turn (180° transfer) at a velocity of only 1.5 km/s, thanks to the large eccentricity of Mercury (0.2). When compared to Messenger, insertion occurs now 3.45 years after the first encounter, but the delta V required for reaching a 5-day orbit is now only 325 m/s, 810 m/s including the DVGA manoeuvres. An additional 1.75 years of cruise time results in a reduction by 900 m/s of the on-board delta V. By implementing a bound orbit launch strategy for the required 3.76 km/s departure velocity, the effective mass in a 5-day orbit around Mercury reaches more than 1100 kg with the Soyuz versions which are considered for 2012. As explained in Section 2.2, windows to Venus repeat nearly identically after 8 years, hence the initial stages for the BepiColombo chemical back-up with a launch in 2012 (Fig. 2.34) are very similar to that of Messenger (launch in 2004).

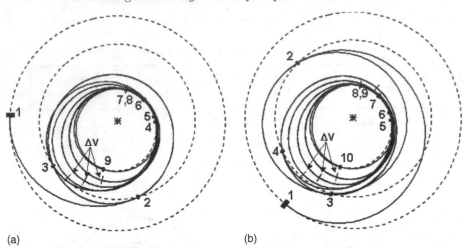

(a)   (b)

FIGURE 2.34. (a) Chemical back up for BepiColombo, launch in March 2012. The initial strategy is identical to that of Messenger for the 2004 window. The DVGA sequence has now three steps (4–7), and a final sequence using a 180° Mercury to Mercury transfer (7–9) reduces the approach velocity down to 1.5 km/s. Cruise time: 6.1 years; launch velocity: 3.76 km/s; on-board delta V: 480 m/s. (b) Chemical back-up for BepiColombo, launch in May 2012. The initial strategy is the same as the related Messenger option. The DVGA sequence with three steps and the final sequence are similar to that of the March 2012 opportunity. Cruise time: 6.25 years; launch velocity: 4.17 km/s; on-board delta V: 615 m/s.

2.4.4. *Planet-planet gravity assists*

In the previous section, we have seen how combining Venus swing-bys with a DVGA sequence provides adequate effective masses for a Mercury orbiter mission, and the interest of Jupiter as a stepping stone for the outer solar system as well as for exotic missions such as Ulysses has been demonstrated. As was shown in Section 2.4.3, a departure velocity of 5.1 km/s and a deep space manoeuvre of 550–580 m/s provides a first solution (the 2-year DVGA) for reaching Jupiter with significant effective mass. Still, the launch capabilities at 5.1 km/s are much lower than that of missions to Venus and Mars, and the deep space manoeuvre eats up 20% of the mass.

A swing-by of Mars or Venus during a round trip from the Earth to the Earth can provide for free the delta V which is required by a DVGA manoeuvre. Mars is most efficiently used in combination with a (3:2) GA strategy, and Venus with a (1:1) GA strategy. Even in poor windows, the launch to Mars (or Venus) requires a departure velocity which is lower than that required for a 2-year DVGA (5.1 km/s). With this planet–planet gravity assist strategy, it is possible to much improve the mass budgets for very ambitious missions, but, as we shall see, the number of opportunities is smaller than for "straightforward" DVGA strategies.

For every Mars window (every 2.1 years), there are several 3:2 Earth–Mars GA strategies. As the spacecraft goes twice around the Sun on an orbit with a period close to 1.5 years, Mars can be encountered at four locations: inbound or outbound, and on the first or second orbit. The cheapest strategy consists of encountering Mars inbound on the second orbit, a few months before the Earth encounter (Fig. 2.35). Launch velocities are in the range of 3.4 km/s and one obtains for free a 9 km/s encounter velocity which is adequate to reach Jupiter or Jupiter family comets. This was the strategy initially selected for the January 2003 launch of the Rosetta mission to comet P/Wirtanen. As this comet was highly inclined, a (2:1) inclined resonant orbit had to be inserted after the

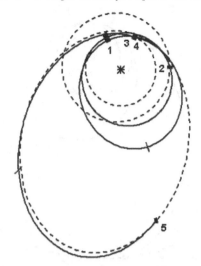

FIGURE 2.35. Initial baseline trajectory for the Rosetta mission. A (3:2) Mars–Earth GA manoeuvre (1–3) increases the relative velocity to 9 km/s. After two swing-bys of the Earth (3–4) separated by a 2:1 resonant orbit, the spacecraft can encounter the comet with a low relative velocity, as their orbits are now very similar. Two deep space manoeuvres are needed, with the main one between the last Earth fly-by and the arrival at P/Wirtanen.

(3:2) Mars–Earth gravity assist (abbreviated as MEGA), hence this strategy is referred to as a MEE strategy where each letter represents a planetary encounter.

For each Venus window (every 1.6 years), there is an effective 1:1 Earth–Venus GA strategy, with a type I transfer from the Earth to Venus followed by a relatively distant swing-by (typically 20,000 km from the center of Venus). The Earth is encountered with an inbound relative velocity of 7.6 km/s after 1 year to 9 km/s after 1.3 years. The semi-major axis is always close to 1. As explained in the presentation of the 1:1 DVGA strategy, velocities larger than 5.5 km/s cannot be rotated enough to become aligned with the velocity vector, and the relative velocity is usually too small to reach Jupiter, hence a (2:1) DVGA (2 years additional cruise time) must be inserted for outer solar system missions, with a deep space manoeuvre requiring up to 250 m/s of delta V, depending on the Jupiter window (see Fig. 2.36).

There are two solutions with Venus which take approximately 2 years:

(1) Type II transfer to Venus followed by a type III transfer back to the Earth (as already explained, a type III transfer is a type I transfer (arc $< 180°$) with a full phasing orbit (360°). This opportunity is not available without delta V on all Venus windows. The resulting relative velocity when coming back to the Earth is higher than 11 km/s, outbound.

(2) Type II transfer to Venus followed by a (2:1) Venus–Venus non-resonant transfer (outbound to inbound), followed by a type I transfer back to the Earth. The relative velocity is outbound, close to 9.4 km/s. It can be much higher if a DVGA is performed on the Venus–Venus leg. This is the option which was selected for the Cassini trajectory to Jupiter then Saturn (see Fig. 2.37).

These strategies are potentially extremely useful. In the best cases, the launch velocity of 3.4 km/s (compared with 5.1 km/s for the "straightforward" 2-year DVGA) and the small on-board delta V requirement (compared with 580 m/s for the DVGA) can improve the mass budget by 40%. An example of the potential of these techniques was given by the search of backups for Rosetta once the launch to Wirtanen in January 2003 had

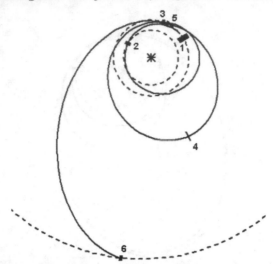

FIGURE 2.36. Example of a Venus–Earth–Earth strategy for reaching Jupiter. The launch (1) is in October 2013 with a departure velocity of 3.68 km/s. Following the Venus swing-by (2), the Earth is encountered in early December 2014 (3) with a relative velocity of 8.9 km/s. A (2:1) DVGA with a manoeuvre of 180 m/s (4) increases the relative velocity to 9.9 km/s at the last Earth encounter (5) so as to reach Jupiter (6) in mid-2019. The total cruise time is 5.7 years, a quite reasonable value for a useful mass of 1160 kg at Jupiter with a Mars Express level Soyuz and a bound orbit launch strategy.

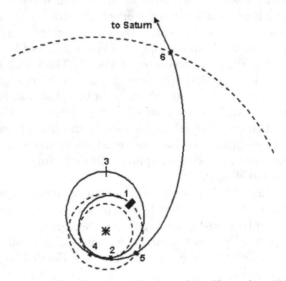

FIGURE 2.37. Mission strategy for Cassini with a launch in November 1997: after a first Venus encounter, a (2:1) Venus–Venus DVGA (2–4) is performed, which increases the relative velocity to 9 km/s. The Earth is encountered (5) less than 2 months after the second Venus encounter (4). A Jupiter swing-by (6) sends the spacecraft to Saturn.

been cancelled due to the problems of Ariane 5. Two solutions using Venus were found with the launches in October 2003 and April 2004 which reached Wirtanen at the same time as the baseline mission (Fig. 2.38). The final transfer orbits are indeed remarkably similar to the baseline when considering the totally different strategies in the inner solar system. Unfortunately, the thermal constraints at 0.7 AU prevented the implementation

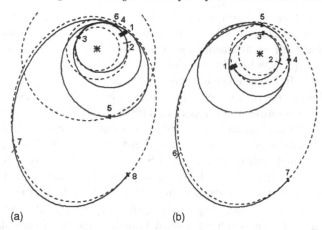

(a)   (b)

FIGURE 2.38. (a) Backup launch to Wirtanen in October 2003. Venus (3) is reached after more than a full orbit an a deep space manoeuvre (2). A large asteroid (8 Flora) can be encountered (5) between the two Earth swing-bys (4, 6) on a 2-year orbit. (b) Backup launch to Wirtanen in April 2014. Venus (3) is reached after a deep space manoeuvre (2). The two Earth swing-bys (4, 5) are, respectively inbound and outbound on a 2-year orbit. Less than 8 years are required to reach Wirtanen.

of these options, and the new selected baseline is a mission to a new comet, Churyumov–Gerasimenko, with a launch in February 2004 and a cruise time of more than 10 years (Fig. 2.39), so that the target is reached in mid-2014 instead of late 2011. This mission constitutes an extreme case in terms of cruise time (more than 10 years) and complexity (DV-EMEE-GA strategy!) for gravity assist strategies.

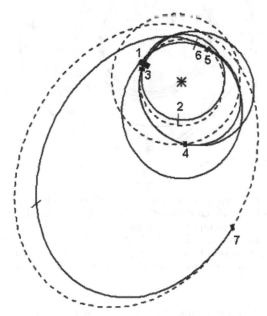

FIGURE 2.39. New baseline of the Rosetta mission to comet Churyumov–Gerasimenko, launched end of February 2004 with a departure velocity of 3.46 km/s. A 1:1 DVGA (1–3) pumps up the relative velocity to 3.86 km/s (3). Mars is reached outbound after more than a full orbit (4). Its pulls down the orbit so that the Earth is encountered more than 90° away from the departure position (5). After a 2-year orbit (5–6), the spacecraft finally departs towards the comet (7).

A particularly challenging example of planet–planet gravity assist techniques is provided by the next generation of Solar Physics mission, which require reaching the close vicinity of the Sun at high ecliptic latitudes. A first solution was discussed in Section 2.4.4 with a swing-by of Jupiter, which is considered by NASA for its solar probe mission, but this requires a relatively large relative velocity at Jupiter (>15 km/s) and the spacecraft visits the inner solar system only once every 5 years. The approach selected by ESA for its Solar Orbiter mission requires shorter periods (100 days), hence a strategy using Venus swing-bys. Even moderate inclinations (>30°) and solar distances (>0.2 AU) require an extremely large relative velocity at Venus of at least 19 km/s. Without a large on-board delta V capability (which will be discussed in Section 2.5), the only way to achieve this with an adequate mass budget is an elaborate sequence of planet–planet gravity assists, progressively pumping up the eccentricity (hence the radial velocity) by braking with the Earth and accelerating with Venus (Fig. 2.40). Quite surprisingly, it is possible to achieve such extremely high relative velocities with a launch velocity of only 3.6 km/s and no on-board delta V. This strategy, which provides a large effective mass (1090 kg with a Mars Express level Soyuz and a direct escape with the Fregat), has been selected for the chemical back-up of Solar Orbiter. The first observations are performed 3.6 years after launch. We will see in Section 2.5 that a solar electric propulsion mission can shorten the cruise time to 2 years.

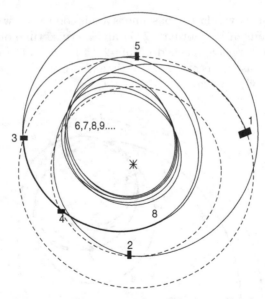

FIGURE 2.40. Chemical back-up strategy for Solar Orbiter with a launch in September 2013 (departure velocity: 3.6 km/s, launch declination: 9°S). A series of alternating Venus swing-bys (2,4) and Earth swing-bys (3,5) pumps up the relative velocity at Venus from 5.6 (2) to 19.5 km/s (6) in 3.5 years without any on-board delta V. There are two complete orbits between swing-by 3 (Earth) and 4 (Venus). A series of Venus swing-bys on a 2:3 resonant orbit (6,7,8,9, ... with 1.2 years between Venus swing-bys) increases the inclination with respect to the solar equator to more than 35°.

Instead of increasing the inclination, the series of swing-bys of Venus can also be used to lower the perihelion. This requires implementing a series of orbits which return to the same position after an integer number of periods of Venus (resonant orbits). As the spacecraft stays in the orbit plane, it is also possible to depart at one of the two intersections with the orbit of Venus (e.g. 2 in Fig. 2.41) and to return at the other

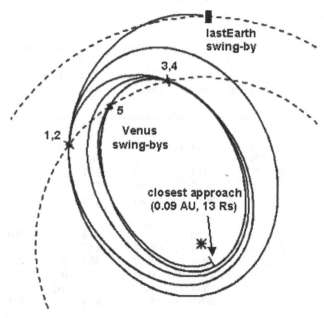

FIGURE 2.41. Strategy to get close to the Sun using Venus swing-bys. The sequence of orbits in resonance 2:3 (swing-bys 1–2), then 1:2 (2 inbound to 3 outbound), then 1:2 (3–4), then 1:2 (4 outbound to 5 inbound) lasts 3 years (6.5 years since launch). It provides approaches of the Sun every 100 days at a minimum distance of 13 Rs for a relative velocity of 19.5 km/s. This minimum distance would decrease to 8 Rs for a relative velocity of 24 km/s, which requires either another Venus–Earth–Venus round trip or ion propulsion. Rs: Solar radius.

intersection (3 in Fig. 2.40). The transfer orbit has then a larger period than the corresponding resonant orbit if one leaves inbound and returns outbound, and a smaller period than the corresponding resonant orbit if one leaves outbound and returns inbound. The higher the relative velocity at Venus, the closer the approximation to the Sun.

Notwithstanding this bewildering range of possibilities, there are limits to the flexibility of planet–planet gravity assist strategies. By definition, they are constrained to Venus and Mars windows, so that the next "delta V free" Solar Orbiter opportunity is in May 2015, and the same applies to the close solar approach scenario. The position of Mars drifts only by 50° between two windows of the same type, hence the position of the target (Jupiter or comet) is much more constrained than with Venus and the opportunities with Mars are 4–5 times less frequent than with Venus.

Table 2.7 indicates the most attractive planet–planet gravity assists opportunities in the 2009–15 time frame, the Rosetta 2003 baseline being given as a reference. In the best cases, the spacecraft departing to Jupiter or a Jupiter family comet has a mass equivalent to that which could be sent to Venus (or Mars). The cruise time of the velocity build-up phase is always 3 years for an Earth–Mars GA, slightly more than 1 year for the shortest Venus opportunities and 2 years for the other Venus opportunities. For Venus opportunities, a 2-year DVGA (with or without delta V) has to be included so as to reach the outer solar system, as the rotation angle would be too large for a single Earth swing-by (the velocity also has to be increased for most 1-year round trips). It may also be required for Mars opportunities as the returning velocity can be too small. The important parameters are the departure velocity and declination, which define the launch capability, the velocity when encountering the Earth, the heliocentric longitude of the return, and the two possible heliocentric longitudes after either a "2−" or a "2+"

TABLE 2.7. Selected Planet–Planet gravity assist opportunities.

Launch date	V departure (km/s)	Launch declination	Return date	V return (km/s)	Heliocentric longitude	After 2 years DVGA (v = 9.4 km/s) 2 − 2 + cost (m/s)		
Mars								
2003/01	3.36	1.5°N	2005/12	9.00	68° (in)	64°	162°	70
2009/07	3.86	18.7°N	2012/04	8.82	201° (in)	197°	292°	95
2011/11	3.96	20.0°N	2014/07	8.80	344° (in)	340°	55°	100
2012/03	3.86	25.0°S	2015/01	9.34	127° (in)	126°	228°	10
2015/11	3.90	23.0°N	2016/01	10.20	119° (out)	6°	119°	0
Venus (1 year)								
2009/01	3.05	18.0°N	2010/01	7.83	113° (in)	101°	198°	250
2009/01	3.40	21.0°N	2010/02	8.94	143° (in)	139°	234°	75
2010/08	3.20	19.2°S	2011/09	8.43	353° (in)	346°	52°	160
2012/03	3.39	1.0°S	2013/04	9.04	219° (in)	216°	308°	60
2013/10	3.66	7.0°N	2014/11	8.85	71° (in)	67°	162°	90
2015/05	3.65	13.0°S	2016/06	8.82	244° (in)	240°	335°	95
Venus (2 years)								
2009/01	3.67	40.0°N	2010/12	11.60	99° (out)	330°	99°	0
2010/09	3.37	44.0°S	2012/07	11.65	300° (out)	174°	300°	0
2012/04	3.20	17.0°N	2014/03	11.14	170° (out)	45°	170°	0
2013/11	3.83	13.0°N	2015/10	12.10	34° (out)	263°	34°	0
2015/06	3.56	24.7°S	2017/05	11.30	234° (out)	109°	234°	0
2× Venus								
2012/03	3.08	24.0°N	2014/02	9.40	130° (out)	64°	130°	0
2015/06	3.65	24.6°S	2016/04	9.30	205° (out)	104°	206°	15

DVGA (see Section 2.4.3) following the planet–planet gravity assist, which is required to go. The inward or outward direction of the relative velocity when coming back to the Earth is mentioned. It always points inward after a "2−" DVGA, outward after a "2+" DVGA. The search for mission opportunities is done by comparing the possible dates and heliocentric longitudes at departure with the positions and dates of launch windows to the target. As an example, the Jupiter VEE opportunity of Fig. 2.35 was built by noting that the date and position of the "2−" possibility for an October 2013 launch matched that of a 2016 launch window to Jupiter.

2.4.5. *Gravity assists in satellite systems*

This discussion of gravity assist strategies would not be complete without addressing gravity assists in satellite systems. Indeed, satellite systems are solar systems in miniature, where the strategies we have just described can be applied. The small orbital periods constitute a major asset, as strategies involving a very large number of periods can now be considered. Apart from Jupiter, the orbital velocities are much lower than for planets, which is beneficial as satellites have smaller gravity assist potentials than planets.

There are three situations in the solar system:

- No satellite with gravity assist potential >200 m/s
  Mercury, Venus, Mars and Uranus

- One satellite with a large gravity assist potential: the relative velocity will remain constant unless DVGA strategies are implemented.

	Period (days)	V orbit (km/s)	V circular (km/s)
Earth/Moon	27.30	1.0	1.68
Saturn/Titan	15.93	5.6	1.94
Neptune/Triton	5.88	4.4	1.02

- The system of **Jupiter**: there are four large satellites, the Galilean satellites, which make it possible to implement satellite–satellite gravity assists similar to the planet–planet gravity assists presented in Section 2.4.5.

	Period (days)	V orbit (km/s)	V circular (km/s)
Io	1.77	17.3	1.82
Europa	3.55	13.7	1.44
Ganymede	7.15	10.9	1.95
Callisto	16.69	8.2	1.67

Even a single satellite system provides very interesting opportunities for improving the observations of a planetary system. A remarkable example is provided by the 4-year orbital tour which has been defined for the Cassini/Huygens mission, beginning in early 2005 with the descent of the Huygens probe in the atmosphere of Titan. There are strong scientific justifications for modifying the orbit: a rotation of the line of apses is essential as the smaller satellites are in tidal lock, hence each satellite needs to be encountered at a different place on its orbit to observe it in different illumination conditions and to obtain a comprehensive coverage. It is also interesting to explore the different regions of the magnetosphere with respect to the direction of the Sun. Similarly, there is strong interest in obtaining high-resolution views of Saturn's atmosphere at different local hours. Excursions to high inclinations are also quite appealing for observing the polar regions of Saturn and the rings from a good vantage point, and to explore high-latitude regions of the magnetosphere.

As already pointed out, Titan is the only satellite with a large gravity assist potential (V_c: 1.94 km/s). The next largest gravity assist potential, that of Rhea, is four times smaller. The orbital period of Titan is 15.9 days, which substitutes for the year in the system of Saturn. The relative velocity to Titan is fixed if delta V is not spent. The optimum choice is to set the relative velocity close to the orbital velocity (5.56 km/s). The maximum delta V provided by a single Titan gravity assist is 1.1 km/s, a large fraction of the orbital velocity.

The basic building blocks of a single satellite tour are presented in Fig. 2.41. The shortest way to rotate the apses is to first reduce the semi-major axis at an inbound encounter, then to increase it back to the original value at an outbound encounter. With the selected encounter velocity (5.5 km/s), the two swing-bys are moderately tight (altitudes of 330 and 460 km, respectively), and the line of apses is rotated clockwise by 40° in 78 days (slightly less than 5 periods of Titan). One can also increase the inclination every 15.9 days, then go back to the orbital plane of Titan, "flip-over" the orbit by

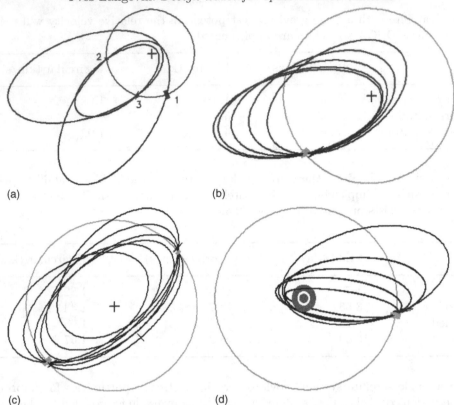

(a)   (b)

(c)   (d)

FIGURE 2.42. (a) "Petal" manoeuvre; the spacecraft leaves Titan inbound on a large orbit (1). After 1.7 orbits, Titan is encountered outbound, with a gravity assist which reduces the orbital velocity (2). After 1.2 orbits, Titan is encountered inbound (3) and the initial orbit is recovered after the swing-by. (b) Sequence of increasingly inclined orbits; the spacecraft encounters Titan every 15.9 days, increasing the inclination at each swing-by, hence decreasing the radial velocity and the eccentricity. The sequence of inclinations is: 20.7°, 36.5°, 46.7°, 53°, 57°, 59°, with 16 days per step up or down. (c) "Flip-over" manoeuvre; the inclination of the orbit is increased until the radial velocity has been eliminated. A 180° transfer becomes possible, then the inclination is reduced step by step. It takes 220 days to go back to the equatorial plane with apses rotated by 180°. (d) High inclination sequence; in such a strategy, the inclination is increased while the period is decreased in a sequence of resonant orbits (1:1, 3:4, 2:3, 3:5, 1:2). The final orbit is nearly polar ($i = 72°$), just staying clear of the edges of the rings.

inserting a 180° transfer between two inclination sequences, and decrease the smallest distance to Saturn while keeping clear of Saturn's rings.

By combining the basic tools presented in Fig. 2.42, it has been possible to build a 4-year tour which provides an extremely thorough exploration of the Saturn system, with 45 Titan fly-bys and seven close fly-bys of the other Saturn satellites (Fig. 2.43). On-board delta V is required only for orbit correction manoeuvres.

The two other single satellite systems in terms of gravity assist potential are the Earth/Moon and Neptune/Triton systems. The gravity assist potential of the Moon is particularly large in relationship with its orbital velocity. It can be used as a turntable for large orbits around the Earth, which can be useful for distributing a fleet of micro-satellites in the Earth magnetosphere with a single launch. Figure 2.44 shows an example of a "flip-over" manoeuvre which rotates the apses by 180° in 14 days.

Once cruise times are reduced in the outer solar system, for example, with Nuclear Electric Propulsion (see Section 2.5), a Cassini/Huygens type mission to the Neptune – Triton

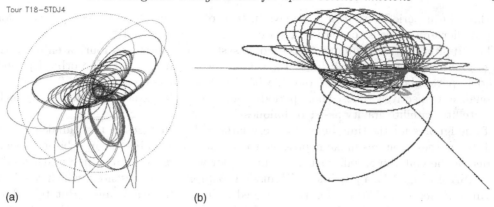

Tour T18−5TDJ4

(a)   (b)

FIGURE 2.43. Baseline Cassini tour in the Saturn system. The view in the equatorial plane (a) and the view as seen from the Sun (b) demonstrate the wide range of regions which will be observed and the comprehensive coverage in local time and latitude of the planet, its ring system and its magnetosphere.

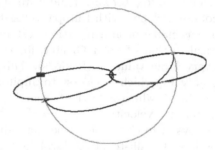

FIGURE 2.44. "Flip-over" manoeuvre in the Earth–Moon system. Thanks to the large gravity assist potential of the Moon with respect to its orbital velocity, only two swing-bys separated by 14 days are needed to rotate the line of apsides by 180°.

system will eventually be programmed, as this system presents very specific science interests: an atmosphere with specific characteristics, a magnetosphere with a large angle between the dipole and the rotation axis, and a large satellite (Triton) in a retrograde orbit which could well be a captured Kuiper Belt Object. Tour methods in this system would be similar to that used for Cassini (Fig. 2.45). Triton fly-bys can be more frequent

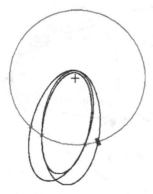

FIGURE 2.45. Petal manoeuvre in the Neptune–Triton system. The gravity assist potential of Triton is twice smaller than that of Titan, hence the rotation of apsides is only 12°, but the period is only 5.9 days, so that this can be repeated every 30 days.

as the orbital period is shorter (5.9 days instead of 15.9 days), which compensates for the smaller gravity assist potential of Triton (half of that of Titan).

The Jupiter system is unique in the Solar system as it provides four satellites with large gravity assists potential: Io, Europa, Ganymede and Callisto. The orbital period of Io is particularly short (1.77 days), with Europa and Ganymede in a 2:1 and 4:1 resonance, respectively. These short periods make the Jupiter system an ideal test bench for satellite–satellite gravity assist techniques.

If one ignores for the time being the very large radiation doses in the radiation belts of Jupiter, these features make it possible to design a remarkably cheap orbiter mission to one of the Galilean Satellite such as Europa, for which such a mission is under study by NASA. In Fig. 2.35, a possible VEE mission to Jupiter was presented, which required 910 m/s of delta V (730 m/s for the powered swing-by from a bound orbit to a 3.68 departure velocity, 180 m/s for a 2 years DVGA manoeuvre) and approached Jupiter at 5.8 km/s after 5.7 years. Due to the very large gravity field of Jupiter, an insertion manoeuvre at 200,000 km from the centre of Jupiter to an orbit with a period of 106 days (60 periods of Io) is relatively cheap (850 m/s). After insertion, the orbit is highly elliptical, and the relative velocity to Io (16.7 km/s) and Europa (15.7 km/s) is very large. One needs to lower the encounter velocity with Europa to less than 2 km/s (the circular velocity) so as to obtain a reasonable orbital insertion cost in terms of delta V. This problem is nearly symmetrical to that of Solar Orbiter, for which the relative velocity had to be pumped up to very large values (>19 km/s). This time, one has to brake with inner satellites (preferably Io, which is closest to Jupiter) and accelerating with outer satellites (preferably Callisto, which is farthest from Jupiter) so as to reduce the eccentricity, hence lower the relative velocity.

With a series of five swing-bys of Io (Fig. 2.46), one can reduce the period to a value smaller than the orbital period of Callisto (16.6 days) in 130 Io periods (232 days). The periods of intermediate orbits can be adjusted to reach Io at an adequate time and position for the transfer to Callisto. Outbound Io swing-bys can be used instead of the inbound swing-bys of the example, which provides further flexibility in positioning the departure to Callisto as outbound swing-bys would be 180° away from the inbound swing-bys on Io's orbit (see Fig. 2.54).

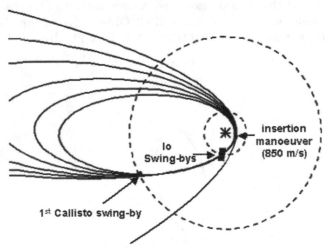

FIGURE 2.46. First phase of a mission strategy to Europa. Four Io swing-bys reduce the orbital period in a sequence of 60, 29, 13 and 10 Io periods. A last Io swing-by sends the spacecraft to Callisto.

Two successive swing-bys of Callisto can then increase the semi-major axis. This satellite–satellite gravity assist manoeuvre (Fig. 2.47) reduces the relative velocity to Io from 16.7 to ~6.5 km/s without any delta V cost apart from orbit correction mano-euvres. A minimum of ~40 Io periods (71 days) are required, depending on the relationship between the orbital positions of Io and Callisto. This double swing-by can be performed with periods bracketing resonance orbits: 1:1 in the example given, 2:1 or 2:3. Similarly to the first phase, inbound or outbound swing-bys can be selected for either Io or Callisto (in the example, the first Callisto swing-by is inbound, the second one is outbound). If the built-in flexibility of the orbit reduction sequence is also taken into account, a reasonably short round trip can be designed for any relationship between the positions of Io and Callisto.

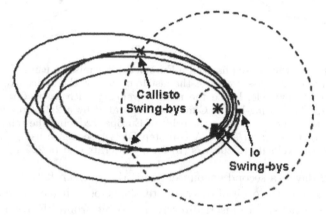

FIGURE 2.47. Reduction of the relative velocity with respect to Io. Two swing-bys of Callisto raise the pericentre to a position slightly inside the orbit of Io. The relative velocity decreases from nearly 17 down to ~6 km/s.

After the last Callisto Swing-by, a new series of Io swing-bys (Fig. 2.48) reduce the orbital period in steps down to 3 Io periods (5.3 days).

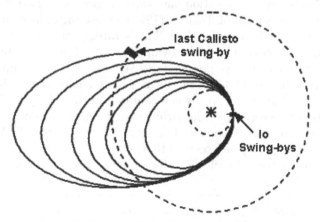

FIGURE 2.48. Second phase of orbital period reduction using Io swing-bys with orbits in the sequence 8, 7, 6, 5, 4, 3.5, 3 Io periods.

A last Io swing-by sends the spacecraft to Ganymede on an orbit with a period of 4 days. As can be seen in Figures 2.48 and 2.49, the orbit is much less elliptical, and the relative velocity to Ganymede is now only ~3 km/s. A first Ganymede swing-by increases the orbit period so as to reach two-third of that of Ganymede (hence 4.8 days). 2 Ganymede

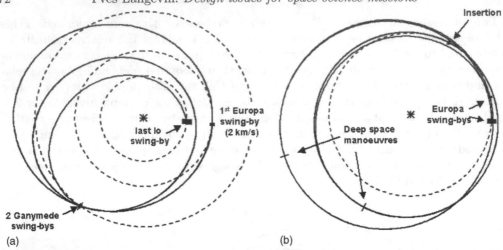

FIGURE 2.49. (a) End of the cruise phase for a Europa orbiter mission using chemical propulsion and satellite–satellite gravity assists. After the last Io swing-by, the apocentre lies only slightly outside the orbit of Ganymede, hence the relative velocity with respect to Ganymede is in the range of 3 km/s. Two swing-bys of Ganymede separated by a 2:3 resonant orbit raise the pericentre from inside the orbit of Io to the orbit of Europa, reducing the relative velocity with respect to Europa down to 2 km/s. (b) A two step DVGA sequence using Europa swing-bys bring down the relative velocity to 800 m/s (b) for a moderate cost of 225 m/s.

period later (14.2 days), a second encounter sends the spacecraft to Europa on an orbit with is nearly tangent to the orbit of this satellite: the velocity has been reduced to 2 km/s.

It would still be expensive to insert directly into orbit around Europa (800 m/s for an orbit with a period of 1 day, even assuming no gravity loss). There are no more outer satellites available, hence the relative velocity cannot be reduced further without spending some delta V. Thanks to the small period of Europa (2 Io periods, or 3.55 days), one can reduce the relative velocity to 0.8 km/s by using a 3:2 DVGA, then a 6:5 DVGA, with an increase in mission time of 9 Europa periods (32 days). The delta V cost of the DVGA manoeuvres is 225 m/s, but the insertion manoeuvre is now only 250 m/s, with less than 1 m/s of gravity losses, a saving of 325 m/s (12% increase in terms of useful mass) well worth the additional month of cruise time. This strategy takes 1.2 years from Jupiter insertion to Europa insertion. The 14 Io swing-bys, 2 Callisto swing-bys, 2 Ganymede swing-bys and 2 Europa swing-bys provide interesting science opportunities on the way with a comprehensive tour of Galilean satellites. The increased cruise time (6.9 years in total) is a very reasonable price to pay when considering that the post Jupiter insertion delta V of 475 m/s all the way to a 1-day orbit around Europa is now quite affordable in terms of its impact on the mass budget. The total cruise time is still shorter than that of Cassini/Huygens (7.3 years) or Rosetta (10.5 years).

The total delta V budget of such a strategy for the Europa mission scenario is the following:

Launch from bound orbit to Venus	730 m/s
2 years DVGA to reach Jupiter	180 m/s
Jupiter Insertion	850 m/s
Final Europa DVGA sequence	225 m/s
Europa insertion	250 m/s
	2235 m/s

New versions of the Soyuz–Fregat (Soyuz T2) launched from Kourou should provide at least 2000 kg of launch capability to a bound orbit. Even if one increases by 10% the delta V budget to take into account the large number of encounters (3 planets and 20 satellites), with correspondingly high-orbit correction costs, the dry mass of the spacecraft in orbit around Europa would be ∼900 kg, and the useful mass ∼790 kg. The surprising conclusion is that with such a strategy it would be possible to send a Mars Express class spacecraft in a 1-day orbit around Europa with a relatively small launcher and a cruise time of 6.9 years. However, a spacecraft of this class would definitely not survive in the extreme radiation environment of the inner regions of the Jupiter System, hence shielding is mandatory for critical subsystems. Furthermore, radioisotope thermo-electric generator (RTGs) are required for power generation, and such subsystems are heavy (several 100 kg). Even with these larger mass requirements, a chemical propulsion mission launched by Ariane 5 is definitely an alternative to the Nuclear Propulsion option which will be discussed in the next section.

2.4.6. *Conclusion: assets and limitations of gravity assist strategies*

The examples in this section demonstrate that gravity assist techniques have been remarkably successful for designing solar system exploration missions. By combining delta V gravity assist and planet–planet gravity assist strategies, one can bring within reach apparently impossible mission scenarios, such as a comprehensive survey of the giant planets followed by an exploration of the Heliosphere all the way to the heliopause (Voyager), a rendezvous and lander mission to a comet (Rosetta), orbiter missions to Mercury (Messenger, BepiColombo) and missions getting very close to the Sun and/or at high ecliptic latitudes (Ulysses, Solar Orbiter and Solar Probe). Even with a "cheap" launcher (Soyuz–Fregat), it is possible to have mass budgets in the range of 800–1000 kg for all those types of missions by accepting long cruise times. One large satellite (Titan) made it possible to design a remarkably comprehensive tour of the system of Saturn for the Cassini/Huygens mission, initiated in early 2005. With four large satellites (Jupiter), the range of possibilities is almost infinite.

Given these remarkable achievements of chemical propulsion when combined with gravity assist strategies, the question which immediately arises is the rationale for developing new propulsion techniques. Three arguments can be put forward:

(1) Cruise times up to 10 years have major drawbacks for a space science program. The nominal procedure for scientific investigations is to define the next step on the basis of previous results. When applied to space science, this can lead to cycles of more than 20 years between investigations of a given set of problems (comets: Giotto in 1986, Rosetta in 2014; Saturn: Voyager in 1983 and Cassini in 2004). Furthermore, science experiments have to be developed 15–20 years ahead of time, while one would like to benefit from the most recent advances in experimentation. Long cruise time is also a programmatic risk as it opens the possibility for another space agency to steal the show by launching a less ambitious, but faster mission to the same target.

(2) The best mission opportunities for gravity assist strategies can be very much tied to a specific launch date, with severe penalties in mass budget and/or cruise time if this launch date cannot be met. The decision not to launch Rosetta in January 2003 eventually led to major consequences for the science, as the target had to be changed to a less active comet with a delay of nearly 3 years for the rendezvous. Similarly, the Messenger mission to Mercury already missed its best launch opportunity (March 2004). If the May 2004 opportunity cannot be met either, the orbit phase will be delayed by nearly 2 years for lack of a similarly favourable mission in terms of mass budget. There is therefore a strong

interest in having more flexibility, with frequent opportunities providing a similar mass budget.

(3) Some important science investigations cannot be performed satisfactorily with chemical propulsion, whatever the cruise duration one would be ready to accept. This is self-obvious for a multiple rendezvous mission to main belt asteroids, which have very little gravity field while the relative velocities are in general higher than 3 km/s: a single insertion in orbit eats up all the capability of an on-board chemical propulsion system. Similarly, a circular polar orbiter at 0.5 AU is not achievable without extremely large on-board delta V capabilities. New propulsion systems are indeed required before such missions can even be considered.

Therefore, for most science mission requirements, the major advantage of the large on-board delta V capabilities provided by low-thrust propulsion methods which will be discussed in the next section is not to make a mission possible (this is in general already the case) or to improve the mass budget (it can be similar with chemical propulsion missions implementing complex gravity assist strategies), but to much reduce the cruise time and to increase the frequency of opportunities, hence the programmatic resiliency of the mission.

2.5. Low thrust propulsion

2.5.1. *Propulsion techniques providing high specific impulse*

In Section 2.4, we have seen that the main incentive for developing new propulsion techniques is the reduction of cruise time, hence the increase of the available delta V on board a spacecraft. From the rocket equation (Section 2.2.2), it is clear that the only way to achieve this is to increase the ejection velocity, hence the momentum provided for each kilogram of ejected mass (the specific impulse). The two main directions which have been considered are to accelerate ions, which can be done very effectively to large velocities, or to use photons, which by definition leave at the highest possible ejection speed, the speed of light.

Two main methods have been developed for accelerating ions. The first approach consists of generating and heating up a plasma and letting it expand into free space ("plasma engines"). This is equivalent in its principle to chemical propulsion, but much higher temperatures, hence ejection velocities, can be achieved. The typical velocity range is from 15 to 20 km/s. There is no built-in lifetime issue. However, instabilities in the plasma (well-known in the development of fusion reactors) can lead to "flame out" events after which the engine has to be restarted. Such events are indeed being observed at weekly intervals on the SMART-1 mission of ESA, which uses a plasma engine.

The second approach is to generate ions by an ion source, which are then accelerated through a high-voltage grid. The ejection velocity can be much higher: in principle, one could consider going up to a large fraction of the speed of light (particle accelerators). Due to the increasing complexity and the energy supply considerations which will be discussed later, ejection velocities achieved by ion thrusters range from 30 to 100 km/s. Ion acceleration is a well controlled process. A fraction of the ions hit the acceleration grid. This reduces the effective impulse and the grid is slowly eroded by ion sputtering, which means that ion thrusters have a built-in lifetime limit. With both approaches, on-board delta V capabilities in excess of 10 km/s can be considered, much exceeding the maximum capability with chemical propulsion (in the range of 4 km/s when maintaining a significant useful mass).

The main issue for any high ejection velocity method is the power supply problem. The conversion of electrical energy into kinetic energy cannot be complete. The best ion propulsion engines have overall efficiencies of at most 70%. The required power for a given thrust is directly proportional to the ejection velocity:

$$\text{Thrust} = dm/dt \; V_{\text{ej}}$$
$$\text{Power} = 0.5 V_{\text{ej}}^2 \; dm/dt$$
$$\text{Watts/Newton} = 0.7 V_{\text{ej}} \; (\text{assuming } 70\% \text{ efficiency})$$

Therefore, higher ejection velocities reduce the mass of propellant which is required for a given delta V, but they require a larger power source so as to provide a similar thrust. The trade-off in terms of mass between the propellant and the power supply has to be considered very carefully, as power requirements are very large if any significant thrust is to be provided. For a plasma engine (ejection velocity 20 km/s), 14 kW are needed for every Newton of thrust. For an ion engine (ejection velocity: 40–100 km/s), the power requirements range from 28 to 70 kW/N. When a large delta V is required (e.g. 10 km/s), ion engines provide better mass budgets than plasma engines, as the reduction in propellant mass outweights the larger power subsystem.

Bi-liquid chemical engines which are used on-board interplanetary spacecraft have thrusts in the range of 400 N. It is clearly impractical to consider such thrust levels with ion engines, as megawatts of power would be required (not even mentioning the huge system mass of thrusters able to accelerate grams of ions during each seconds). All high specific impulse methods are therefore limited to much lower thrust levels (typically in the range of 0.1–5 N). 20–40 m/s are provided every day, and months are needed to build up delta Vs of several km/s, which leads to very specific mission design issues as discussed in Chapter 3. The many months of thrusting which may be needed for large delta Vs become an issue when the lifetime of thrusters is taken in consideration.

Two power supply approaches can be considered: solar energy and nuclear energy. Solar energy is only available in significant amounts in the inner solar system. Assuming a high photocell conversion efficiency of 20%, 280 W/m^2 can be obtained at 1 AU. A Solar Electric Propulsion (SEP) system with an installed power of 14 kW, providing 1 N of thrust with a plasma engine, requires 50 m^2 of solar panels. Such a solar generator already masses several 100 kg and represents a significant fraction of the mass of the propulsion stage. The typical launch mass ranges from 1500 to 3000 kg for thrust levels of 0.2–0.6 N. At 10 AU (mission to Saturn), 5000 m^2 would be required, with very large system masses, hence SEP cannot be considered for missions to the outer solar system. SEP is now operational, with the success of the DS-1 test mission of NASA and the on-going missions of ESA (SMART-1) and Japan (Hayabusa – Muses C).

Nuclear Electric Propulsion works independently of the heliocentric distance. It is the obvious choice for missions to the outer solar system. There are however major constraints on NEP missions: No nuclear reaction can be maintained below a critical mass of fissile material, hence nuclear power generators cannot be miniaturized. The smallest systems which have been flown in space by the Soviet Union and the US provide several 100 kW. Extensive shielding as well as a long boom is needed to protect the electronics of the probe from ionising radiation. As a result, a NEP stage has a typical mass of 5 tons. It does not make sense to carry a science payload of a few 10 kg with such a stage, hence NEP missions are by design very ambitious (and very costly). NASA is considering very seriously a NEP mission to the Jupiter system (JIMO) for the 2013–20 time frame, which could orbit successively all four Galilean satellites, in particular Europa and its buried ocean.

Using photons for propulsion is potentially very attractive, and such missions have been considered since the early 1970s (e.g. the extremely ambitious rendezvous mission to comet P/Halley studied by NASA). The most effective strategy is to directly use the momentum of solar photons (radiation pressure). With a perfectly reflective surface ("solar sail") perpendicular to the direction of the Sun, one can get a total thrust which is at most twice the radiation pressure. However, this thrust would be in the radial direction, hence it could not be used for modifying the orbit. With the optimum strategy of Fig. 2.50, the useful thrust is equivalent to the radiation pressure. There is no "propellant usage", hence solar sail propulsion could make possible missions which require very large delta V (50–100 km/s) such a the Halley rendezvous or a circular polar orbiter of the Sun.

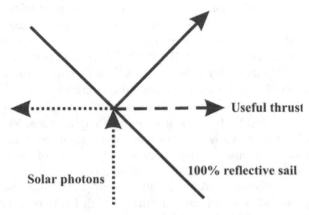

FIGURE 2.50. Optimum configuration for a solar sail, in which the sail is set at an angle of 45° to the direction of the Sun. The total thrust is then 1.41 times the radiation pressure, and the useful thrust (along the orbital velocity) is equal to the radiation pressure.

The comparison between power and thrust is even more unfavourable for solar sails than for ion or plasma engines. At relativistic speeds, the relationship between momentum energy is:

$$E = pc$$

Hence 300,000 kW of solar energy are needed for each Newton of thrust. At 1 AU, each m^2 receives 1 kW of solar energy (taking into account the 45° tilt), hence 100,000 m^2 are required for a thrust of 0.3 N, equivalent to that provided by a SEP stage of 500 kg. The mass available for the solar sail and its (very large) supporting structure is therefore in the range of 5–10 g/m^2. The challenge is even more daunting when considering that the sail must be reflective over a wide range of wavelengths (at least from 0.2 to 3 μm, which correspond to more than 80% of the total solar energy), and that the coating must retain these optical qualities over many years of cruise. Contrarily to ion propulsion, solar sail propulsion is yet far from the operational stage, no test mission being yet programmed by major space agencies.

It is interesting to discuss whether other types of "photon drives" can be considered out of science fiction novels. It is clear that ejection velocities close to the speed of light are required for considering interstellar missions with reasonable cruise time: as an example, a one way robotic mission to alpha centauri in 20 years would require a total delta V of at least 0.4 c. Even if photons are massless, the rocket equation still applies (with c as the ejection velocity), as energy generation requires consuming some mass:

$$E = pc = mc^2$$

hence:

$$m\,\mathrm{d}v = c\,\mathrm{d}m$$

then:

$$\Delta V = c\log(m_\mathrm{i}/m_\mathrm{f})$$

A delta V of 40% of the speed of light therefore requires that 33% of the initial mass is converted into energy, then in a light beam. A "practical" way of achieving this could be to bring on 16% of the mass as anti-electrons, then to find some way of orienting the two 512 keV photons resulting from annihilation.

2.5.2. *The specific problems of mission design with low thrust*

The main asset of high specific impulse is to increase the on-board delta V capability. It is not possible to take full advantage of this asset due to the low-thrust capability. This applies both to changes in the orbital energy and to changes in the orbital plane. The most important parameter for low-thrust strategies is the mass load, in kg/N of thrust, which defines how many m/s of delta V can be accumulated every day. Typical mass loads for SEP missions range from 4000 to 10,000 kg/N, which correspond to delta V capabilities of 8–20 m/s per day, hence from 3.3 to 8 km/s per year of thrusting.

Contrarily to chemical propulsion, an optimal Hohman transfer cannot be implemented with low thrust. Instead of two impulsional burns (one at departure, the other at arrival), the transfer between two orbits is performed as a spiral with the thrust direction close to the orbital velocity. The delta V budget on such a spiral is very straightforward; On a circular orbit, the kinetic energy is equal and opposite to the total energy. Once integrated over a spiral arc, a delta V modifies the total energy by $V\cdot\mathrm{d}V$, hence the kinetic energy by $-V\cdot\mathrm{d}V$, and the change in the orbital velocity is equal and opposite to the delta V. This means that the delta V required between two circular orbits is simply the difference between the two orbital velocities.

With this strategy, there are losses with respect to the optimal transfer, but this penalty is not very large: for a transfer from the Earth to Venus, as an example, it amounts to less than 1% of the total delta V (5.2 km/s). These heliocentric "gravity losses" (see Section 2.2) increase for more eccentric orbits. In such cases, the optimum thrust arcs are located close to the apses, so as to be as close as possible to a Hohman transfer.

The problem is much more severe when considering plane changes. A plane change maneuver is perpendicular to the orbit plane, and it is optimally performed at one of the two intersections of the initial and final orbit planes (relative nodes), and the farther from the sun the better: a plane change is a rotation of the orbital momentum ($\boldsymbol{M} = \boldsymbol{R} \times \boldsymbol{V}$), hence $\mathrm{d}\boldsymbol{M} = \boldsymbol{R} \times \mathrm{d}\boldsymbol{V}$.

With very long thrust arcs (e.g. 120°), the thrust out of the orbit plane is not performed at the optimal position. Thrusting out of plane decreases the effect on orbital momentum and energy (in plane component). The definition of an optimum thrusting strategy for combining in orbit changes with plane rotation is a very complex mathematical problem. Fortunately, a simple heuristic approach gives results which are close to the optimum. It can be derived by balancing along the orbit the positive impact on a plane change of increasing the out of plane thrust angle and the negative impact on in orbit maneuvers. If α is the out of plane thrust angle:

- efficiency: $E = R\sin\alpha\cos\Phi$, where Φ is the angle to the relative node
- in-plane thrust: $P = \cos\alpha$
- marginal effect of increasing α:
  $\mathrm{d}E = R\cos\alpha\cos\Phi\,\mathrm{d}\alpha$
  $\mathrm{d}P = -\sin\alpha\,\mathrm{d}\alpha$

If one attributes a weight K to plane rotation, it is possible to derive the angle which spreads the cost as effectively as possible by maximizing the total $(K \cdot E) + P$. This leads to $K \cdot dE + dP = 0$, hence:

$$\alpha = \operatorname{atan}(KR\cos\Phi) \tag{2.7}$$

A good out of plane thrust strategy can therefore be defined from only two parameters: the direction in which one wants to change the orbit plane (the direction of the relative node, which defines Φ) and the weight K given to the plane change with respect to in orbit maneuvers. When applied to very complex mission strategies such as that of the BepiColombo mission to Mercury, this heuristic thrust strategy came within 2% of a full optimization, hence it is perfectly adequate for evaluating different strategies with low thrust.

An example of a thrust strategy is given in Fig. 2.51 for a direct spiral mission to Mercury. As expected from Equation 2.7, the optimum out of plane thrust angle increases away from the Sun and close to the selected line of relative nodes. The projection of the orbit plane on the ecliptic and the evolution of the orbit plane are given in Fig. 2.52.

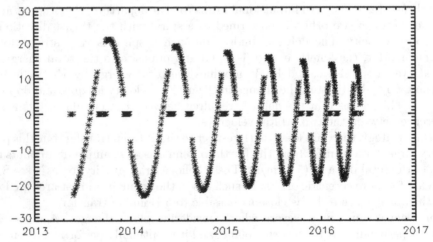

FIGURE 2.51. Angle between the thrust direction and the orbit plane as a function of time for a direct transfer to Mercury, as derived from the approximation in Equation 2.7.

As seen in Fig. 2.52, the thrust arcs extend over the apses, so as to cheaply increase the eccentricity. They also extend over the nodes, so as to cheaply modify the orbit plane. The evolution of the orbit plane takes into account the direction of the launch (dashed line) and the plane evolution towards the plane of Mercury using the approximation of Equation 2.5.2. Early arcs of the spiral are more effective as the spacecraft is farther from the Sun. This trajectory, with a launch in April 2013, requires slightly more than 3 years until the rendezvous with Mercury after a launch at 1.4 km/s ("lunar launch", see Section 2.5.3). The delta V is close to 18 km/s, and the thrust time exceeds 22,000 h with a maximum thrust of 0.4 N, which becomes a problem in terms of thrusters lifetime (qualification tests having been made up to 10,000 h). Such a trajectory was considered in the early stages of the BepiColombo mission to Mercury. It was found out that the mass budget was marginal even with an Ariane 5 (G), hence this "brute force" approach was abandoned in favour of mixed strategies including gravity assists.

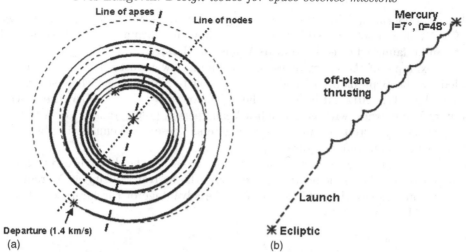

FIGURE 2.52. Direct spiral strategy from a lunar launch (1.4 km/s) to a rendezvous with Mercury. The evolution of the orbit is presented on the left, and that of the orbit plane on the right. Thick solid lines correspond to thrust arcs, which constitute a major fraction of the travel time (3 years). The dashed line on the right indicates the plane change which is provided by the launch direction, which was selected close to the line of nodes of Mercury.

2.5.3. *Combining low thrust with gravity assists*

The main drawback of low-thrust strategies is the time it takes to accumulate a significant delta V. Combining low thrust with gravity assists is therefore very attractive: a single gravity assist provides a large velocity change in a matter of days, and the large delta V capability provided by low-thrust engines makes it possible to adjust the timing before the next encounter. The combination is particularly powerful for the DVGA strategies in Section 2.4.3, as the required deep space maneuvers can easily be provided by low thrust during the months long cruise arcs. The leverage effect of the delta V gravity assist, reduces the required thrust time.

The first example of such an approach is the launch strategy using the Moon, then a 1:1 DVGA, which was presented in Fig. 2.28. After launching on a bound orbit, a lunar swing-by can provide a velocity of 1.4 km/s pointing inward in the ecliptic (the inclination of the lunar plane is only 5°). A maneuver at perihelion is then used to pump up the eccentricity, hence the encounter velocity when returning at the Earth is increased, with a leverage factor ranging from 2.4 to 3. This launch strategy was not very attractive with chemical propulsion, as it was only marginally more effective than a bound orbit launch strategy, in which the same delta V was applied close to the Earth, while the cruise time was increased by 1.2 years. With low-thrust, its advantages are overwhelming, as the loss of efficiency for a maneuver distributed over a long cruise phase is not very large, while a low thrust propulsion system would provide only a few m/s close to the Earth. By spending 0.7–1.25 km/s of low thrust delta V, it is possible to return to the Earth with velocities of 3.5–5.1 km/s, a velocity which is adequate either to reach Mars and Venus with a wide range of encounter velocities or to initiate a 2:1 DVGA strategy (for outer solar system missions). Such maneuvers do not require more than 120 days even for the largest mass loads which are considered for SEP missions, and these thrust arcs can be accommodated in the 1.2 years long cruise phase with acceptable penalties. There is very little penalty in terms of cruise time over a direct strategy, as the velocity change of up to 3.5 km/s provided by the 1:1 DVGA strategy would have required between 5 months and 1 year of continuous thrusting, depending on the mass load. The only case in which a 1:1 DVGA should not be used is that of a mission with such a favourable mass budget that it can be launched directly towards Venus (or Mars).

A first example of the effectiveness of combining low thrust with gravity assists is provided by the BepiColombo Rendezvous mission to Mercury, which is scheduled by ESA for a launch in 2012. After a lunar launch followed by a 1:1 DVGA, the spacecraft encounters Venus twice, with either half a Venus period (180° transfer) or a full Venus period (1:1 resonant orbit) between the encounters. These two Venus swing-bys save up to 7 km/s when compared to a direct spiral.

The 1:1 DVGA strategy was very effective for increasing the departure velocity. Such a strategy can also be implemented for reducing the arrival velocity at Mercury. This end of mission strategy (Fig. 2.53) can also take advantage of the large eccentricity of the orbit of Mercury.

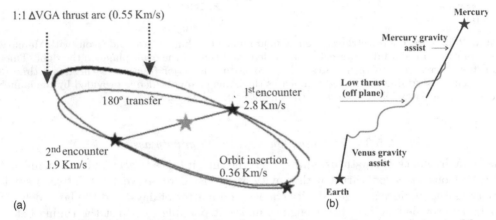

FIGURE 2.53. (a) End of mission strategy for a low thrust rendezvous mission to Mercury (BepiColombo). The 180° transfer followed by a 1:1 DVGA reduce the encounter velocity from 2.8–0.36 km/s with a delta V expenditure of only 0.55 km/s, a saving of 1.9 km/s which requires only 0.4 years of additional cruise time. (b) The plane evolution required for a rendezvous with Mercury is obtained by combining planetary gravity assists (black lines) and out of plane thrusting (grey arcs).

After leaving Venus, a spiral makes it possible to encounter Mercury with a relative velocity of 2.8 km/s close to perihelion. A 180° transfer from perihelion to aphelion reduces the relative velocity to 1.9 km/s, thanks to the ratio of more than 1.5 between the heliocentric distances of Mercury at perihelion (0.308 AU) and aphelion (0.466 AU). This part of the strategy was already implemented for the chemical back-up (Fig. 2.34). A relative velocity of 1.9 km/s can be rotated by 90° back to the orbit plane, initiating a 1:1 DVGA which reduces it to 360 m/s. For such a low encounter velocity, the velocity close to Mercury is only 16 m/s above the local escape velocity. The evolution of the orbit plane shows how Venus and Mercury gravity assists also markedly reduce the plane change which has to be provided by the low thrust arcs.

As demonstrated in Fig. 2.54, a solution can be found for each Venus window, hence with a launch opportunity every 1.6 years. Thanks to the flexibility which is provided by the large on-board delta V capability, the overall characteristics are very similar for each window, with a delta V budget of 7.2–7.7 km/s and a cruise phase of 3.1–3.8 years. The combined benefits of gravity assists by the Earth, Venus and Mercury have reduced the delta V budget (and the associated thrust times) by more than half when compared to a direct spiral strategy. This much improved mass budget made it possible to consider a

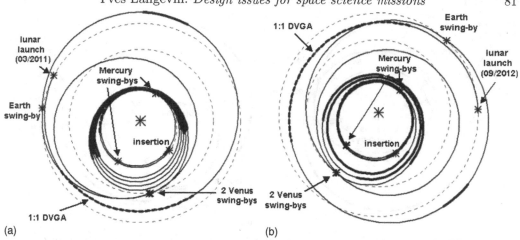

FIGURE 2.54. BepiColombo mission options with a launch in March 2011 (a) and September 2012 (b). The early stages are specific to each window while the end of mission strategy is very similar.

launch with a much cheaper launcher (Soyuz instead of Ariane 5). The cruise time is only marginally longer than that of a direct spiral (3.1 years). It is much shorter than the 6 years of the best chemical trajectory with a similar mass budget (launch in March 2012). The advantage becomes overwhelming for less favourable windows, as a launch in 2013 would require 8.1 years for a chemical mission. This is not just a theoretical situation: the recent decision of NASA to delay the launch of Messenger from March 2004 (nearly identical to the 2012 opportunity) to August 2004 (with a return to the Earth in 2005, hence a situation similar to 2013) resulted in delaying by 2 years the arrival date at Mercury (2011 instead of 2009). These drastic consequences of a delay by a few months in the preparation of a mission demonstrate the advantage of SEP missions in terms of programmatic flexibility.

A second example of the advantages of using SEP for inner solar system missions is provided by the Solar Orbiter mission. As discussed in Section 2.4.5, the first mission stages are dedicated to pumping the relative velocity to Venus up to 19 km/s, at which stage a series of Venus swing-bys can increase the inclination of the orbit to more than 30°.

As only Venus is involved, the different windows are nearly equivalent for SEP missions, and the differences for chemical missions are smaller than for missions involving more than one planet other than the Earth. The main asset of SEP is therefore to shorten the cruise time from 3.5 years (chemical mission) down to 1.8 years, as demonstrated by Fig. 2.55.

The flexibility provided by SEP makes it possible to consider a wide range of mission strategies, such as those presented in Fig. 2.56, which would be completely out of reach for a chemical mission. This further increases the range of possible launch dates, launch conditions and on-board delta V requirements. A list of opportunities for Solar Orbiter is given in Table 2.8. It demonstrates the remarkable programmatic flexibility of low thrust missions, even when combined with planetary gravity assists, as there are now five possible launch periods with a wide range of mass budgets distributed over the period of 2.3 years which has been targeted by ESA for the launch of Solar Orbiter.

The emphasis on SEP for the future Solar System exploration missions of ESA has led this agency to define a technology mission for testing SEP, SMART-1. This 360 kg spacecraft has been successfully launched in late September 2003 on a Geostationary Transfer Orbit as a companion satellite on a commercial launch of Ariane 5. The plasma

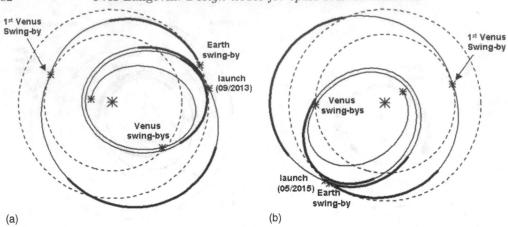

FIGURE 2.55. Short mission options for Solar Orbiter with launches in September 2013 (a) and May 2015 (b). A 1:1 Earth–Venus–Earth gravity assist strategy complemented by aphelion thrusting (thick arc) increases the encounter velocity with the Earth up to more than 9 km/s. In a second phase, a DVGA strategy on a type IV Earth–Venus transfer provides the required encounter velocity with Venus of more than 19 km/s. The cruise phase from the Earth to the first Venus swing-by at 19 km/s is only 1.8 years.

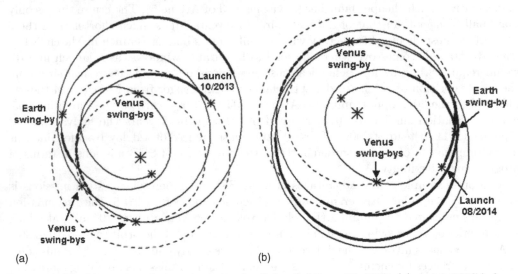

FIGURE 2.56. Alternate mission strategies for Solar Orbiter with SEP. (a) A 2:1 DVGA from Venus to Venus is implemented so as to increase the relative velocity. It is followed by a Venus–Earth–Venus gravity assist. (b) A DVGA approach is applied to a type IV transfer from the Earth to Venus (1.4 orbits around the Sun), again followed by a Venus–Earth–Venus gravity assist.

engine has a thrust of 70 mN, hence the mass load of 5000 kg/N is fully representative of future missions. Since launch, the SEP system has been used to enlarge the orbit by nearly continuous thrusting (Fig. 2.57(a)), getting clear of the radiation belts of the Earth by early January, then further enlarging the orbit, which had a period of nearly 30 hours by mid-march (to be compared with 11 h for the GTO).

At the end of 2004, SMART-1 has benefited from a series of distant gravity assists by the Moon, which has drastically increased the perigee. This has made possible a capture by the lunar gravity field. SMART-1 has spiralled down to a polar orbit with its pericentre over the South pole. With its two spirals combined with lunar gravity assists

TABLE 2.8. Mission opportunities for Solar Orbiter from January 2013 to April 2015.

Launch date	Launch (km/s)	Cruise time (years)	delta V (km/s)	Mass budget (kg, Soyuz "T")
01/2013	1.40	4.20	2.61	1790
01/2013	1.40	3.66	4.24	1720
09/2013	2.61	2.93	3.23	1400
10/2013	2.98	2.13	4.09	1310
11/2013	2.41	3.44	1.14	1500
03/2014	1.40	3.26	4.25	1719
03/2014	1.40	3.88	3.08	1820
08/2014	1.40	3.69	4.40	1713
04/2015	2.70	2.16	4.12	1360

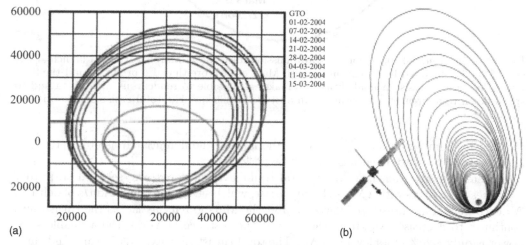

(a)   (b)

FIGURE 2.57. (a) Orbital evolution of the SMART-1 spacecraft over the first 6 months of the mission. Notwithstanding some operating problems due to radiation, the plasma engine has been operated for a large fraction of the time until February, so as to escape the radiation belts of the Earth, which extend up to 20,000 km. After a pause due to long eclipses, the spacecraft has now resumed thrusting close to perigee, so as to increase the apogee until it reaches the vicinity of the Moon. (b) After being captured by the Moon following a series of weak lunar gravity assists, SMART-1 has spiralled down to its operational orbit.

and its total delta V of 3.4 km/s (in the same range as Solar Orbiter and nearly half of that required for BepiColombo), SMART-1 constitutes a very representative test for inner solar system missions. Its operating environment has actually been more severe than interplanetary missions, with the large flux of high-energy particles in the radiation belts of the Earth. The successful completion of the SMART-1 mission constitutes a very positive development for future ESA missions relying on SEP.

As indicated in Section 2.5.1, SEP becomes less attractive as one moves away from the Sun, due to the rapid decrease of the available solar energy. This decrease is actually somewhat slower than the expected inverse square law, as the temperature of the panels also decreases, which improves their efficiency. Nevertheless, the practical limit for SEP missions is the main belt of asteroids. NASA has selected a Discovery mission, Dawn, which uses SEP to perform a rendezvous with 4 Vesta, the largest differentiated asteroid

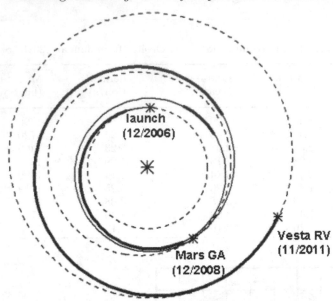

FIGURE 2.58. Nominal option for the Dawn mission until the Vesta rendezvous. A launch at the end of 2006 is followed by an encounter with Mars after more than 1.5 orbits around the Sun. After this gravity assist, a final thrust arc makes it possible to rendezvous with Vesta and to spiral down to a circular polar orbit at an altitude of 300 km. The total cruise duration is less than 5 years.

(560 km in diameter), then with 1 Ceres, the largest undifferentiated asteroid (1000 km in diameter), if an extended mission can be performed.

Not surprisingly, even if a direct spiral could be considered, the selected trajectory combines thrust arcs with a Mars Gravity assist, which saves more than 2 km/s of delta V. An extremely long thrust arc is needed after the Mars gravity assist, as the thrust level continues to decrease away from the Sun: at the distance of Vesta, the available solar power is only $\sim$20% of its level at 1 AU. The total thrust time exceeds 3 years, and three identical thrusters have to be carried so as to overcome the lifetime limitations. The total delta V ($<$7 km/s) is comparable to that required for BepiColombo.

From Jupiter outward, Nuclear Electric Propulsion is the prime candidate for missions implementing advanced propulsion methods. NEP missions are by design very ambitious affairs, with a very large spacecraft. The main perspective is that offered by the JIMO study of NASA, which should perform an orbital survey of three Galilean satellites (Callisto, Ganymede and Europa) and possibly Io as well. Europa and its buried ocean is the prime target of the mission. With its large mass (up to 10 tons, with at least 6 tons for the nuclear reactor and nuclear propulsion stage), a JIMO type mission can implement a "brute force" spiral approach both for leaving the Earth and for spiralling down into the system of Jupiter. Given the large available power, it makes sense to use the highest possible ejection velocity (e.g. 100 km/s). Thrust levels of 5 N are considered, which provide relatively low-mass loads of 2000 kg/N. It should be noted that at Jupiter, the acceleration level provided by the engine (5×10^{-4} m/s^2) is twice larger than solar gravity (2.4×10^{-4} m/s^2), making it possible to depart markedly from Keplerian orbits. Even with such performances, NEP missions can benefit from combining low thrust with gravity assist techniques.

If one compares a direct spiral approach with a combination of a lunar launch and a 1:1 DVGA (Fig. 2.59), the savings in terms of delta V ($\sim$5 km/s), correspond to only

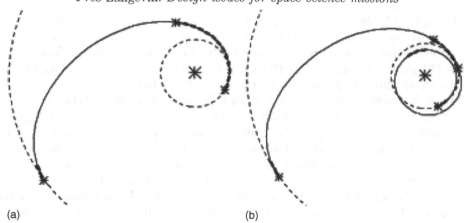

(a)   (b)

FIGURE 2.59. (a) Options for NEP missions to Jupiter. A direct spiral is shorter by 1 year than a transfer using a lunar launch. (b) A 1:1 DVGA, but the delta V requirement is increased by up to 5 km/s. Solid thick lines correspond to braking, dashed thick lines correspond to acceleration.

5% of the mass (assuming an ejection velocity of 100 km/s). However, when applied to the very large total mass of a JIMO type spacecraft (8–10 tons), the improvement in the mass budget represents 400–500 kg, which would make it possible to deploy an additional orbiter around one of the Galilean satellites.

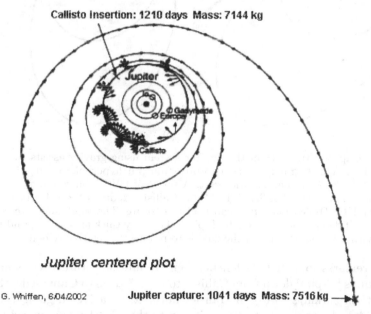

FIGURE 2.60. Direct spiral around Jupiter leading to a capture in orbit around Callisto, from a preliminary study of JIMO by NASA; 5 km/s of delta V are required from the capture by Jupiter to the capture by Callisto.

The relatively large accelerations provided by NEP systems are quite effective for matching the velocity of Jupiter, as demonstrated by the small thrust arc close to arrival in Fig. 2.59. Once close to Jupiter, the gravity field of this giant planet much exceeds the NEP acceleration (by a factor of 70 at the distance of Callisto). It is still possible to implement a capture strategy that is more effective than a slow spiral: 5 km/s are needed from the capture by Jupiter to the capture by Callisto, compared to 8.2 km/s for a near circular spiral (as explained in Section 2.5.2, the cost from infinity down to a circular orbit is then equal to the circular velocity at this orbit, here 8.2 km/s). This strategy requires 170 days. It is far more effective (and not much longer) to use gravity assists by the Galilean satellites themselves during the capture sequence.

With a first capture from heliocentric orbit followed by a series of DVGA on resonant orbits (Fig. 2.61), it is possible to limit the cost of capture to a value of 1.3 (capture by Ganymede) – 1.4 km/s (capture by Callisto). This is similar to the approach velocity, hence the savings for a capture by Callisto is in the range of 5 km/s. The improvement on the delta V budget would reach 7.7 km/s for a capture by Ganymede, as the local circular orbital velocity is 10.9 km/s, 2.7 km/s larger than that of Callisto. The time required ranges from 200 (capture by Ganymede) to 240 days (capture by Callisto), while one should add 15–20 days to the 170 days of the direct spiral capture so as to match the 1.3 km/s relative velocity which is the starting condition for Fig. 2.61. The improvement of the mass budget by 500 kg or more is definitely worth the slight delay.

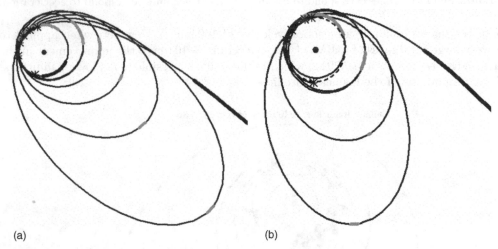

(a)   (b)

FIGURE 2.61. Capture strategies in the Jupiter system using gravity assists by Ganymede (a) and Callisto (b). A first gravity assist captures from an hyperbolic orbit with an approach velocity of 1.2–1.4 km/s into an elliptic orbit. A series of DVGA on resonant orbits (11:1, 5:1, 3:1, 2:1, 3:2 for Ganymede, 6:1, 3:1, 2:1, 3:2 for Callisto) reduces the velocity down to a level at which a final 1:1 DVGA maneuver can lead to capture. The solid thick arc corresponds to the final stages of deceleration relative to Jupiter. The grey thick arcs correspond to the DVGA maneuvers, with a forward thrust as the goal is to reduce the relative velocity.

The same remarks apply to transfers between satellites: a "brute force" spiral requires a total of 9.1 km/s to spiral down from Callisto to Io (2.7 km/s to Ganymede, then 2.8 km/s to Europa, then 3.6 km/s to Io). A transfer initiated by a 1:1 outgoing DVGA, then a second DVGA to bring the relative velocity up to the 1.5–2 km/s required for a transfer orbit, then a third DVGA to lower the relative velocity, then an ingoing 1:1 DVGA to be captured by the next Galilean satellite can reduce the delta V requirements by a factor of more than 3, with a similar travel time of a few months. The implementation of strategies combining NEP thrust and satellite gravity assists can reduce the cost of capturing in the Jupiter system, then transferring between satellites from more than 10 to less than 2 km/s.

The circular velocities around the Galilean satellites at a safe altitude of 1.1 radii are 1.6, 1.85, 1.4 and 1.7 km/s for Callisto, Ganymede, Europa and Io, respectively. Apart from the transfers, a tour including Callisto, Ganymede and Europa requires 8.3 km/s for spiralling in, then out of orbit around Callisto, then Ganymede, and finally spiralling down to a circular orbit around Europa. There are no more tricks to be played once in orbit around a Galilean satellite, and these costs are unavoidable. However, if JIMO runs into mass problems, a DVGA capture by Ganymede, followed by a spiral in and out of orbit around Ganymede followed by a DVGA transfer to Europa, then a spiral down to circular orbit would be an attractive alternative: the delta V cost would then be less than 6 km/s after capture by Jupiter, to be compared with nearly 19 km/s with a scenario relying on a spiral capture to Callisto then spiral transfers from Callisto to Ganymede to Europa.

As a conclusion of this section on high specific impulse propulsion methods, the examples of BepiColombo, Solar Orbiter, Dawn or JIMO demonstrate that low thrust propulsion should whenever possible be combined with gravity assist techniques. The negative impact on cruise time and programmatic flexibility is marginal, while the improved mass and thrust time budgets can make the difference between a viable mission and a runaway monster which will cost itself out of the program.

This is a somewhat surprising general conclusion: new propulsion techniques could have been expected to simplify mission analysis by providing such large on-board capabilities that one would not need to rely on complex mission strategies. However, reducing requirements on on-board systems and launchers remains mandatory, in particular for science missions which always push these constraints to their limits. The range of available strategies which have to be assessed by mission analysis is far wider with low thrust propulsion than with chemical propulsion, which will make it very difficult indeed to be sure that the optimum strategy for achieving a given science objective has been uncovered.

REFERENCES

BELBRUNO, E. & AMATA, G.B. *August 1996 Low Energy Transfer to Mars and the Moon Using Fuzzy Boundary Theory*, ALENIA doc. SD-RP-AI-0202, Final Report of ESA Contract.

FARQUHAR, R. Halo orbits and lunar swing by missions of the 1990s, *Acta Astronautica*, **24**, 227–234.

HECHLER, M. & COBOS, J. *Herschel, Planck and GAIA Orbit Design*. 7th International Conference on Libration Point Orbits and Application Parador d'Aiguablava Girona, Spain 10–14 June, 2002. http://europa.ieec.fcr.es/libpoint/presentations/hechler.pdf

3. Instrumentation in X-ray Astronomy

By X. BARCONS

Instituto de Física de Cantabria (Consejo Superior de Investigaciones Científicas-Universidad de Cantabria), Santander, Spain

In this set of lectures I discuss instrumentation for astronomical X-ray observatories. In particular I first briefly outline the physical processes that are observed in cosmic X-ray sources, and then I discuss X-ray telescopes and X-ray detectors. An overview of the development of X-ray Astronomy, since its beginnings in the 1960s to today's missions, follows. The lectures end with a look into the future with special emphasis in the next decade or two.

3.1. Motivation: astrophysics in the high-energy domain

Observing the Universe at X-ray wavelengths is not an easy task. X-rays do not reach the Earth's surface and therefore all X-ray observatories need to be space borne. X-rays are not particularly easy to collect: normal incidence to a mirror, no matter how reflecting it is, fails to reflect X-rays in any sizeable amount. In addition, there are not very many X-rays in cosmic sources. Each one, however, carries a lot of energy and can only be produced in a very extreme environment.

This is exactly the science driver for X-ray Astronomy. Since the very early days of X-ray Astronomy (1962, see Section 3.4), it was realized that X-ray observations produced in no way redundant information to other wavelengths. Totally inconspicuous optical sources turn out to be the strongest X-ray emitters. As an example the X-ray to optical flux ratio is around 10^{-4} for a normal galaxy like our own, but it can be 1 or even 10 for an active galaxy.

What appears to do the trick of producing significant X-ray emission is the presence of strong gravity in some sense. The majority of the extragalactic X-ray sources known (Active Galactic Nuclei, AGN) obtain their energy from accretion onto a supermassive black hole (i.e. from a very deep gravitational potential well). Clusters of galaxies emit X-rays because large amounts of gas are trapped in the shallow but very extended potential well of the same dark matter that binds the galaxies into the cluster. Accreting binaries are also a further manifest of the same process that powers AGNs, where the accreting material is provided by the companion star onto a compact one (white dwarf, neutron star or black hole). Stars with very active coronae (possibly caused by both gravity and magnetic fields) are also strong X-ray emitters.

With current highly sensitive X-ray observatories, X-ray emission is found among all classes of astronomical objects, from solar system objects to the farthest galaxies and quasi-stellar objects (QSOs). The Universe is filled with X-ray radiation, the X-ray background (XRB), which contains about 10^{-4} times the energy of the cosmic microwave background. The XRB comes from the integrated emission of all cosmic X-ray sources, most of them AGN, of which a very large fraction are absorbed and therefore inconspicuous at virtually all wavelengths other than hard X-rays and possibly the far infrared.

Prior to the launch of the current generation of X-ray observatories, National Aeronautics and Space Administration's (NASA) *Chandra*, European Space Agency's (ESA) *XMM-Newton* to be joined very soon by Japan Aerospace Exploration Agency's (JAXA) *ASTRO-E2*, X-ray Astronomy was driven by serendipity. Each new mission reported many more unexpected discoveries, than deeper knowledge of previously known sources. To some extent the situation is now reversing. *Chandra* and *XMM-Newton* are delivering their share of unexpected new phenomena, but the amount of *astrophysics* that is being carried out with these observatories on previously known sources is now fractionally much larger than in former missions. In the 0.1–10 keV band, X-ray Astronomy appears now to have cleaned many of the surprises, paralleling the very high fraction of the XRB resolved (90% below 2 keV and 50% up to 10 keV).

X-ray Astronomy still has a long way to go to match observing facilities at other wavebands. *Chandra* is the first observatory to achieve sub-arcsecond angular resolution, but with limited collecting area. The highest spectral resolution is achieved with wavelength-dispersion spectrographs, but the spectral resolution has not yet reached $E/\Delta E \sim 1000$. To the amazement of optical Astronomers, this is called *high* spectral resolution in X-ray Astronomy. The *thermal* spectroscopy limit (i.e. the capability of resolving lines thermally broadened only) is still in the distant (but perhaps foreseeable) future. The largest conceived (not even planned) X-ray telescopes call for a collecting area of $\sim$100 m^2, similar to the largest existing ground-based optical facilities, but at the moment our largest X-ray observatory *XMM-Newton* has the same effective area than an amateur's society optical telescope.

This presentation is an attempt to put some order and logics in a set of post-graduate lectures that I have collected from previous years, on instrumentation for X-ray Astronomy. To understand what we need to observe and what are the requirements on instrumentation, I discuss the physical processes that are relevant to cosmic X-ray emission (Section 3.2). In Section 3.3 I present and discuss the various approaches to X-ray telescope optics. Section 3.4 is devoted to detectors and instruments. Section 3.5 presents a personal view on the development of X-ray Astronomy, with emphasis on the current missions, their capabilities and shortfalls. I finish with a look at the future, including the various planned or understudy missions by the big space agencies.

Throughout this contribution, I tend to discuss in more detail ESA or ESA-related missions, as this is the topic of this Winter School.

3.2. Physical processes in high-energy astrophysics

To design and specify the requirements of a high-energy astrophysics instrument, it is mandatory to know the physical processes that will be observed and what are their spectral signatures. Continuum processes are very nicely reviewed and explained in terms of basic physics by Longair (1992), while details are deeply exposed in Rybicki and Lightman (1979). The physics and astrophysics of partially ionized plasmas (leading to discrete spectral features) are very well reviewed by Mewe (1999) and Liedahl (1999). Please consult these references for details, as the account given here is nothing but a very coarse summary.

3.2.1. *Synchrotron radiation*

The basic ingredients of this important continuum process at many wavelengths are strong magnetic fields and relativistic electrons. The Lorentz force exerted by a magnetic field B on a moving electron, makes it spin around the direction of the magnetic field with a *gyro* frequency $\omega_B = eB/(2\gamma mc)$ ($\omega = 2\pi\nu$), where γ is the electron's Lorentz factor. Since the electron is accelerated due to its curved trajectory it emits radiation (*cyclotron* radiation) which is mostly monochromatic at the gyro frequency. These electrons can also absorb radiation at that frequency.

When the electron moves close to the speed of light c, γ becomes large and then several effects happen, that qualitatively change the above monochromatic spectrum. The first one is that higher-order harmonics of the gyro frequency become more important. The case of the isolated neutron star 1E1207.4-5209 (Bignami *et al.*, 2003) shows a nice detection of 3 (maybe 4) cyclotron line absorption harmonics, enabling the measurement of the star's magnetic field. In the very large γ limit, the harmonics superpose to each other and instead of a resolved discrete spectrum, a continuum spectrum is emitted. The second phenomenon happening is relativistic beaming, and therefore we (observers) receive highly amplified emission, but only when the electron moves towards us, within a cone of opening angle $\sim 2/\gamma$. The third relativistic effect is the Doppler effect which boosts the frequency where we detect the radiation by a factor γ^2 with respect to the emitted one. These effects conspire so that the emission spectrum becomes a very broad distribution centered around a characteristic frequency $\omega_c = 3eB/(4\pi mc)\gamma^2 \sin\Theta$, where Θ is the angle between the direction of the line of sight with respect to the magnetic field. The total power radiated by synchrotron emission of one electron is:

$$\frac{dE}{dt} = \frac{4}{3}\sigma_T c \left(\frac{v}{c}\right) \gamma^2 U_B \tag{3.1}$$

where $\sigma_T = 6.65 \times 10^{-25}$ cm^2 is the Thomson scattering cross-section and $U_B = B^2/(8\pi)$ is the energy density of the magnetic field.

If we finally assume an isotropic distribution of electrons and that their energies follow a power-law distribution (see later for a justification on this) $N(\gamma) \propto \gamma^{-p}$, then the emitted spectrum radiated by synchrotron radiation is $dE/d\omega \propto \omega^{-\alpha}$, where the energy spectral index (as used in X-ray Astronomy, please note the minus sign difference with radio astronomy) is $\alpha = (p-1)/2$. Therefore a power-law distribution in the electron's energy $mc^2\gamma$ leads to a power-law in the synchrotron emission spectrum.

Indeed, a power-law spectrum cannot indefinitely continue to low frequencies. The same electrons that produce synchrotron emission can ultimately absorb that radiation if there is enough electron column along the line of sight. That gives rise to a "synchrotron self-absorbed" spectrum, whereby the power-law exhibited at high frequencies turns into a $\omega^{5/2}$ spectrum at low frequencies. The turnaround frequency depends ultimately on the electron column density and its specific location can make a big difference concerning the radio properties of the source.

3.2.2. *Bremsstrahlung*

In any ionized plasma, electrons moving close to ions deviate their trajectories (mostly due to Coulomb electrostatic force) and therefore emit "braking" radiation (or *bremsstrahlung*). For this processes to be relatively efficient, high temperatures (say $T > 10^6$ K) are needed in addition to having partially or totally ionized atoms. At temperatures exceeding 5×10^9 K, electron–electron bremsstrahlung becomes as important as electron–ion bremsstrahlung and corrections to the estimates below need to be applied.

The shape of the spectrum and the radiated power through this process can be obtained by considering a single electron moving towards an ion at rest with velocity v and with impact parameter b. Very simple arguments lead to the existence of a cutoff frequency $\omega_0 = v/2b$, above which there is very little power emitted. Specifically, the radiated power drops as:

$$\frac{dE}{d\omega} \propto \exp\left(-\frac{\hbar\omega}{\omega_0}\right),$$

flattening to a constant at frequencies $\omega < \omega_0$.

For a thermal electron gas at temperature T, the emission spectrum is:

$$\frac{dE}{dV d\omega} \propto Z^2 n_i n_e T^{-1/2} g(\omega, T) \exp\left(-\frac{\hbar\omega}{kT}\right) \tag{3.2}$$

where n_i is the ion density (of effective atomic number Z), n_e is the electron number density and $g(\omega, T)$ is a correction called *Gaunt* factor. The total amount of power emitted per unit volume is:

$$\frac{dE}{dt\,dV} = 1.4310^{-41} Z^2 T^{1/2} n_i n_e g(T)\,\text{erg/cm}^3/\text{s} \tag{3.3}$$

Note that unless the plasma is very dense, bremsstrahlung (being a two-body process) is usually very inefficient in cooling it. Bremsstrahlung emission is relevant when no other channels are available (e.g. no strong magnetic fields).

3.2.3. *Compton scattering*

Compton scattering is the elastic scattering between electrons and photons. In the classical interpretation of radiation, the energy exchange between both is small and it is called the *Thomson* effect. In general there is energy exchanged between electrons and photons, direct Compton referring to photons transferring energy to electrons, and inverse Compton to the opposite energy flow.

The total power scattered via Thomson effect by an electron at rest is:

$$\frac{dE}{dt} = c\sigma_T U_\gamma \tag{3.4}$$

where σ_T is the Thomson cross-section and U_γ is the incident photon density. In a proper relativistic approach, the electron can move at very high speed after the scattering if the incoming photon energy is high. This modifies the cross-section which is now smaller (the Klein–Nishina cross-section), which for comparison is only $\sim \sigma_T/3$ at $\sim 1\,\text{MeV}$. This increasing inefficiency at high energies turns out to be a general property of Compton scattering.

However, looking at photons as particles, the elastic collision between a photon and an electron can have energy transfer between both, as far as the total energy is preserved. For a target electron at rest the energy of the outgoing photon E_{out} can be related to the energy of the incoming photon E_{in} and the scattering angle θ by:

$$E_{\text{out}} = \frac{E_{\text{in}}}{1 + E_{\text{in}}/mc^2 (1 - \cos\theta)} \tag{3.5}$$

In the opposite limit, when the electron is initially moving at relativistic speeds $\gamma \gg 1$, we have:

$$E_{\text{out}} \sim \gamma^2 E_{\text{in}} \tag{3.6}$$

Note in this last case that energy is very efficiently transferred from electrons to photons. There is indeed a limit to that energy exchange which is the electrons kinetic energy $mc^2\gamma$. The total power transferred from the electron to the photons is:

$$\frac{dE}{dt} = \frac{4}{3}\sigma_T c U_\gamma \left(\frac{v}{c}\right)^2 \gamma^2 \tag{3.7}$$

This shows that the influence of Compton scattering on the emission spectrum grows very rapidly ($\propto \gamma^2$) when there are relativistic electrons in the source.

If we now think of a thermal electron gas at a temperature T, the average energy dumped from the electrons to the photons in each collision is:

$$\frac{\Delta E}{E} = \frac{4kT - E}{mc^2} \tag{3.8}$$

which is valid for both direct and inverse Compton effect. Note that Compton scattering always tends to "equilibrate" the temperatures of the electron and the photon populations. The depth of this process (Compton depth) is $\tau_T \approx \sigma_T N_e$, where we assume low photon energies (far from the Klein–Nishina regime) and an electron column density

$N_e = \int \mathrm{d}x n_e$. The number of Compton scatters experienced by a photon before it leaves the source is $\sim \tau_T$ if $\tau_T < 1$. But if the Compton depth is large, then photons follow a "random walk" trajectory instead of an approximately straight line and therefore they undergo $\sim \tau_T^2$ collisions before they leave the source. We define the Comptonization parameter for the thermal gas as:

$$y = \frac{kT}{mc^2} \times \text{Number of collisions} \qquad (3.9)$$

After undergoing all collisions, the energy of the input photon is modified to $E_{\mathrm{out}} = E_{\mathrm{in}} e^{4y}$.

When the Compton parameter y is large, the incident spectrum is significantly modified before the photons leave the source and it is said to be *Comptonized*. The effects of Comptonization can be studied by using the much celebrated Kompane'ets equation if the Compton depth is large (note that this is a Fokker–Planck diffusion-like equation and therefore it can only be guaranteed to work in the random walk limit). If the spectrum is only weakly Comptonized, then a more careful treatment needs to be conducted. Comptonized radiation always ends up in a Bose–Einstein-like spectral distribution at high energies (only a black-body if electrons and photons have reached real thermal equilibrium). This produces a very characteristic hump decaying with a Wien law at high energies, which is seen in Comptonized sources at X-ray and γ-ray energies.

Since, once again, Compton scattering is an elastic process, all the energy gained by photons is lost by electrons. For relativistic electrons, the cooling rate is (see Equation 3.7) $\dot{\gamma} \propto \gamma^2$, so the cooling time is $t_{\mathrm{cool}} = \gamma/\dot{\gamma} = t_{\mathrm{esc}} 3\pi/(l\gamma)$. Therefore, in a compact source Compton cooling is likely to play an important role (see Section 3.2.4 for the definition of the compactness parameter l).

X-ray spectroscopy has been for decades that of power-laws. When more sensitive, higher spectral resolution observatories have become available, deviations from power-laws have been found and attributed to various processes, but still an underlying power-law is unavoidable. How is a power-law spectrum produced? Compton cooling provides a simple physical example on how to produce an electron power-law distribution, regardless of the injected distribution of electron energies. Let $N(\gamma)$ be the number density of electrons with Lorentz factor γ, and $Q(\gamma)$ the rate of electrons injected at that Lorentz factor. The balance equation:

$$\frac{\partial N(\gamma)}{\partial t} + \frac{\partial}{\partial \gamma} \left(\dot{\gamma} N(\gamma) \right) = Q(\gamma)$$

has the following steady solution:

$$N_{\mathrm{st}}(\gamma) = \frac{1}{\dot{\gamma}} \int_{\gamma}^{\infty} \mathrm{d}\gamma\, Q(\gamma) \qquad (3.10)$$

where $\dot{\gamma}$ is the Compton cooling. If electrons are injected at a single energy γ_{max}, then the steady Compton cooled distribution gives a power-law $N_{\mathrm{st}}(\gamma) \propto \gamma^{-2}$. If synchrotron radiation dominates, the energy spectral index would be $\alpha = 0.5$. If, on the contrary, a power-law distribution of electrons is injected $Q(\gamma) \propto \gamma^{-b}$, then the Compton cooled electrons will be distributed as $N_{\mathrm{st}}(\gamma) \propto \gamma^{-1-b}$ if $b > 1$ or $N_{\mathrm{st}}(\gamma) \propto \gamma^{-2}$ for $b < 1$ (as in the monoenergetic injection). If electrons radiate via synchrotron radiation, they will produce a power-law continuum with $\alpha = 0.5$ for $b < 1$ or $\alpha = b/2$ for $b > 1$.

3.2.4. *Electron--positron pairs*

A very friendly review on electron–positron pairs and the observational effects of their existence in astrophysical plasmas can be found in Svensson (1990). Electron–positron pairs can be created in photon–photon collisions if photons have enough energy and if they have a reasonable chance to collide. The first condition is an *energy* one, and requires $E_1 E_2 > (mc^2)^2$ for the energies of the colliding photons E_1 and E_2. Note that pairs can be created out of the interaction between very high-energy photons and, say, microwaves.

The cross-section for pair creation $\sigma_{e^- \, e^-}$ peaks at $(E_1 E_2)^2 \sim 1\,\mathrm{MeV}$ and has a value of about $0.5\sigma_T$ (σ_T is the Thomson scattering cross-section). This is the energy where photons are most likely to create pairs. But for this process to play any role, the density of the photon gas n_γ needs to be large enough for binary collisions to happen. The optical depth for photon–photon interactions is $\tau_{\gamma\gamma} = n_\gamma \sigma_{e^- \, e^-} R$, where R is the size of the source. Since most of the action takes place at photon energies $\sim mc^2$, we estimate n_γ at that energy from the luminosity of the source L. Defining the *compactness* parameter as:

$$l = \frac{L\sigma_T}{Rmc^3} \tag{3.11}$$

we find $\tau_{\gamma\gamma} \approx l/60$. This means that for pairs to be efficiently created, the compactness parameter l must be large (close to ~ 100). This parameter can be estimated from observations, where a lower bound on the source size R could be obtained from variability. Many AGNs that have been found and studied at γ-ray energies indicate that they might be compact and since γ-ray photons are observed, electron–positron pairs might play a role.

There are a number of effects that pairs can create in a plasma. The first one is obviously an electron–positron annihilation line at 511 keV. This has been observed in a number of sources. A further effect, which can have an impact on the X-ray continuum, is known as *pair cascades*. These can be understood as a sequence of Compton upscatterings of photons off high-energy electrons, followed by a photon–photon pair creation. Each one of the pair particles will share about half of the energy of the primary energetic electron. If they have enough energy to Compton upscatter a further photon the cycle can continue. If it occurs many times, then we have a saturated pair cascade. It can be easily seen that a far from saturated pair cascade started by an injection of mono-energetic electrons leads to an energy spectral index of 0.5, but as saturation progresses with the compactness of the source the energy spectral index approaches 1. This has been proposed as a basic process to explain the underlying power-law observed in accreting sources.

3.2.5. *Atoms and ions*

Astrophysical plasmas very often have temperatures where most of the H and He atoms are fully ionized, but the remaining elements are only partly ionized (coronae of active stars, supernova remnants, galaxies and clusters of galaxies, etc.). This causes a number of atomic processes to play an important role in the structure and the emission spectrum of these plasmas. A comprehensive review of the structure and spectra of astrophysical plasmas can be found in Mewe (1999) for coronal plasmas and Liedahl (1999) for nebular (photoionized) plasmas.

The most important processes that lead to emission at temperatures $T < 10^8\,\mathrm{K}$ are:

- *Thermal bremsstrahlung*: This process dominates at the highest temperatures, where all species are fully ionized.
- *Bound-bound emission lines*: These give rise to (permitted or not) emission lines, which might be dominant at temperatures $T < 5 \times 10^7\,\mathrm{K}$
- *Radiation recombination continua*: Result from the capture of a free electron to a bound state. The process is important at temperatures $T < 10^7\,\mathrm{K}$.
- *Di-electronic recombination lines*: Capture of a free electron giving rise to a doubly-excited state and therefore to two emission lines.
- *Two-photon continuum*: Simultaneous emission of two photons from a meta-stable state.

The structure of astrophysical plasmas depends on their chemical composition, density and intensity of any ionizing field. Two complementary simplified models are commonly used to describe the ionization state of atoms and ions in optically thin or moderately thick plasmas: the *coronal plasma*, where collisions are the main cause of the ionization, and the *nebular plasma*, where the ionization is driven by photo-ionization from an external field. Examples of the former include stellar coronae, supernova remnants and intra-cluster gas. Examples of the latter are AGN and planetary nebulae. The borderline between both extreme cases is dictated by the value of the ionization parameter:

$$\xi = \frac{L}{n_e r^2} \tag{3.12}$$

where L is the luminosity of the ionizing source, r the distance to it and n_e the electron density. For small values of the ionization parameter (usually measured in units of erg cm/s) collisions dominate the ionization, but when this parameter is large, the plasma is said to be *over-ionized* with respect to its temperature (indeed due to the effects of the ionizing field).

Once the ionization structure of a plasma has been found, then the emission spectrum can be computed. The presence or absence of emission lines and other features can be used to *diagnose* the plasma, that is to constrain the existence or absence of specific ion species and therefore the temperature and ionization parameter of the plasma. There are a number of diagnostics that can be used at X-ray photon energies to measure the physical parameters of the plasma.

Measuring the plasma density directly can be done using the so-called He-like triplet of 2-electron ions, which constitutes a very useful diagnostic in X-ray astronomical plasmas. In a 2-electron ion, the first excited states are the $1s2s\ ^1S$, $1s2s\ ^3S$, $1s2p\ ^3P$ and $1s2p\ ^1P$. Transitions to the ground state can occur only from the last 3 excited states to the ground state $1s^2\ ^1S$, giving rise to the so-called forbidden (F, magnetic dipole), inter-combination (I, magnetic quadrupole and electric dipole) and resonance (R, electric dipole) lines. Since the excitation from the ^3S to the ^3P state is driven by collisions, the ratio between the intensities of the F and I lines is a function of the plasma density: the larger the density, the more populated the ^3P state with respect to the ^3S state, and therefore the smaller the ratio between the F and I line intensities. The expected behaviour of the line ratio with the density is $F/I \propto (1 + n/n_{\mathrm{crit}})^{-1}$, where n is the plasma number density and n_{crit} is a critical density that can be computed from atomic physics parameters. In order to be able to conduct this diagnostic, the triplet needs to be resolved by the spectrograph, posing a demanding requirement on this type of X-ray instrumentation. For guidance, the OVII line energies are R: 21.60 Å, I: 21.80 Å and F: 22.10 Å.

3.2.6. *Atomic absorption*

Electrons in atoms can also absorb high-energy (X-ray) radiation. There are two processes relevant:

• *Bound-bound absorption*: This is the opposite process to bound-bound emission and gives rise to resonance absorption lines. Note that only permitted (electric dipole) transitions are relevant for this process.

• *Bound-free absorption*: This is the opposite process to free-bound recombination; that is, a photon is absorbed by an atom and its energy used to free a bound electron, the remaining energy being taken away by the electron.

The latter process is very commonly observed in many X-ray sources, even at very moderate spectral resolution. The bound-free absorption cross-section is $\propto E^{-3}$, at $E > E_0$, where E is the photon energy and E_0 is the ionization energy of that particular electron.

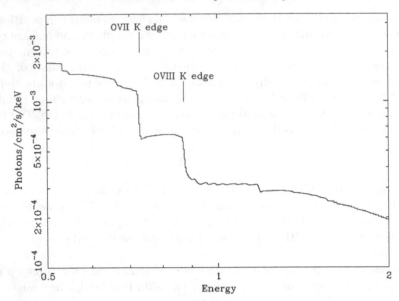

FIGURE 3.1. Illustration of the effects of a moderately ionized absorber with $N_H = 10^{22}/\text{cm}^2$ with the most prominent absorption edges marked.

The shape of the features observed is shown in Fig. 3.1. The most important element for absorption is oxygen, and the absorption "edge" K-shell energies are at 0.529 keV for OI, 0.739 keV for OVII and 0.874 for OVIII. The Fe K absorption edge energies range from 7.1 keV for neutral Fe to 9 keV for hydrogen-line Fe.

3.3. X-ray telescope optics

Obtaining images with high-energy photons needs a completely different strategy compared to the optical or radio light collecting devices. Normal incidence of X-rays onto a metal surface results primarily in either absorption or transmission but only to a very small extent in reflection. Several techniques have been devised along the years to form images under these circumstances. Some of them consist of multiplexing the sky signal and then deconvolving it onto real images (collimators, coded masks). For soft X-rays, grazing incidence optics provides a way to obtain direct images by focusing the incoming X-rays into the focal plane. These systems are reviewed in this chapter with their trade-offs, since the imaging system is an integral part of high-energy observatory payloads.

3.3.1. *X-ray collimators*

The first multiplexing imaging devices to be used in X-ray Astronomy were the collimators. These consist (see Fig. 3.2) of a set of metallic walls that limit the path of X-ray photons to the detector when they come from off-axis directions. For perfectly absorbing walls, the Field Of View (FOV) of a collimator is limited to angles $\theta < \theta_{\max}$, where $\tan \theta_{\max} = a/h$, and it is therefore dictated by the aspect ratio of the walls. Indeed it is very difficult to achieve FOV of less than a degree, as in that case the walls would have to be prohibitively high.

The collimator angular response is approximately triangular with the maximum along the walls (no photons are then absorbed), going down to zero at the edges of the FOV. Deviations from this ideal behaviour occur as there is some probability for the off-axis

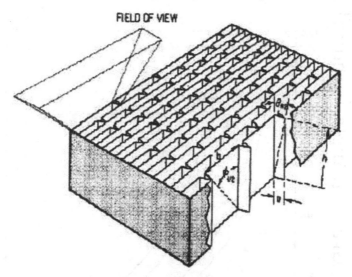

FIELD OF VIEW

FIGURE 3.2. Scheme of the collimator of the large-area proportional counter (LAC) on board *Ginga*, after Turner *et al.* (1989).

X-rays to penetrate the walls and ultimately landing in the detector. This probability increases with photon energy. The effective area is $A_{\text{eff}} = N \times a^2$, where N is the number of collimator cells. Note that if the angular resolution needs to be improved at the expense of reducing a, the number of collimator walls will need to be increased accordingly to keep the same value of A_{eff}.

The angular resolution of a collimator depends on which mode it operates. If in pointing mode, then the angular resolution is the FOV itself. If, as usual, the collimator is operated in scanning mode, then the multiplexing comes into play and the resolution can be improved along the scan direction by "deconvolving" the obtained sequence of observations.

One way to improve the angular resolution of collimators is the "modulation collimator", operated in scan mode. In that device, the walls were substituted by two sets of blocking metal rods, one placed just in front of the detector and the other one at a distance d in front of it. The rods are separated by a distance a among themselves. When a source enters the FOV, during the scan the signal appears and disappears as a consequence of the shadowing produced by the modulation collimator. In this way, the *HEAO-1* A3 modulation collimator was able to locate the position of bright sources with an accuracy of $\sim$1 arcmin along the scan direction (see Section 3.4). By scanning the same source with different roll angles, the source could be positioned down to that accuracy. Collimators are still in operation (e.g. in the *Rossi X-ray timing experiment XTE* mission), but since their major limitation in sensitivity is source confusion, they are only used to study bright enough sources.

3.3.2. *X-ray grazing incidence optics*

Incidence almost parallel to the reflecting surface has proven to be the most efficient way to produce real X-ray images by focusing the X-rays. The goal here is to achieve total reflection, by aligning the direction from the incoming X-rays with the reflecting surface to an angle of less than the critical angle θ_{cr}. If we write the refraction index of the reflector as $n = 1 - \delta - i\beta$, then at X-ray energies both δ and β are very small. The critical total reflection angle is $\theta_{\text{cr}} = \sqrt{2\delta}$, which is a *decreasing* function of the energy of

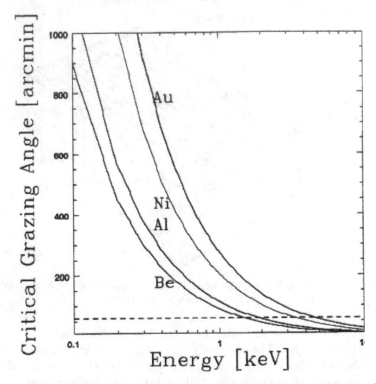

FIGURE 3.3. Critical angle for various reflecting materials as a function of X-ray photon energy.

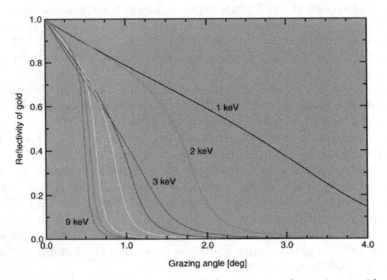

FIGURE 3.4. Reflectivity of Au as a function of photon energy for various incidence angles.

the incident photon as well as a linearly growing function of the atomic number of the reflector. For the most used gold coating, the total reflection angle is of the order of 1° at a photon energy of 1 keV. The dependence of the critical grazing incidence angle on photon energy is shown in Fig. 3.3 for a variety of coatings and photon energies from 0.1 to 10 keV. Note that this fixes the FOV of this reflecting shell, as photons coming from more off-axis angles will just not be reflected. To further emphasize this, Fig. 3.4

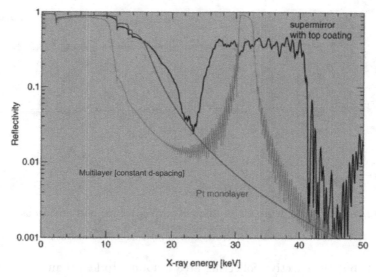

FIGURE 3.5. Reflectivity of a single Pt reflecting layer, of single-spacing multilayers (displaying a broad peak) and of multilayers with multiple-spacing.

shows the reflectivity for gold as a function of photon energy and for different grazing angles. For a maximum FOV of 0.5° the reflectivity drops dramatically at photon energies ~10 keV. Note also that vignetting is an intrinsic property of grazing incidence optics, as the reflectivity is higher for lower grazing incidence angles (i.e. for sources closer to the optical axis).

One way that has been devised to improve the reflectivity at higher energies is by coating the reflecting surface with multilayers. These consist of depositing a number of layers with various refraction indices, alternating a highly reflecting one with a highly transmissive one. In this way, multiple-reflections of X-ray photons can occur, giving the high-energy radiation a chance to reflect in one of the layers. If the layers are deposited all with the same thickness, the probability is maximized around the wavelength that fulfils the Bragg condition. A similar effect on a much broader wavelength range can be obtained by the "supermirror" technology, where the layers have variable thicknesses (see Fig. 3.5). This technique is shown to work up to 40–50 keV, therefore expanding considerably the use of grazing incidence optics to hard X-rays.

Having proper X-ray images also requires that the optical design defines a converging image. To first order this actually needs that every photon undergoes two reflections. Designs used include the Kirkpatrick–Baez one, based on cylinders, and the Wolter-type telescopes, based in reflection on two conical surfaces. In particular Wolter-I type optics uses reflection on an hyperboloid followed by a second reflection on a paraboloid (see Fig. 3.6 for a schematic view of the light-path in both optical systems).

In the design of an X-ray telescope there are a number of trade-offs that need to be taken into account. Indeed the goal is to achieve a very large effective area, high angular resolution and a large FOV. In grazing incidence optics the surface utility $A_{\text{eff}}/A_{\text{geom}}$ (where A_{eff} is the effective area and A_{geom} the geometric area of the resulting surface) is $R^2(\theta_{\text{cr}}) \sim \theta_{\text{cr}}$, which is a small number and decreasing with increasing photon energy. To reach a sufficiently large photon collecting capability, mirror shell pairs are usually nested inside each other and focused to the same focal plane. For example, the mirror shells in the three *XMM-Newton* X-ray telescopes add to an area comparable to that of

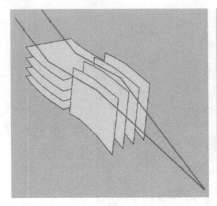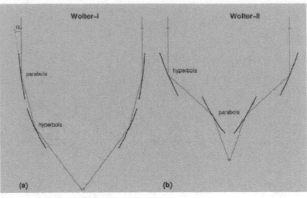

FIGURE 3.6. Scheme of the Kirkpatrick–Baez X-ray optics (a) and of various Wolter-type X-ray optics (b).

a tennis court, but the effective collecting area amounts to less than half square metre. The focal length also needs to be long given the small total reflection angles at X-ray energies. Again high-energy photons are only deflected very small angles and therefore need longer focal lengths for a constant effective area. The FOV is basically limited to the critical total reflection angle θ_{cr}, which means that it decreases with increasing photon energy. That builds in energy-dependent vignetting in all X-ray reflecting telescopes, as sensitivity to higher-energy photons is progressively lost at larger off-axis angles.

The quality of the image (e.g. measured in terms of the point-spread function, PSF) deserves separate discussion. To start with, reflecting surfaces need to have very small roughness to prevent diffusion as opposed to reflection. Polishing is usually required to leave root mean square (rms) residuals of <5 Å. But what ultimately determines the size of the PSF is the stiffness of the mirrors in the standard design. Three different approaches to mirror fabrication have been adopted:

• *Thin foil mirrors*: These are the thinnest shells (∼100 μm), and therefore many of them can be nested gaining in effective area. Angular resolutions achieved are of the order of 1 to a few arcmin. Given this, a conical approximation for the shape of the mirrors is adopted, as opposed to the quadratic Wolter shapes. This technique has been used in *ASCA*, the *Beppo SAX* concentrators and *ASTRO-E*.

• *Replicas*: These are the next in thickness, with up to 300 μm, so nesting a large number of them can still be done. The mirror shells are built by replicating on polished mandrels. Angular resolutions achieved so far are in of the order of 10–15 arcsec, although there might be room for improvement. The best example is *XMM-Newton*, where the shells are of electroformed Ni.

• *Monoliths*: These are thick and heavy mirrors, each one needs to be polished individually and difficult to nest many of them. The angular resolution achieved is extremely good (0.5″), to the point that each mirror pair has to be carefully aligned to ensure that it focuses in the same place as the other nested mirrors. Examples of this are the X-ray telescopes in *ROSAT* and *Chandra*. In the latter case, zerodur was used.

The above sequence follows closely the tendency of improving the PSF as the stiffness parameter $((E/\rho)^{1/2}$, where E is Young's module and ρ, mass density) increases. The mass per geometrical area (in kg/cm^2) needed in these different techniques goes from <0.1 for the foils, to ∼0.2 for the replicas and to 0.5–1.5 for the monoliths. Mission mass constraints are therefore a strong limiting factor on how much effective area can be achieved in monolithic mirror (and therefore best PSF) X-ray telescopes.

Besides the most common design of large mirror shells, micro-channel plate (MCP) optics is also gaining importance in X-ray Astronomy optics. This consists of very large numbers of very small (mm or sub-mm size) pores, each one of them replicating a Kirkpatrick–Baez or a Wolter optics type design. The pores need to be manufactured in such a way that all of them focus X-rays in the same place of the focal plane. One advantage of the MCP optics is that if the pores are fabricated on a curved surface, then a very large solid angle of sky can be observed simultaneously. This is the concept of the "lobster eye" optics, which was designed for the *LOBSTER International Space Station* (*LOBSTER-ISS*) payload to be installed in the External Columbus Payload Facility in the ISS, to monitor variable X-ray sources in the sky (see Section 3.6).

3.3.3. *Wavelength-dispersion optics*

Diffraction is the physical phenomenon used to disperse X-ray photons into different directions as a function of wavelength. These photons are then detected by a position-sensitive X-ray detector, as is usually done (e.g. in optical dispersive spectrographs). There are two main classes of wavelength spectrographs, those based on gratings and those based on Bragg crystals.

3.3.3.1. *Gratings*

In grating spectrographs, the dispersion angle θ is related to the incidence angle χ as a function of wavelength λ according to:

$$m\lambda = d(\sin\theta - \sin\chi) \tag{3.13}$$

where d is the grating spacing and m, the order of the spectrum, which can be positive in transmission gratings and negative in reflection gratings. The resolving power is $\Delta\theta/\Delta\lambda = m/d/\cos\theta$, which for near-normal dispersion ($\theta \sim 0$) is approximately constant at all wavelengths. Grating spectrographs can therefore be used for a variety of photon energies, as far as dispersion is efficient enough. The spectral resolution $\lambda/\Delta\lambda$ improves when the d-spacing decreases, and it improves at higher spectral orders, indeed at the expense of smaller efficiency.

The problem with the wavelength-dispersed X-rays is that they are no longer focused in general. To first order this can be solved following the concept of the Rowland circle, which means that the gratings should be curved with a curvature radius $L = 2R$. If both the source (entrance slit), the grating and the detector, are placed around a circle of radius R, then to first approximation there are no optical distortions. In practice the detector is placed next to the telescope system, and the detectors at the same distance from it than the focal point of the telescope. *Chandra* has two transmission gratings, which to first spectral orders yield spectral resolutions of 0.02–0.03 Å.

XMM-Newton instead has two reflection grating spectrographs (RGS). In these instruments there is a zeroth-order spectrum, which is the direct undispersed image that reaches the EPIC-MOS CCD detectors. The dispersion power $\Delta\theta/\Delta\lambda$ and the spectral resolution are no longer independent of wavelength:

$$R = \frac{\lambda}{\Delta\lambda} = \frac{\cos\theta - \cos\chi}{\sin\chi\Delta\chi} \tag{3.14}$$

but it rather increases at smaller incidence angles. In principle this value could be made arbitrarily high by going to very small incidence angles, but the problem is that the efficiency is then very small. The RGS have a resolution of 0.06 Å for $m = -1$ and they can reach lower photon energies than the transmission gratings on *Chandra*.

3.3.3.2. *Bragg crystals*

Bragg crystal spectrographs can in principle reach very high resolutions ($R > 1000$). They are based on the Bragg effect, which uses constructive interference of diffracted X-rays in various atomic layers in a perfectly periodic crystal. The Bragg condition $2d \sin \theta = m\lambda$ delimits a very narrow wavelength range where Bragg diffraction works, for a given crystal and a given angle. The way they work in practice is by scanning different dispersion angles θ to sample a wide range of wavelengths, of course at the expense of exposure time. The resolving power $\Delta\theta/\Delta\lambda = m/2/d/\sin\theta$ can reach several thousands. The BCS instrument on board the *Einstein Observatory* was one of the few Bragg crystal spectrographs ever operating.

3.3.4. *Coded masks*

Even with multilayers, grazing incidence mirrors fail to focus photons of energy $100 \, \mathrm{keV}$ or more. Spatial multiplexing in the form of coded masks has been often used to produce images in the hard X-ray and γ-ray domains. Multiplexing implies that a mask and a position-sensitive detector are needed, that a reliable image reconstruction algorithm has to be applied to the raw data and that the noise is spread over all of the FOV and not just at the source's positions.

The size of the coded mask (made of holes and opaque zones) must be matched to the detector for optimal use of the multiplexing technique. The opaque zones must be really thick to hard X-ray radiation to prevent the encoding system being energy dependent. In the simplest version of the mask-detector system they have the same size, but for optimum configuration the mask is either two times larger or two times smaller than the detector. This last option leaves partially illuminated areas at the borders of the detector. Both the γ-ray imaging instrument IBIS, the γ-ray spectroscopy instrument SPI and the X-ray monitor JEM-X on *INTEGRAL* use coded masks as imaging mechanisms. Image reconstruction works well for bright enough sources, but it leaves correlated noise (similar to the aperture synthesis radio maps) which ultimately defines the sensitivity of the image. Ghost sources may also appear in corners of the detectors after image reconstruction.

3.4. Detectors and instruments

After X-ray photons have passed through the imaging device, an efficient detector is needed. Detecting high-energy photons is indeed an increasingly hard task, as detecting means stopping them and matter becomes progressively transparent at higher energies. This is indeed the main limiting factor in very high-energy observatories in space, as the amount of mass needed to stop the very high-energy photons is prohibitively large. Fortunately, novel-detector developments now permit the detection of X-rays of up to many tens of keV with high efficiency and relatively low mass.

The ideal detector should be able to detect photons on a one-by-one basis and characterize, at least, the following features for each one of them:

• *Position in some detector coordinates* (X, Y) convertible to sky coordinates (α, δ), after the attitude solution of the spacecraft has been provided. The angular resolution of the observatory should not be limited by the "pixel" size, but by the PSF of the focusing telescope.

• *Arrival time of every individual photon*: This is very important to study timing of variable sources. In particular for solar-mass Galactic X-ray binaries, variability is expected on scales down to $\sim 1 \, \mu s$. These timescales become progressively larger when the mass scale of the objects under study is larger.

● *The energy of every individual photon*: Since the photon is ultimately stopped, all its energy is deposited in the detector and it should be possible to measure it. The precision that is needed depends on the strength of the feature that needs to be measured. Obtaining the continuum needs only low resolution $E/\Delta E \sim 10$; detecting the most prominent emission or absorption features requires $E/\Delta E \sim 100$; detecting narrow emission or absorption lines requires $E/\Delta E \sim 1000$ and resolving these features would require $E/\Delta E \sim 10000$.

● *Polarization state*: High-energy light is polarized either by atomic processes (some atomic transitions produce polarized photons), by scattering or by magnetic fields in synchrotron radiation.

On top of that the ideal detector should operate at very wide-energy bands, be highly efficient, be stable over long operating periods, not be too affected by the high-radiation environment expected in orbit and have a very low background noise (leading to photon-limited detection, rather than background-limited detection). To achieve the latter, detectors often have a combined hardware and software background rejection mechanism.

In the following, I discuss the most popular detector devices used in X-ray Space Astronomy, emphasizing the main strengths and weaknesses regarding the above requirements. A very good reference for details on many of these devices can be found in Fraser (1989).

3.4.1. *Proportional counters*

Gas-filled Proportional Counters (PC) have been the standard X-ray energy detector for many years. They offer a very wide dynamic range, from a few tenths of a keV to several tens of keV, depending on the gas pressure. Their peak efficiency reaches values close to 100% and are amongst the most stable detectors in orbit on scales of years. Their time response can be as low as $\sim 1\,\mu s$. Background rejection can be made extremely efficient, exceeding 99%. The drawbacks of PCs are their very low spectral resolution $E/\Delta E \sim 2$–4 and their limited spatial resolution, which has never been better than a fraction of 1 arcmin.

The way PCs work is illustrated in Fig. 3.7. They consist of a gas chamber filled with noble gases, usually a mixture or Ar and Xe at one or several atmospheres (gas pressure

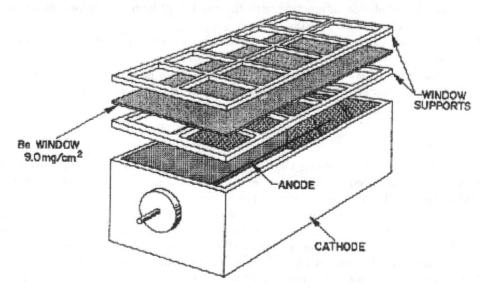

FIGURE 3.7. Scheme of a PC.

determines the dynamic range). When an X-ray enters the chamber it photoionizes one atom, releasing a small number of electrons. These are accelerated by a mesh of anodes which is inserted inside the chamber, producing a cloud of secondary electrons. The voltage of these anodes determines the final number of electrons that reaches the cathode, which is related to the energy deposited by the detected photon.

The electron pulse is then detected and analysed. By examining the pulse spatial shape, high-energy particles can be discriminated from X-ray photons. The former leave a trail of ionized atoms in the chamber and therefore produce a very different spatial signal.

The pulse height can be related to the energy of the detected photon. If E is the incoming photon energy, w the ionization energy of the atoms in the gas (26.2 eV for Ar), the number of primary electrons stripped by the X-ray photon is $N = E/w$. Fluctuations in this number are $\sigma_N^2 = FN$ where F is the Fano factor (~ 0.17 for these gases). Note that fluctuations are well below Poisson statistics, as the number of primary electrons released is approximately fixed by energy conservation. The primary electrons are then accelerated by the voltage and a large number of electrons finally reach the cathode. The resulting number, proportional to the energy of the incoming photon, is $P = GN$ where G is the gain, typically of the order $G \sim 10^3$–10^5. The spectral resolution is limited by the noise in the amplified number of electrons, and ultimately by the Fano factor, that is:

$$\frac{\Delta E}{E} = 2.36 \frac{\sigma_P}{P} \sim 0.35 E^{-1/2} \qquad (3.15)$$

Care has to be taken to work in a regime where the bias voltage is large enough to produce a large number of electrons but does not saturate the cascade (i.e. in the proportional counter regime). Note the rather low-energy resolution, which allows the measurement of the broad-band continuum shape of the sources, but other emission or absorption features can only be detected if they are very broad.

The efficiency of a gas-filled PC is the result of two factors: the probability that an X-ray photon is absorbed by the gas and the probability that the X-ray photon is absorbed by the entrance window and therefore lost. Peak efficiencies occur typically around a few keV, where the entrance window is mostly transparent to radiation and the probability of the X-ray photon interacting with the gas is still high. Quantum efficiency at these energies can be as high as 80%.

There are many examples of missions that have used (and use) gas-filled PC as X-ray detectors. Among the most recent ones the *Ginga* Large-Area Proportional Counter (LAC) (Turner *et al.*, 1989), which enabled the detection of Fe broad emission lines, the *ROSAT* Position-Sensitive PC (PSPC, Pfeffermann and Briel, 1986), and the *Rossi XTE* array (PCA).

3.4.2. *MCP detectors*

MCP detectors are built to reach the highest spatial resolutions available in current X-ray missions. Compared to PCs, this is achieved mostly at the expense of a very high gain, and therefore by the loss of spectral resolution.

MCPs (see Fig. 3.8) consist of several layers of very narrow (diameter typically between 2 and 12 μm) metal tubes which act as photo-multipliers. Incoming X-ray photons are expected to release a primary electron from the photocatode. To facilitate this, MCP tubes are either curved or oriented in a slightly slanted fashion with respect to the optical axis. Subsequent collisions of the electron(s) with the tube walls, which are maintained at a voltage $V \sim 1.5$–4 kV, produce a very large burst of secondary electrons. The cascade

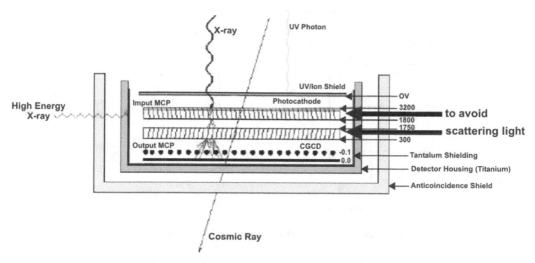

FIGURE 3.8. The MCP detector HRC on *Chandra*.

of electrons is finally detected and used to centroid the position of the incoming X-ray photon.

Since the emphasis here is in the spatial resolution, very high values of the gain $(G \sim 10^7 - 10^8)$ are needed to improve the statistics, which means that the detector will work close to the saturated regime. This implies that the spectral resolution is very low $\Delta E/E \sim 1$. Quantum efficiency of MCP detectors can exceed 30% in the soft X-ray regime, but it drops rapidly at energies of a few keV. A further problem is the higher background level with respect to PCs, due to the stronger difficulty to identify non-X-ray events.

Examples of MCP detectors which have been flown in X-ray missions include the high-resolution imager (HRI) on board the *Einstein Observatory*, the HRI on board *ROSAT* and the high-resolution camera (HRC) on board *Chandra*. Devices very similar to these are also used for bio-imaging.

3.4.3. *Charge-coupled devices*

Charge-Coupled Devices (CCDs) are semiconductor-based devices which revolutionized optical Astronomy. Their use in the X-ray band has provided a leap forward during the last decade. CCDs provide a spatial resolution comparable to (or better than) those of PSPC, but their spectral resolution is of the order of $\sim$10 times better. CCDs do not need consumables (gas) and although there is in-orbit degradation due to the high-radiation environment, methods have been devised to keep these detectors in good shape for many years. Limitations with respect to PCs are most significant in the time-resolution domain, but the next generation of CCDs might actually overcome this problem.

The basic physics behind the detection of X-rays in CCDs is the excitation of electron-hole pairs in a semiconductor. When an X-ray is absorbed, a number of electrons from the valence band jump to the conduction band. The charge deposited in the conduction band can therefore be read out and it is actually proportional to the energy of the incoming X-ray photon. The major advantage over PCs is the better spectral resolution and this comes mainly from the fact that the energy of the gap (now $w \sim 3\,\mathrm{eV}$ for Si or Ge) is much lower and therefore the energy of a single X-ray photon is now broken into more pieces and can be more accurately measured. The Fano factor is 3–5 times smaller in CCDs than in gas-filled PCs.

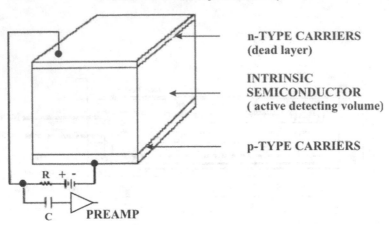

FIGURE 3.9. Sketch of a simple pn CCD detector.

Indeed CCDs need to work at (not too) low temperatures to suppress thermal excitation of electrons from the valence to the conduction band. However, a simple semiconductor aimed at detecting X-rays would detect only thermal leakage currents even at
liquid nitrogen temperatures (note that the aim is to detect the small signal produced
by a single X-ray photon). To prevent this, CCD detectors are constructed with different
layers of variedly doped semiconductors. In particular p-doped (with an excess of holes
in the valence band) and n-doped (with an excess of electrons in the conduction band)
semiconductors are used. A simple version of a CCD working in the X-ray regime consists of a p-layer, followed by a deep intrinsic semiconductor zone and then a n-layer is
sketched in Fig. 3.9. The intermediate layer acts as the real absorber here, where upon
arrival of an X-ray photon a number of electron-hole pairs are created and move towards
the electrodes following the action of the bias voltage. The depth of this *depletion layer*
$d \propto (\rho_S V)$ (where ρ_S is the resistivity of the semiconductor and V, the applied voltage)
has to be large enough if quantum efficiencies close to 100% have to be achieved at photon
energies higher than a few keV.

The quantum efficiency of a CCD is given by:

$$Q = \prod_m \exp(-\mu_m t_m)[1 - \exp(-\mu_S d)] \qquad (3.16)$$

where μ_S and μ_m are the linear absorption coefficients in the depletion layer and in
any layers that X-rays need to cross before reaching the active area, and t_m are their
thicknesses. The quantum efficiency can reach values in excess to 95% with a layer of a
few mm thickness of the most used semiconductors.

The energy (spectral) resolution is limited by three effects: the Fano noise (as in PCs),
the loss of charge during collection and the amplifier plus electronics noise. The last two
are collected in the standard "read noise" R (in units of electrons):

$$\Delta E = 2.36w(FE/w + R^2)^{1/2} \qquad (3.17)$$

Typical values for $\Delta E/E$ reach 20–50 in modern CCD X-ray detectors. Traps and
absorption in non-active areas all act to degrade the spectral resolution.

The way CCDs are operated in X-ray Astronomy is by reading out very frequently,
in order to avoid the "pile-up" effect, that is, measuring the effect of more than one
incoming X-ray photon together. This would cause the detector to assign a single count

with a higher energy to several low-energy photons. Strategies to limit pile-up in real X-ray observatories all turn around the reduction of the count rate per pixel. Electrons are transferred pixel-to-pixel by a moving potential well and the current is finally read. The reading time must be small enough compared to the exposure time or otherwise the device would be extremely inefficient. Even though, for bright enough sources it might happen that during readout more photons are detected, leaving a trail of out-of-time events. Charge Transfer Efficiency (CTE) is another issue to worry about, as the high-radiation environment can cause this to decrease significantly with time. The next generation of CCDs will feature Active Pixel Sensors (APS), enabling readout without pixel-to-pixel charge transfer and will therefore resolve most of these issues, besides providing a much better timing capability.

X-ray Astronomy CCD detectors are often "front illuminated", which means that the X-rays enter the depletion layer through the electrode structure. Unfortunately this suppresses the lowest-energy X-ray photons which are absorbed before they reach the depletion layer. As a way to solve this, "back-illuminated" CCDs are also operated in X-ray observatories, where the substrate is thinned and then the CCD is illuminated from the back. The drawback of this, as demonstrated by operating CCDs, is the higher background created by particles.

Controlling and reducing the particle background is a very important issue in the use of CCDs as X-ray detectors. Charged particles tend to deposit their energy in various pixels, as opposed to X-rays which tend to leave a more concentrated "pattern", and therefore a software distinction between both types of events can be used. However, their difference becomes narrower at higher photon energies where X-ray photons might leave a signal in various neighbouring pixels. This rejection mechanism (pattern filtering) has an efficiency of up to 90% in rejecting non-X-ray events, to be compared with the >99% efficiency reached by PCs.

The first X-ray orbiting observatory featuring CCDs as X-ray detectors was *ASCA*, where two out of the four X-ray telescopes were equipped with CCDs in their focal planes. Having an imaging spectroscopic device with a resolution 10 times higher than PCs, really introduced a qualitative difference in X-ray Astronomy. Both *Chandra* and *XMM-Newton* have CCDs on board.

3.4.4. *X-ray micro-calorimeters based on semiconductors*

X-ray micro-calorimeters constitute the first attempt to produce a non-dispersive imaging spectrograph of spectral resolution of ~500–1000. The basic principle is remarkably simple: X-rays absorbed by the bolometer heat it up (i.e. phonons are excited by the X-ray photon), and its resistivity increases consequently. The excess energy is subsequently deposited to a thermal bath through a weak link. The increase in the temperature is $\Delta T \propto E/C$, where E is the energy of the photon and C, the heat capacity. The relaxation time is typically ~1 ms, which has to be long enough to be detectable and as short as possible to prevent a second photon being absorbed during this process. This trade-off is mostly controlled by the link between the absorber and the thermal bath.

The energy of the absorbed photon is measured by the integrated excess voltage. The energy spectral resolution in a single pixel micro-calorimeter is mainly limited by the thermal phonons, and this scales as $\Delta E \propto T^{5/2}$. The operation temperature for these detectors has to be of the order of ~100 mK; that is, they need to work in cryogenic conditions which is always challenging in space. Building imaging arrays of these detectors poses an extra technological difficulty in preventing the leakage of heat in neighbouring pixels.

The first X-ray micro-calorimeters to be flown were those on board *ASTRO-E*, whose launch unfortunately failed. These had a spectral resolution of the order of 10–12 eV. The new calorimeter to be flown on *ASTRO-E2* has been considerably improved with respect to its predecessor, both in an improved spatial layout and in the spectral resolution that is now close to 6 eV.

3.4.5. *X-ray imaging spectrographs based on superconductors*

One option to gain spectral resolution with respect to the semiconductor-based micro-calorimeters is to use Transition Edge Sensors (TESs). These consist of a superconducting metal operated at the critical temperature which separates the normal from the super-conducting phase. The device acts as a bolometer, in the sense that the excess heat deposited by an absorbed X-ray rises the temperature enough to place the metal in the non-superconductive phase. The sharp rise in the resistance of the metal is readily de-tected and used to measure the energy of the X-ray photon to high precision (a few eV). The excess energy is transferred to a thermal bath through a weak link. Operation temperatures of 70–100 mK require cryogenic conditions. Spectral resolutions achieved are of the order of 4 eV at 6 keV and 2 eV at 0.25–3 keV. NIST in the USA and SRON in Europe are at the lead in the investigations of these devices.

Superconducting Tunneling Junctions (STJs) offer an alternative detector technology, especially at energies <3 keV. The basic detector consists of two superconducting metals separated by an insulating barrier, where thermal Josephson current is suppressed by a magnetic field (~100 Gauss). When an X-ray is absorbed it breaks a very large number of Cooper pairs that can then establish a measurable current. The theoretical energy resolution is mostly dictated by the gap energy; that is, by the number of quanta created by a single photon (millions for a 1 keV photon). So far, Ta/Al layers have achieved spectral resolutions of 2 eV, but there is substantial room for improvement by testing different materials and geometries. STJs do not scale up well (as many of the cryogenic detectors) and therefore covering large areas is a challenge. Arrays of STJs scale up the amount of cabling by N^2 making it an impractical option for any sizeable number of pixels. Matrix readout also appears impractical in space, as one dead pixel will kill an entire row and an entire column. A more interesting approach has been put forward by ESTEC in the form of Distributed Readout Imaging Devices (DROIDs), where a large superconducting absorber of X-rays transports the broken quasi-particles to tunneling readout ends. One-dimensional DROIDs appear very promising.

3.4.6. *X-ray polarimeters*

Classical X-ray polarimeters were designed to take advantage of the polarization depend-ence of Bragg diffraction, maximal at 45°. To measure polarization, the Bragg crystal need to be rotated continuously. This was the design for the Stellar Rotating X-ray Polarimeter (SXRP) on board the spectrum–X–Γ mission.

A novel design for an X-ray polarimeter has been put forward by Costa *et al.* (2001). It is based on a micro-pattern detector, and it uses the fact that when an X-ray produces photoelectric effect, the resulting photoelectron trajectory depends on the polarization on the incoming photon. The photoelectron track needs to be measured in the very vicinity of its original point, as later the polarization information is lost. The design of this X-ray polarimeter consists in a very fine pixel detector, so the photoelectron track can be detected with high accuracy.

3.5. The development of X-ray Astronomy

3.5.1. *The early days*

X-ray Astronomy was officially born on June 1962. Prior to that date the only cosmic X-ray source known was the Sun itself. It was also known that solar flares resulted in an increased X-ray emission and in important disruptions in the Earth's ionosphere which resulted in communications problems. But assuming that the Sun was a typical star, its X-ray to optical flux ratio is so small that there was no hope to detect any other cosmic sources in X-rays prior to that date. Riccardo Giacconi and his colleagues launched a rocket with three mica detectors to detect solar X-rays scattered in the surface of the Moon – that was probably the only other cosmic source that could be seen in X-rays (Tucker and Giacconi, 1985 provide a very detailed account of what happened in those early days, including personal points of view).

The rocket was launched successfully and flew over 80 km of altitude for several minutes. One of the three counters failed but the other two were operational. The results were astonishing (Giacconi *et al.*, 1962): the Moon was not seen, but a very strong signal came from an unidentified direction in the sky. That strong signal was in fact the very first extra-solar X-ray source to be discovered (named Sco X-1), and was later identified with an optically inconspicuous star which turned out to be a low-mass X-ray binary. The second big discovery reported from that pioneering flight was the detection of an approximately isotropic X-ray emissivity coming from everywhere in the sky: the cosmic XRB. Note that the XRB was discovered 3 years before than the cosmic microwave background radiation, but its origin took a bit longer to be understood (see, e.g. Fabian and Barcons, 1992 for a view in retrospect of the studies and models for the XRB). It is also remarkable that the Moon was first seen on the 29 June 1990 by Schmitt *et al.* (1991) with the *ROSAT* observatory (equipped with an X-ray telescope). The very same image of the Moon showed shadowing of the background X-ray radiation in the Moon's dark side; that is, it also showed that the XRB arose at least beyond the Moon.

In the decade that followed this pioneering X-ray observation, many other rockets and balloons were flown to detect X-rays from the outside of the solar system. The 1962 rocket had indeed opened a new window in the study of the Universe. Although it took some time to understand why some stars (and galaxies) produced so many X-rays compared to our Sun and to our Galaxy, it was clear that (1) X-ray observations were worth pursuing and (2) the information gathered by X-ray instruments was in no way redundant to optical observations.

3.5.2. *Collimators*

The 1970s saw the first orbiting observatories, whose operations lasted for years instead of minutes. *UHURU* set a milestone on that front. It was launched on the 12 December 1970 from the Italian San Marco platform in Kenya. Being Kenya's independence day, the satellite was named after the swahili word for *freedom*. The mission operated with success until March 1973.

UHURU carried two sets of gas-filled PCs sensitive to X-rays from 2–20 keV, with an effective area of $0.08 \, m^2$. Two different collimators were placed in front of each set of PCs: one with a FOV of $5° \times 5°$ (to study isolated sources) and the other one with $0.5° \times 0.5°$ to position with higher accuracy the brightest sources. The full *UHURU* payload weighted 56 kg!

The satellite was put in an almost-equatorial low-Earth orbit (520–560 km), with an orbital period of 96 min, and spinning every 12 min. The major legacy of that mission is the fourth *UHURU* catalogue, containing 339 X-ray sources distributed over the whole sky (Forman *et al.*, 1978). This catalogue provided a first comprehensive view of the contents of the X-ray sky: binary stars and supernova remnants in the galaxy, and active galaxies and clusters in the extragalactic sky. Qualitatively this is still a good picture of the overall X-ray source populations.

A number of orbiting X-ray observatories, where the X-ray optics was based in collimators, were launched since then. *HEAO-1* was amongst the most celebrated ones, as it carried out the, to date, most sensitive all-sky survey in hard X-rays. *HEAO-1* was launched on the 12 August 1979 and it scanned the full sky more than twice. Its payload consisted of four experiments, which covered the energy range from 0.2 keV to 10 MeV: the A1 experiment (large-area sky survey experiment) with collimators of $1° \times 4°$ and $1° \times 0.5°$ was sensitive to 0.25–25 keV photons; the A3 experiment, which was a modulation collimator, was mostly used to position A1 sources with ~arcmin accuracy; the A2 cosmic X-ray experiment provided the most sensitive measurement of the XRB surface brightness and spectrum to date in the 2–60 keV band; the A4 experiment extended this measurement to 10 MeV using phoswich detectors. The most important legacies of the *HEAO-1* mission include the measurement of the spectrum of the XRB (Marshall *et al.*, 1980) and the first complete high-galactic latitude catalogue of hard X-ray sources (Piccinotti *et al.*, 1982). The *HEAO-1* A2 all-sky maps have been extensively used to study the large-scale structure of the X-ray sky.

With the launch of missions equipped with X-ray grazing incidence telescopes (see next section), and prior to the operation of detectors based on CCDs, PCs with collimated FOV were used mostly for broad-band spectral studies and/or time variability. The Japan/UK mission *Ginga* (Japanese word for Galaxy, launched on 5 February 1987 and operational until 1 November 1991) featured one of such instruments (the LAC), with which many important discoveries in the spectral domain were achieved: discovery of an Fe emission line and Fe absorption edges due to cold matter in Seyfert galaxies (Pounds *et al.*, 1990), as well as the discovery of cyclotron absorption lines in X-ray pulsars.

In more recent times, NASA's *Rossi XTE* observatory (launched on the 30 December 1995 and, as of March 2004, still in operation) was designed and built to conduct timing studies of X-ray sources. The payload consists of a suite of collimated FOV PCs (PCA) sensitive in the 2–10 keV band, with a ~1 µs time resolution and a very large effective area (0.65 m^2), the High-Energy X-ray Timing Experiment (HEXTE) sensitive to 15–250 keV photons and an all-sky monitor operating in the 2–10 keV band.

The ultimate limitation of all collimated FOV instruments is source confusion created by having several sources within the same resolution element (~1°). This, in practice, limits studies to bright sources, which are far apart enough from each other and that dominate on much fainter sources. The study of fainter sources requires proper grazing incidence imaging.

3.5.3. *The telescopes*

The first real X-ray mission with proper imaging (non-multiplexing) capability was NASA's *Einstein Observatory*. It was launched on the 12 November 1978 and ended operations on April 1981. The X-ray telescope adopted the Wolter-I type design, with stiff mirrors resulting in a very good intrinsic angular resolution (~arcsec). Its relatively short focal length meant that only soft X-ray photons ($E < 4$ keV) could be focused onto the focal plane, where there was a choice of four instruments: the imaging PC (IPC) a position-sensitive gas-filled PC with angular resolution $>1'$ and effective area ~0.01 m^2 and spectral resolution $E/\Delta E \sim 1$–2 which was the *de facto* prime instrument; a MCP HRI which achieved ~2″ angular resolution with no spectral resolution and 5–20 times smaller effective area (plus higher background); the solid state spectrometer which provided CCD-type spectral resolution $E/\Delta E \sim 3$–20 over a FOV of 6′; and a Bragg-based focal plane crystal spectrometer which delivered very high resolution $E/\Delta E \sim 100$–1000 spectroscopy but with a very small effective area ($\sim 1\,\mathrm{cm}^2$). A monitor PC co-aligned with the X-ray telescope and sensitive to photons of up to 20 keV and an objective grating spectrometer, delivering spectroscopy of $E/\Delta E \sim 50$ in conjunction with the HRI completed the payload.

Einstein was the first true X-ray astronomical facility in many respects. Even at its moderate spatial resolution, it resolved the emissivity of all types of extended sources like Supernova remnants, galaxies and clusters of galaxies. It detected a large number of active coronal stars in our galaxy, which created a new class of galactic X-ray source population. It resolved, for the first time, a significant fraction of the soft XRB into individual sources. And as a major legacy, the *Einstein Observatory* left us with the Extended Einstein Medium Sensitivity Survey (Gioia *et al.*, 1990; Stocke *et al.*, 1991) a sample of >800 identified X-ray sources at high galactic latitude – still the largest sample of serendipitous X-ray sources completely identified – which formed the basis for studies of the QSO population, luminosity function, evolution, etc.

In May 1983 ESA launched *EXOSAT*, which operated until April 1986. It combined two soft X-ray Wolter-I telescopes (0.05–2 keV) with two collimated FOV proportional counters extending the response to 20 and 50 keV, respectively. In the focal plane of the X-ray telescopes, there was a Channel Multiplier Array (CMA), with a spatial resolution of 18″ and a position-sensitive detector. The peak effective area was $\sim 10\,\mathrm{cm}^2$. Two transmission gratings could also be used in conjunction with the CMA detectors. *EXOSAT* performed pioneering variability studies of galactic X-ray binaries – discovering Quasi-Periodic Oscillations (QPOs) – and of active galaxies. It also measured broad-band spectral properties of accreting sources and obtained the first high-energy grating spectra at low X-ray photon energy.

Perhaps the most celebrated soft X-ray observatory was the German–UK–USA observatory *ROSAT*, launched on the 1 June 1990 and operational until February 1999. *ROSAT* carried a stiff mirror short focal length Wolter-I telescope, covering a 2° FOV. On its focal plane *ROSAT* had two PSPC, one of them was burned during an erroneous pointing to the Sun and a HRI similar to that of the *Einstein Observatory*. The PSPC provided a spatial resolution of 20–30″ with a moderate spectral resolution of $E/\Delta E \sim$ 3–4. The HRI had a much better angular resolution (4″), but had its sensitivity limited, among other reasons, by its higher background. *ROSAT* also had an extreme ultraviolet telescope, the Wide-Field Imager (WFI), which was also burned out following the erroneous pointing to the Sun.

The *ROSAT* PSPC was used to conduct an all-sky survey (mostly during the first 6 months of the mission) which is one of the most important legacies of the project (Voges *et al.*, 1999). Besides providing all-sky maps in various soft X-ray bands, the *ROSAT* all sky survey delivered the largest catalogue of X-ray sources with $\sim 50{,}000$ entries. Indeed during its 10 years of pointed observations, *ROSAT* studied all types of sources: comets, stars, X-ray binaries, supernova remnants, galaxies, clusters, AGNs and the XRB. In all fields, it provided major breakthroughs in our understanding of the X-ray Universe. As an example, the *ROSAT* deep survey in the Lockman Hole resolved 70% of the soft XRB into sources, most of which are AGN.

The next step was to have a longer focal length grazing incidence X-ray telescope sensitive to X-rays from 0.5 to 10 keV. Besides the shuttle-borne BBXRT, the Japan–USA *ASCA* mission was the first X-ray observatory with this capability (Tanaka, Inoue and Holt, 1993). *ASCA* (launched on February 1993 and operational until March 2001) consisted of four thin-foil mirror telescopes (resolution $\sim 3'$), two of them with gas scintillation PCs and another two, for the first time, with CCDs on their respective focal planes. The improvement in spectral resolution with respect to previous missions (both in hard and soft X-rays) was of paramount importance to crack many important astrophysical problems in high-energy astrophysics. Among the most important discoveries was the detection of strong gravity effects in the Fe Kα emission line in Seyfert galaxies, in particular in MCG-3-30-15 (Tanaka *et al.*, 1995).

The *Beppo SAX* observatory (named after Italian astronomer Giuseppe Occhiliani) of the Italian Space Agency (ASI) and the Netherlands Agency for Aerospace Programs (NIVR) was put into an equatorial low-Earth orbit on the 30 April 1996. It re-entered the atmosphere exactly 7 years after launch and successful scientific operations. Without any doubt, the major strength of *Beppo SAX* was its combination of several instruments, which operated simultaneously to provide significant sensitivity over a very broad energy band: from 0.1 to 300 keV.

Beppo SAX had a suite of co-aligned instruments. The low-energy concentrators (LECs) and the three medium-energy concentrators (MECs) consisted of thin-foil X-ray reflecting telescopes with gas scintillation PCs in their focal planes. The telescopes delivered 1′–2′ angular resolution and the imaging detectors provided spectral resolution of the order of $E/\Delta E \sim 10$. A high-pressure gas scintillation proportional counter (HPGSPC) with a collimated FOV of $\sim 1.1°$ extended the sensitivity to 120 keV. At even higher energies the phoswich-detector system (PDS) (15–300 keV) delivered a spectroscopic resolution $E/\Delta E \sim 7$ over a collimated FOV of $\sim 1.4°$. The combination of these co-aligned instruments, and in particular of the LECs, MECs and PDS, provided a coherent view of high-energy bright sources where their continuum could be studied over more than three decades of energy. In particular a number of Compton-thick AGN (i.e. with column densities in excess of $10^{24.5}/cm^2$) were observed by *Beppo SAX*, and the Compton reflection "hump" at high energies was studied in detail.

Pointing in the diametrically opposite direction, *Beppo SAX* had two identical Wide-Field Cameras (WFCs), where a coded-mask optics delivered 5′ positional accuracy for the sources in a (fully coded) FOV of 20° × 20°. The WFCs operated PSPC sensitive to 2–28 keV photons. Their main goal was to detect and study X-ray transient sources. But perhaps one of the major contributions of the WFCs, and even of *Beppo SAX* to Astronomy, was due to the use of the WFC anti-coincidence shields as high-energy detectors. In particular, they operated as a very successful GRB monitor in the energy range from 60–600 keV. It must be noted that the positional accuracy delivered by the WFCs for GRBs (5′) allowed a follow-up of the GRB afterglow with an unprecedent efficiency without any further refinement. This had a lot to do with the identification of the origin of GRBs.

3.5.4. *Chandra*

At the end of the 20th century, both NASA, ESA and Institute of Space and Astronautical Science ISAS/JAXA (Japan) had their major X-ray observatories ready for launch. Delays and coincidences lead to both *Chandra* (NASA) and *XMM-Newton* (ESA) being launched in 1999, with *ASTRO-E* (ISAS) being launched in February 2000. Unfortunately the last one failed, and a re-make of it (*ASTRO-E2*) is now planned for launch in 2005.

Chandra† was launched on 23 July 1999 into a highly eccentric orbit (64 h) by the space Shuttle. If anything defines *Chandra* is its superb imaging capability. Its monolithic mirrors can deliver an on-axis spatial resolution of $\sim 0.5″$, for the first time matching

† http://cxc.harvard.edu

and challenging similar angular resolutions at optical wavelengths from ground-based facilities. It has a peak effective area of $\sim$0.08 m^2.

On its focal plane, *Chandra* has two types of detectors available: the CCD-based ACIS (advanced camera for imaging spectroscopy) and the MCP-based HRC (a renewed version of the previous similarly named instruments on *Einstein* and *ROSAT*).

ACIS comes in two formats: ACIS-I (mostly for imaging), which covers a square FOV of $16' \times 16'$; and ACIS-S (mostly for dispersive spectroscopy) which consists of six aligned CCD chips covering a FOV $\sim 8' \times 48'$. The four ACIS-I chips and a further four of the ACIS-S are front illuminated with a spectral resolution of $E/\Delta E \sim 20$–50. The remaining two CCDs of the ACIS-S are back illuminated, providing an improved sensitivity at low energies, but with more modest energy spectral resolution ($E/\Delta E \sim 10$–35). Soft protons reflected onto the *Chandra* mirrors produced some charge transfer inefficiency on the front-illuminated CCDs during the first weeks of scientific operations. After identifying the problem, the CCDs are never on the focal plane when *Chandra* approaches the Earth's radiation belts.

HRC also has two settings: HRC-I, covering a square FOV of $30'$ and with a pixel size of $0.5''$ to match the telescope angular resolution, and HRC-S which has a rectangular shape covering $7' \times 97'$ and mostly used for dispersive spectroscopy.

Two grating assemblies can be placed below the telescope to wavelength-disperse X-ray photons. The high-energy transmission grating (HETG), operating at energies 0.5–10 keV, but most efficient at energies >1–2 keV, can deliver spectroscopy with a resolution of $E/\Delta E \sim 400$ in the first order. It is often used with the ACIS-S detector, which can additionally separate the various spectral orders by measuring the (approximate, at CCD-like resolution) energy of every individual X-ray photon detected. The Low-Energy Transmission Grating (LETG) operates in the 0.08–6 keV energy band and delivers similar spectral resolution at softer energies. The LETG-dispersed X-rays can be detected either by ACIS-S or HRC-S, the latter having a higher background but better efficiency at the very soft end.

Fig. 3.10 displays the effective area of the various *Chandra* instruments as a function of photon energy. *Chandra* has already completed 4 years of science operations. Its superb

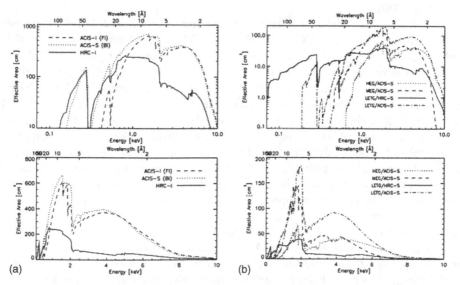

FIGURE 3.10. Effective area of the various *Chandra* instruments without gratings (a) and with gratings (b).

angular resolution, along with the high spectral resolution of its gratings combined with a moderate photon collecting capability, has indeed revolutionized our views of the X-ray Universe. For instance, *Chandra* surveys of the deep X-ray Universe have now resolved 90% of the XRB at 1 keV into individual sources (Mushotzky *et al.*, 2000). Highly obscured X-ray sources, which were mostly missed by previous soft X-ray missions, are now routinely discovered in *Chandra* observations. An account of some highlights of the *Chandra* mission is presented in Weisskopf *et al.* (2002).

3.5.5. *XMM-Newton*

ESA's *XMM-Newton* X-ray observatory* was launched on the 10 December 1999 by an Ariane V from Kourou (French Guiana) onto a highly eccentric orbit (48 h). *XMM-Newton* is the largest X-ray Astronomy mission ever launched, weighing over 3.5 tons. Its major strength in comparison to *Chandra* is its effective area which is 3–10 times higher (at 1 and 7 keV, respectively), besides the simultaneous operation of various instruments (including imaging detectors, dispersive spectrographs and an optical/UV monitor). Its major drawback is its poorer spatial resolution, which is 12″–15″. All this makes *XMM-Newton* a superb X-ray spectroscopy instrument (except at very faint fluxes where it is confusion limited), the most sensitive X-ray instrument at energies >2 keV and the most powerful imaging spectrograph for extended sources.

XMM-Newton consists of three co-aligned X-ray telescopes, each one with 58 mirror pairs with Wolter-I type optics (Jansen *et al.*, 2001). The mirrors are Au-coated Ni shells which were fabricated via replication on mandrels. One of the telescopes focuses the X-rays on a single CCD-based spectroscopic imaging detector named EPIC-pn (Strüder *et al.*, 2001). The remaining two have reflection gratings which wavelength-disperse about 40% of the X-rays, with a further 40–50% being focused on two versions of another class of CCD-based spectroscopic imaging detectors: the EPIC-MOS (1 and 2, Turner *et al.*, 2001). The wavelength-dispersed X-rays are then collected by MOS-type CCDs, constituting the two RGSs (den Herder *et al.*, 2001). Further, a co-aligned optical/UV small aperture telescope, the Optical Monitor (OM) equipped with a UV-intensified counting detector, offers imaging in the optical and UV as well as grism spectroscopy (Mason *et al.*, 2001).

The EPIC cameras deliver spectroscopic imaging over the 0.2–12 keV band with a spectral resolution typical of CCDs ($E/\Delta E \sim 20$–50). A very important asset of EPIC (and of *XMM-Newton*) is that the cameras cover a region of approximately $30' \times 30'$. The degradation of the PSF is moderate over the whole EPIC FOV region, and almost unnoticeable within a region of radius 8'. As it can be seen from Fig. 3.11, the EPIC-pn camera reaches an effective area approaching $0.2\,\mathrm{m}^2$, with each one of the MOSs having about half of this. Since the EPIC-pn is a back-illuminated CCD device, is more sensitive to soft X-ray photons than the MOSs, which have, however, a comparatively lower background. The sensitivity of the EPIC cameras is limited by source confusion at a flux level below $\sim 10^{-15}\,\mathrm{erg/cm}^2/\mathrm{s}$ in the 0.5–2 keV band. This limit is reached with exposures of around a few hundreds of kilo-seconds. Due to the decrease of effective area with growing photon energy, confusion is not a real limitation >2 keV even for megasecond long exposures.

The RGS operate in the wavelength range from 5 to 35 Å (0.3–2.5 keV). The spectral resolution is of the order of 0.06 Å for first-order spectroscopy (and twice better for second-order spectroscopy). The various orders are separated by the CCD detectors in the RGS using the pulse-height amplitude measuring capability of the detectors. The

---

* http://xmm.vilspa.esa.es

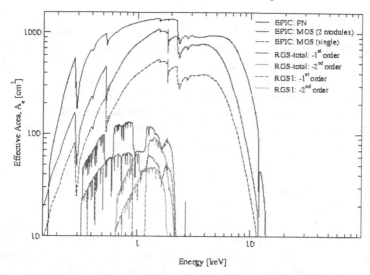

FIGURE 3.11. Effective area of the various instruments on board *XMM-Newton*.

sensitivity of the RGS is limited by photon counting, with a source of flux $\sim 10^{-11}$ erg/cm^2/s being well detectable for a 100-ks-long exposure.

The OM telescope has a PSF of 1.3–2.5″ (depending on filter) over a FOV of $\sim 17' \times 17'$. It is equipped with a suite of optical and UV filters covering the range from 1600 to 6600 Å. Its sensitivity is such that it can reach typical AB magnitudes of the order of 23.5^{mag} in 1000 s of integration. It is also equipped with optical and a UV grisms.

XMM-Newton Science Operations are such that all instruments operate simultaneously. The EPIC cameras have windowing modes (and even a timing mode) which are used to mitigate pile-up effects from bright sources. About 20% of the observing time suffers of high-background flares in the X-ray CCDs (possibly due to the spacecraft crossing the Earth's magneto-tail). These passages make the data obtained of very little use for the study of faint or extended sources.

As in many other ESA missions, the Mission Operations Centre is located in Darmstadt (Germany). The Science Operations Centre is located in the European Space Astronomy Centre near Madrid (Spain). The instrument data downloaded via telemetry is then combined with attitude and housekeeping information from the spacecraft and an observation data file (ODF) is produced. There are about 3000 ODFs produced during the first 4 years of science operations.

The ODFs are then sent the *XMM-Newton* Survey Science Centre (SSC)*, a consortium of 10 European Institutes lead by Leicester University. The SSC team was appointed by ESA following an Announcement of Opportunity and its activities are funded by the national agencies. One of the tasks of the SSC is to pipeline-process all the *XMM-Newton* data. Besides producing calibrated event lists, images and other standard X-ray Astronomy products, the SSC pipeline searches for all sources in the EPIC FOV and then it looks for catalogued counterparts of these sources in a large collection of catalogues. This is all supplied to the observer. All data products are screened by the SSC before they are sent back to the SOC for delivery to the observer and ingestion in the *XMM-Newton* Science Archive (XSA)†.

* http://xmmssc-www.star.le.ac.uk
† http://xmm.vilspa.esa.es/xsa

The XSA has a central role in the data delivery to the community. It contains all the ODFs and the associated pipeline data products. The XSA is designed to protect the proprietary rights of the guest observers (which download their data from the archive when ready), and offers a user-friendly interface to search for public archived data. Since version 2.5, it can also perform on-the-fly data processing; for example, filtering of high-background episodes, extraction of images, spectra and time series, etc.

The time elapsed since the observation is conducted and the observer can access the calibrated and screened data products now ranges from 20 to 30 days for the vast majority of the observations, the remainder having processing and/or instrument problems. A major re-processing of all ODFs is planned for 2004. In addition to the pipeline processing, the SSC also collaborates with the SOC in producing and maintaining the standard *XMM-Newton* Science Analysis Software (SAS).

The third SSC task is to exploit the serendipitous sky survey conducted by *XMM-Newton*. The potential of this survey is enormous: every *XMM-Newton* observation with one of the EPIC cameras in full window mode finds between 30 to 150 new X-ray sources. These sources are collected in the *XMM-Newton* source catalogue, the first version of which (1XMM) is available since April 2003 (both from the science archive at SOC and from the SSC web pages). The 1XMM catalogue contains around 30,000 X-ray sources, almost as many as the *ROSAT* all-sky survey. Every source has been visually inspected, a quality flag has been assigned to it and a large collection of data items referring to it (position, X-ray spectral information, possible candidate counterparts, etc.) is included in the catalogue. The second version of the catalogue (2XMM) is likely to be released by the end of 2005, and it will contain between 100,000 and 150,000 X-ray sources.

Besides the catalogue, the SSC is also conducting an identification programme of the serendipitous X-ray sources (Watson *et al.*, 2001). This consists of three steps: (a) a full spectroscopic identification of representative samples of sources, in and out of the galactic plane and at various flux levels; (b) an imaging programme whereby many fields targeted by *XMM-Newton* are observed in a number of optical filters, most of them with the Sloan filters g', r' and i' (typical limiting magnitudes $r' \sim 23$) and also in the infrared; (c) a statistical identification procedure where with the use of X-ray and optical imaging information a likely identification can be assigned to serendipitous X-ray sources without optical spectroscopy. Significant progress has been made with point (a), point (b) is close to completion and point (c) has started recently. All this information (including the optical data) will ultimately reside in the XSA.

XMM-Newton is making important contributions to many fields of X-ray Astronomy. At the time of writing these proceedings there are close to 500 papers in referred journals using *XMM-Newton* data. An earlier summary of science with *XMM-Newton* can be found in Barcons and Negueruela (2003).

3.5.6. *ASTRO-E2*

ASTRO-E2 is the recovery mission from the JAXA/NASA mission *ASTRO-E* whose launch failed on the 10 February 2000. Its launch to a circular low-Earth orbit is currently planned for mid 2005 by a M-V rocket. The most novel contribution of *ASTRO-E2* will be its ability to conduct spatially resolved non-dispersive spectroscopy with a resolution approaching $E/\Delta E \sim 1000$ at $E \sim 6$ keV.

ASTRO-E2 has five co-aligned thin-foil mirror telescopes with an angular resolution of 1.5'. Four of them will have CCD-based detectors on their focal planes (similar to the ACIS instrument in *Chandra*), delivering moderate resolution $E/\Delta E \sim 20$–50 imaging spectroscopy over the 0.4–10 keV X-ray band and with a FOV of $\sim 20' \times 20'$. The fifth telescope (called X-ray spectrometer, XRS) will have an imaging detector based on a

TABLE 3.1. Comparison of the main parameters of the *Rossi XTE*, *Chandra* ASTRO-E2, and *XMM-Newton* observatories.

Mission	Effective area (m^2)	Angular resolution	Spectral resolution $E/\Delta E$	Time resolution
RXTE	0.65	1° FWHM	5.5	1 μs
Chandra	0.08	0.5″ HEW	20–50 (non-dispersive) 300–1000 (dispersive)	0.18 s
XMM-Newton	0.4	12″–15″ HEW	20–50 (non-dispersive) 200–400 (dispersive)	0.1 s–7 μs
ASTRO-E2	0.05	1.5′	1000 (non-dispersive)	100 ms

semiconductor micro-calorimeter. The XRS has an energy resolution of about 6 eV, (i.e. 10 times better than CCDs) but with the same very high quantum efficiency and with imaging capability. A collimated FOV (1–3°) hard X-ray detector sensitive to 10–800 keV photons complements the payload.

The XRS is really the *ASTRO-E2* primary instrument and represents the first step towards a new generation of high spectral resolution imaging spectrometers. Its imaging capability is still limited: 32 pixels only, each pixel of 625 μm covering a FOV of $3' \times 3'$. Its effective area at 6 keV is about 150 cm^2.

3.5.7. *Outlook*

The present situation is such that there are three main X-ray observatories in orbit: one for timing studies (*Rossi XTE*), one where imaging is the major strength (*Chandra*) and a further one whose major strength is in the spectroscopy (*XMM-Newton*). Table 3.1 summarizes the key parameters of these three observatories.

Complementing the last two, ESA launched the *INTEGRAL* γ-ray observatory on the 17 October 2002. *INTEGRAL* has two γ-ray instruments: a spectrograph (SPI) and a wide-field imager (IBIS, 12° fully coded). Both of these instruments consist of coded-mask optics and detectors with various technologies. An X-ray monitor (JEM-X) provides response from a 3 to 30 keV, in the form of a coded-mask telescope with a micro-strip PC as detector. A further monitor (the OM camera, OMC) completes the payload. Volume 411 of *Astronomy and Astrophysics* is dedicated to the first results from *INTEGRAL*. The *INTEGRAL* Science Data Centre, located in the Geneva Observatory, takes care of receiving the data from the instruments, to process and archive them, as well as of preparing the necessary software for the analysis.

3.6. The future of X-ray Astronomy

The future of X-ray Astronomy missions in space is obviously dictated by both the science drivers and the technological possibilities. Within the horizon of a decade or so, the international road map of X-ray Astronomy goes in at least two directions: (1) Wide area survey telescopes, to either map frequently the whole sky, or to map the structure of the Universe in X-rays; (2) Deep-sky X-ray observatories improving on the existing *Chandra* and *XMM-Newton* missions. Figure 3.12 summarizes the requirements on several features of the new observatories from a number of science goals.

Further away into the future, the prospects of having really huge X-ray observatories, equivalent to 10–50 m optical ground-based telescopes might lead to worldwide missions.

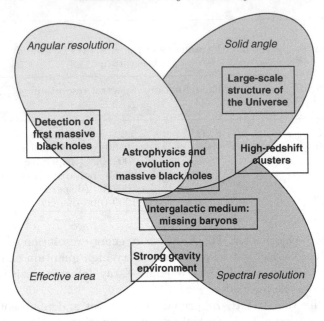

FIGURE 3.12. Requirements of various scientific goals in terms of angular resolution, solid angle, effective area and spectral resolution. Various missions will emphasize different regions in this cartoon.

Exploring other frontiers, like spatial resolution via X-ray interferometry, might also lead to missions in the foreseeable future.

3.6.1. *Surveying the X-ray sky*

LOBSTER-ISS is a mission under study by ESA to place and operate an all-sky X-ray monitor in the Columbus External Payload Facility of the ISS. The mission is lead by the Space Research Centre of the University of Leicester and with collaborations from NASA/Goddard Space Flight Center. The primary goal of *LOBSTER-ISS* is to study the variable X-ray Universe, which includes X-ray flashes, γ-ray bursts, active galactic nuclei, compact binaries and stellar flares.

LOBSTER-ISS will be equipped with a novel "lobster-eye" (i.e. curved) X-ray telescope which uses MCP optics, covering a very wide strip of 180° × 30° (see Fig. 3.13). It will use PSPC as detectors. In one 90-min ISS orbit, *LOBSTER-ISS* will cover 80% of the sky.

The current performance of the X-ray optics yields 9' resolution, but the goal is to achieve 4'. The sensitivity will be of 10^{-12} erg/cm^2/s in the 0.5–3.5 keV band. *LOBSTER-ISS* will deliver an unprecedent view of the variability of the whole X-ray sky, for many thousands of sources.

ROSITA is a German-lead mission to study the local obscured X-ray Universe, by conducting a sensitive all-sky survey in the energy band 0.5–11 keV. Besides going significantly deeper than the *ROSAT* all-sky survey in the soft X-ray band, *ROSITA* will discover a large fraction of mildly to heavily absorbed X-ray sources which will show up only in hard X-rays (soft X-rays being photoelectrically absorbed). *ROSITA* will also discover many (>150,000) distant AGN.

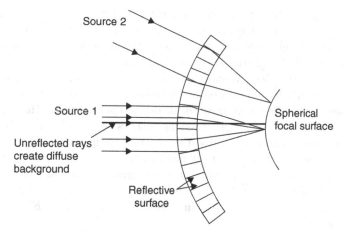

FIGURE 3.13. Design of the lobster-eye optics.

ROSITA is a re-make of the *ABRIXAS* mission, lost due to a failure in the batteries. It comprises seven replicated X-ray telescopes to cover a FOV of ~1°. X-rays are then focused onto a (set of) CCD detectors. These will probably incorporate some of the recent advances in APS on the design of the operating EPIC-pn camera on *XMM-Newton*. The angular resolution of the telescopes will be 30–60″. After a 3-year operation, *ROSITA* will reach a sensitivity of $2 \times 10^{-12} \, \text{erg/cm}^2/\text{s}$ in the 0.5–2 keV band, and $4 \times 10^{-13} \, \text{erg/cm}^2/\text{s}$ in the 2–11 keV band.

DUO (Dark Universe Observatory) is a proposal for a NASA Small Explorer mission (SMEX) which aims at measuring the cosmological parameters with unprecedent precision via observations of X-ray clusters. The main astrophysical tools to be used by *DUO* are the number counts of clusters as a function of redshift and also the cluster spatial distribution; both are very closely linked to the amount of dark matter and dark energy, and to the dark energy pressure. The precision that will be achieved in the *DUO* measurement of these parameters is comparable (or better) than the figures to be achieved by the *Planck* surveyor of the microwave background radiation, but they are "orthogonal" in parameter space, so complementary information is provided by the two missions.

Technically, *DUO* is a free-flying derivative of *ROSITA*. *DUO* is to conduct a wide-area shallow survey on the Sloan Digital Sky Survey (SDSS) region looking for extended emission from clusters of galaxies, most of which will have either spectroscopic or photometric redshifts from the SDSS. It will also conduct a deeper survey onto a separate region to map the more distant cluster population.

Other missions, particularly those looking for GRBs (like *SWIFT* or *Agile*) will also have the capability of mapping large parts of the sky to an unknown extent.

3.6.2. *The next generation of X-ray observatories*

The great success of both *Chandra* and *XMM-Newton* in discovering and analysing local and distant sources is paving the way towards the definition of the next generation of X-ray deep-sky missions. Clearly more effective area is needed to discover and analyse the most distant sources, or to collect more photons for high spectral resolution analyses. Going deep also requires good spatial resolution (to escape from confusion). It is clear that the ideal deep-sky observatory will need good angular resolution, large collecting area, high efficiency and high spectral resolution.

NASA is approaching this via a sequence of two missions: *Constellation-X**, mostly a spectroscopy mission, and *Generation-X*† aiming at order of magnitude improvements in spatial resolution, collecting area and spectral resolution. ESA and JAXA are concentrated on the XEUS‡ (X-ray Evolving Universe Spectroscopy) programme.

In addition, the $E \sim 10 - 50 \, \text{keV}$ energy range is well behind in exploration compared to photon energies $<10 \, \text{keV}$. The coded-mask telescopes do not provide enough angular resolution for sensitive enough studies. It is not surprising that there are a number of initiatives (some as part of larger missions, others as stand-alone observatories) to further our knowledge in the hard X-ray domain, but using reflecting optics rather than coded masks. It is important to note that most of the energy content of the Universe in the X-ray band resides at $\sim$30 keV, where the integrated emission from all sources (the XRB) peaks. The JAXA mission *NEXT*, the recently proposed CNES mission *Symbol-X* or specific instruments and capabilities in *Constellation-X* and *XEUS* are part of the activities in this field.

3.6.2.1. *Constellation-X*

As indicated by the name itself *Constellation-X* will consist of a number of spacecrafts operated in conjunction (i.e. observing towards the same direction). The current baseline mission concept envisages four free-flying elements in the Lagrange 2 (L2) Sun–Earth region. *Constellation-X* is part of the *Beyond Einstein* NASA programme, and will follow the ESA–NASA gravitational wave observatory Laser Interferometer Space Antenna (*LISA*). Prior to the newly announced NASA plan *The Vision for Space Exploration*, the launch(es) of *Constellation-X* were foreseen to happen before 2015, but now the situation regarding the launch date is uncertain.

The observatory will feature a spectroscopy X-ray telescope operating in the 0.25–10 keV band, with a minimum of 1.5 m^2 of peak effective area, and covering a FOV of 2.5′ with an angular resolution of 15″. Part of the incoming X-rays ($E < 2 \, \text{keV}$) will be wavelength-dispersed by reflection gratings, in a similar way as it is done by *XMM-Newton*. When combined with CCD detectors, the grating spectrometer will deliver a spectral resolution of several hundreds to 1000 in $E/\Delta E$, the final number depending on the angular resolution achieved. The remaining X-rays focused by these telescopes will reach an imaging spectrograph with a spectral resolution of 2–4 eV, which will probably be based on TESs. This will be not only the most sensitive but also best spectral resolution instrument in the energy range 1–10 keV (see Fig. 3.14).

Constellation-X will also have a co-aligned hard X-ray telescope system which will collect X-rays in the 10–40 keV range with a minimal effective area of 0.15 m^2 at 40 keV, with a minimum angular resolution of 1′ over an 8′ FOV. The spectral resolution delivered in this hard-energy range will be >10, and probably accomplished by CdZnTe detectors.

Constellation-X will be an extremely powerful tool to conduct spectroscopic observations of sources as faint as the faintest ones discovered by *XMM-Newton*. That will therefore include low to intermediate redshift X-ray sources (mostly AGNs), and all the local sources, except for the very extended sources. It will also provide deep insight into non-thermal processes occurring in clusters of galaxies and AGN, as well as the study of many heavily absorbed and Compton-thick sources, thanks to its high-energy capability.

* http://constellation.gsfc.nasa.gov
† http://generation.gsfc.nasa.gov
‡ http://www.rssd.esa.int/XEUS

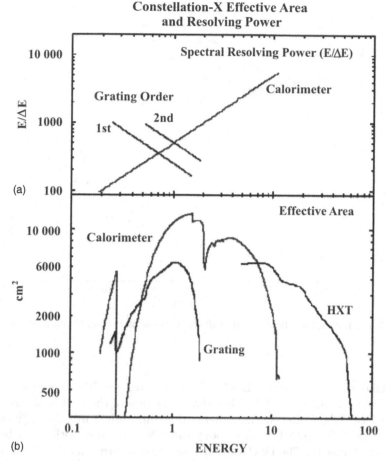

FIGURE 3.14. (a) Spectral resolution of the grating spectrograph and of the calorimeter in *Constellation-X*. (b) Effective area of the various *Constellation-X* instruments. HXT: hard X-ray telescope.

3.6.2.2. *XEUS*

XEUS is an ESA–JAXA initiative for a very powerful X-ray observing facility in space. It is currently under pre-phase A study by ESA, with the aim of being included in the *Cosmic Vision* science programme. The key science goals of *XEUS* are as follows (see Arnaud *et al.*, 2000; Barcons 2005):

• Study matter under extreme conditions, either in strong gravity fields near black holes or at supra-nuclear densities inside Neutron Stars.

• Study the assembly of baryons into groups and clusters along cosmic history.

• Study the growth of supermassive black holes in the centres of galaxies, and its relation to galaxy and star formation.

All these science requirements lead to the need for a large collecting area ($10\,\mathrm{m}^2$ at $1\,\mathrm{keV}$), high spatial resolution (minimum $5''$ with a goal of $2''$), high spectral resolution ($\Delta E \sim 1\,\mathrm{eV}$ at $1\,\mathrm{keV}$, and $5\,\mathrm{eV}$ at $6\,\mathrm{keV}$) X-ray mission, with sensitivity from 0.1 to $40\,\mathrm{keV}$. The sensitivity of *XEUS* will be at least $4 \times 10^{-18}\,\mathrm{erg/cm^2/s}$.

The baseline concept for the mission (see Bleeker *et al.*, 2000) consists of two spacecraft, one with the mirror assembly (the mirror spacecraft, MSC) and another one with the

PerXEUS

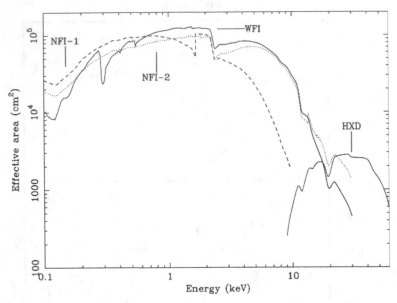

FIGURE 3.15. Estimated effective areas of the various instruments planned for *XEUS*.

detectors (the detector spacecraft, DSC). The focal length is 50 m, enough to provide at least $2\,m^2$ of effective area at 6 keV. Multi-layer coatings of the central shells can extend the sensitivity of *XEUS* into the hard X-ray domain, with a minimal effective area of $\sim 0.2\,m^2$ at 30 keV. Technical details about the telescope and optics can be found in Aschenbach *et al.* (2001). The DSC and MSC operate in formation flying (to a precision of $\sim 1\,mm^3$), the DSC being the active partner.

The baseline *XEUS* payload is defined to contain a suite of instruments to fully exploit its main capabilities and deliver the expected science goals (Barret *et al.*, 2003):

• A WFI, based on CCDs with AXS covering a FOV of $5' \times 5'$, and with an energy spectral resolution of around 120 eV.

• A Narrow-Field Instrument (NFI-1) operating at soft X-ray energies (0.05–3 keV), with energy spectral resolution of ~ 1 eV, and covering a $1' \times 1'$ FOV. The baseline for this instrument is an STJ-based cryogenic detector.

• A Narrow-Field Instrument (NFI-2) operating at harder energies (1–10 keV), with energy spectral resolution of 5 eV, and covering a $1' \times 1'$ FOV. The baseline for this instrument is a TES-based cryogenic array.

• A Hard X-ray Camera (HXC), sensitive to 5–80 keV photons, with an energy resolution of 0.5–1 keV. The detector will be probably placed below the WFI.

• An Extended Field Imager (EFI), expanding the WFI FOV up to $15'$. This is mostly to do serendipitous science.

• A High-Time Resolution Spectrograph (HTRS) designed to obtain spectroscopy of the brightest Galactic sources, and capable of handling up to $\sim 10^6$ counts per second. The baseline for this instrument is a small array of silicon drift detectors.

• An X-ray polarimeter, based on a micro-pattern detector.

The XEUS baseline mission concept foresees operations at L2 following direct injection by a single launcher.

3.6.3. *Far into the future: X-ray interferometry*

In addition to large extensions of the current X-ray observatories, a new frontier in X-ray Astronomy might open with X-ray interferometry. There is a mission concept, the micro arcsecond X-ray imaging mission (MAXIM*), whose goal would be to image the immediate vicinity of supermassive black holes in X-rays. That will require $\sim\mu$arcsec spatial resolution, which can only be achieved via interferometry from space.

X-ray interferometry requires two steps: (a) Achieving the diffraction limit in X-rays and (b) combining the signals from various X-ray telescopes and achieve fringes. Fringes at the level of $100\,\mu$arcsec have already been achieved in the Laboratory (Cash *et al.*, 2000), but going down to 1μarcsec is a real challenge (even more from space). However, progress is expected in this front, as the science goal of imaging the event horizon of a black hole is scientifically very attractive.

Acknowledgements

I am grateful to the many colleagues that have taught me all I have tried to present in this talk. This includes the X-ray Astronomy group at the Instituto de Física de Cantabria, the *XMM-Newton* SSC, the XEUS Science Advisory Group, the EURECA team, and many other people with whom I had the pleasure to collaborate through the years. I am indebted to Valentín Martínez-Pillet and his colleagues at IAC for having organised this wonderful Winter School and for his warm hospitality. Partial financial support for this work was provided by the Spanish Ministerio de Educación y Ciencia, under grant ESP2003 00812.

REFERENCES

ARNAUD, M. ET AL. 2000 ESA SP-1238, The XEUS Science Case.

ASCHENBACH, B. ET AL. 2001 ESA SP-1253, The XEUS telescope.

BARCONS, X. & NEGUERUELA, I. 2003 Boletín de la SEA, **10**, 19 (also available from http://sea.am.ub.es).

BARCONS, X. 2005. The Energetic Universe. Proceedings of the 39th ESLAB Symposium. ESA-SP (in the press).

BARRET, D. ET AL. 2003 ESA SP-1273, The XEUS Instruments.

BIGNAMI, G.F., CARAVEO, P.A., DE LUCA, A. & MEREGHETTI, S. 2003 *Nature*, **423**, 725.

BLEEKER, J.A.M. ET AL. 2000 ESA SP-1242, The XEUS Mission Summary.

BRINKMANN, A.C. ET AL. 1987 *Astrophys. Lett.*, **26**, 73.

CANIZARES, C.R. ET AL. 1987 *Astrophys. Lett.*, **26**, 87.

CASH, W., SHIPLEY, A., OSTERMAN, S. & JOY, M. 2000 *Nature*, **407**, 160.

COSTA, E. ET AL. 2001 *Nature*, **411**, 662.

DEN HERDER, J.W. ET AL. 2001 *Astron. Arstrophys.*, **365**, L7.

FABIAN, A.C. & BARCONS, X. 1992 *Ann. Rev. Astron. Astrophys.*, **30**, 429.

FORMAN, W., JONES, C., COMINSKY, L., JULIEN, P., MURRAY, S., PETERS, G., TANANBAUM, H. & GIACCONI, R. 1978 *Astrophys. J. Suppl. Ser.*, **38**, 357.

FRASER, G.W. 1989 *X-ray Detectors in Astronomy*. Cambridge University Press, Cambridge, UK.

GIACCONI, R., GURSKY, H., ROSSI, B. & PAOLINI, F. 1962 *Phys. Rev. Lett.*, **9**, 439.

GIOIA, I.M., MACCACARO, T., SCHILD, R.E., WOLTER, A., STOCKE, J.T., MORRIS, S.L. & HENRY, J.P. 1990 *Astrophys. J. Suppl. Ser.*, **72**, 567.

*http://maxim.gsfc.nasa.gov

JANSEN, F.A. ET AL. 2001 *Astron. Astrophys.*, **365**, L1.

LIEDAHL, D.A. 1999 The X-ray spectral properties of photoionised plasmas and transient plasmas. In: *X-ray Spectroscopy in Astrophysics*, J. Van Paradijs & J.A.M. Bleeker (Eds.), pp. 189–266.

LONGAIR, M.S. 1997 *High Energy Astrophysics, Volume 1: Particles, Photons and Their Detection.* Cambridge University Press.

MARSHALL, F.E., BOLDT, E.A., HOLT, S.S., MILLER, R.B., MUSHOTZKY, R.F., ROSE, L.A., ROTHSCHILD, R.E. & SERLEMITSOS, P.J. 1980 *Astrophys. J. Suppl.*, **235**, 4.

MASON, K.O. ET AL. 2001 *Astron. Astrophys.*, **365**, L36.

MEWE, R. 1999 Atomic Physics of hot plasmas. In: *X-ray Spectroscopy in Astrophysics*, J. Van Paradijs & J.A.M. Bleeker (Eds.), pp. 109–182.

MURRAY, S.S. ET AL. 1987 *Astrophys. Lett.*, **26**, 113.

MUSHOTZKY, R.F., COWIE, L.L., BARGER, A.J. & ARNAUD, K.A. 2000 *Nature*, **404**, 459.

NOUSEK, J. ET AL. 1987 *Astrophys. Lett.*, **26**, 35.

PFEFFERMANN, E. & BRIEL, U.G. 1986 *SPIE*, **597**, 208.

PICCINOTTI, G., MUSHOTZKY, R.F., BOLDT, E.A., HOLT, S.S., MARSHALL, F.E., SERLEMITSOS, P.J. & SHAFER, R.A. 1982 *Astrophys. J. Suppl.*, **253**, 485.

RYBICKI, G.B. & LIGHTMAN, A.P. 1979 Radiative processes in Astrophysics. Wiley and Sons, Inc, New York.

SCHMITT, J.H.M.M., SNOWDEN, S.L., ASCHENBACH, B., HASINGER, G., PFEFFERMANN, E., PREDEHL, P. & TRÜMPER, J. 1991 *Nature*, **349**, 583.

STOCKE, J.T., MORRIS, S.L., GIOIA, I.M., MACCACARO, T., SCHILD, R., WOLTER, A., FLEMING, T.A. & HENRY, J.P. 1991 *Astrophys. J. Suppl. Ser.*, **76**, 813.

STRÜDER, L. ET AL. 2001 *Astron. Astrophys.*, **365**, L18.

SVENSSON, R. 1990 In: *Physical processes in hot cosmic plasmas*, W. Brinkmann, A.C. Fabian, F. Giovanelli (Eds.), Kluwer Dordrecht, p. 357.

TANAKA, Y., INOUE, H. & HOLT, S. 1993 *Pub. Astron. Soc. Jap.*, **46**, L37.

TANAKA, Y. ET AL. 1995 *Nature*, **375**, 659.

TUCKER, W. & GIACCONI, R. 1985 *The X-ray Universe.* Harvard University Press, Cambridge, MA, USA.

TURNER, M.J.L. ET AL. 1989 *Pub. Astron. Soc. Jap.*, **41**, 345.

TURNER, M.J.L. ET AL. 2001 *Astron. Astrophys.*, **365**, L27.

VOGES, W. ET AL. 1999 *Astron. Astrophys.*, **349**, 389.

WATSON, M.G. ET AL. 2001 *Astron. Astrophys.*, **365**, L51.

WEISSKOPF, M., BRINKMANN, B., CANIZARES, C., GARMIRE, G., MURRAY, S. & VAN SPEYBROEK, L.P. 2002 *Pub. Astron. Soc. Pac.*, **114**, 791.

WILLINGALE, R.W. 1999 New developments in X-ray optics. In: *X-ray Spectroscopy in Astrophysics*, J. Van Paradijs & J.A.M. Bleeker (Eds.), pp. 435–475.

4. EUV and UV imaging and spectroscopy from space

By RICHARD A. HARRISON

Rutherford Appleton Laboratory, Chilton, Didcot, Oxfordshire, UK

Observations of the ultraviolet (UV) and extreme-ultraviolet (EUV) universe provide us with the tools to examine the atomic, ionic and molecular properties of many phenomena, including the Sun, planetary atmospheres, comets, stars, interstellar and intergalactic gas and dust, and extragalactic objects. This chapter takes the reader from the dawn of the UV space age, through to the modern instruments operating on missions such as the Solar and Heliospheric Observatory, and the Hubble Space Telescope. We examine the properties of the UV region of the electromagnetic spectrum and explore the reasons for utilizing it for space research. This includes a detailed discussion of the basic processes which lead to EUV/UV radiation from space plasmas, and an introduction to the "EUV/UV Toolbox", which allows us to diagnose so much from the emission we detect. Frequent reference is made to recent and ongoing missions and results. However, such a review would not be complete without a glance at the future strategy for EUV/UV space research. This highlights some new techniques, and a range of upcoming missions, though the emphasis of the near-future space programme in this region of the electromagnetic spectrum is more on solar physics than non-solar astronomy; there are many exciting developments in solar EUV/UV research, but the lack of mission opportunities for astronomy in general is a concern.

4.1. Introduction

The dawn of space astronomy occurred on 10 October 1946 with the observation of the solar ultraviolet (UV) spectrum, from a V-2 sounding rocket launched from the US White Sands Missile Range. This observation, shown in Fig. 4.1, identified, for the first time spectral features down to 2200 Å (220 nm). In Fig. 4.1, we compare this early result with a modern UV solar observation, in this case, made using the Solar and Heliospheric Observatory (SOHO) spacecraft, where high-resolution images are produced from selected UV wavelengths. Whilst the details of the images of Fig. 4.1 are described later, they demonstrate immediately the tremendous advances made in UV astronomical observation – in this case, solar astronomy – and give a brief insight to the potential rewards that such observations can provide.

In fact, what we will see in this chapter is that the UV range is unique in providing tools for probing astrophysical plasmas and this has provided many fundamental insights to the nature of the Universe.

We start by revisiting the most basic definitions of the UV and extreme-UV (EUV) range, and explore the reasons behind our need to perform EUV/UV observations from space. As already noted, it was in this wavelength range that we first embarked on space-based remote sensing astronomy. For this reason, we review the early missions and the opening of the EUV/UV space window. Before exploring the details of modern EUV/UV instrumentation, it is necessary to examine the basic processes leading to radiation in this wavelength region, from astrophysical sources, and to the methods required for determining plasma diagnostic information on those sources. We complete this chapter by outlining some of the future missions and techniques which are promising exciting results for the future of this field.

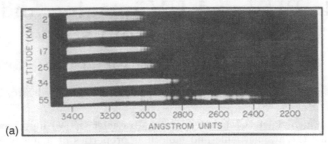

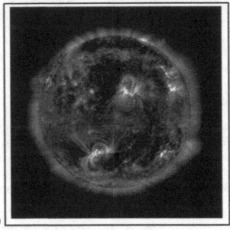

FIGURE 4.1. The image that heralded the dawn of space astronomy and a modern UV observation of the Sun. (a) Solar UV spectra taken by Tousey and co-workers during 10 October 1946 V-2 rocket flight from White Sands Missile Range (courtesy US Naval Research Laboratory). (b) An image of the Sun taken in an iron emission line at 171 Å, using the EUV imaging telescope (EIT) instrument aboard the SOHO spacecraft (courtesy SOHO EIT Team).

It has to be stressed that this is a rather selective and somewhat biased review. We cannot review the details of all EUV/UV missions and observations in a few tens of pages, and the author, as a solar physicist, has to admit that some bias has crept in. Nevertheless, this chapter should serve to give a flavour for the worldwide efforts in EUV/UV space astronomy and the rewards this field is providing.

4.2. The electromagnetic spectrum and the EUV/UV region

4.2.1. *The definition of EUV/UV*

The electromagnetic spectrum is, of course, the familiar spectrum of all electromagnetic waves ordered by frequency, wavelength or energy. As shown in Table 4.1, this spectrum extends from the long-wavelength radio waves to the short-wavelength gamma-rays and includes the visible range to which are eyes are suited.

The portion of the spectrum at shorter wavelengths than the visible, starting from 4×10^{-7} m is known as the ultraviolet, or UV. Unfortunately, the broad range of wavelengths and energies means that people become comfortable with different units, depending on their experiences. Thus, the 4×10^{-7} m UV boundary can be given as 0.4 microns (μm), 400 nm, 4000 Å, 3 eV or 7.5×10^{14} Hz.

Whereas this long-wavelength boundary for the UV is pretty much defined by the sensitivity of the human eye, the shorter-wavelength boundary, between the UV and

TABLE 4.1. The electromagnetic spectrum.

	Wavelength (m)	Frequency (Hz)	Energy (eV)
Radio	>0.1	$<3 \times 10^9$	$<10^{-5}$
Microwaves	10^{-4} to 0.1	3×10^9 to 3×10^{12}	10^{-5} to 0.01
Infrared	10^{-4} to 7×10^{-7}	3×10^{12} to 4.3×10^{14}	0.01 to 2
Visible	7×10^{-7} to 4×10^{-7}	4.3×10^{14} to 7.5×10^{14}	2 to 3
Ultraviolet	4×10^{-7} to 10^{-8}	7.5×10^{14} to 3×10^{16}	3 to 100
X-rays	10^{-8} to 10^{-11}	3×10^{16} to 3×10^{19}	100 to 10^5
Gamma-rays	$<10^{-11}$	$>3 \times 10^{19}$	$>10^5$

TABLE 4.2. Commonly used UV divisions in astronomy.

	Wavelength (m)	Wavelength (Å)
Near-UV	3–4×10^{-7}	3000–4000
Middle-UV	2–3×10^{-7}	2000–3000
Far-UV	1–2×10^{-7}	1000–2000
EUV	0.1–1×10^{-7}	100–1000

X-rays is rather subjective and often down to opinion. A reasonable figure to use is a wavelength of order 10^{-8} m, which is 0.01 μm, 10 nm, 100 Å, 100 eV or 3×10^{16} Hz.

However, to complicate the issue further, the UV region suffers from a plethora of subdivisions which are not always sharply defined and are used sporadically or by different research communities, and one hears terms such as vacuum-UV (VUV), near-UV, far-UV, EUV, X-ray-UV (XUV) and middle-UV. Perhaps the most commonly used divisions in astronomy are those given in Table 4.2, and these are the divisions we will use in this chapter.

This wavelength region provides a unique set of tools for astronomy. However, before exploring the emission and absorption processes that constitute the UV range, let us consider the most basic requirements for making astronomical observations in this range.

4.2.2. *Transmission of the Earth's atmosphere: the need to go into space*

The Earth's atmosphere is opaque to much of the electromagnetic spectrum. The degree of transparency is dependent on altitude – how much of the atmosphere is above the observer – but, in effect, there are only two basic windows for which we can use ground-based observations for astronomical purposes. These are in the visible region and the radio region and they are clearly illustrated in Fig. 4.2.

The radio region of 1 cm to 30 m provides the classical window for radio astronomy. At longer wavelengths, the ionosphere becomes opaque to radio waves. At the shorter-wavelength end of the window, observations can be made to less than 1 mm if strong atmospheric molecular absorption bands are avoided.

The visible range falls between 0.3 and 1 μm, that is, 300 and 1000 nm (3000 and 10,000 Å) and, of course, this is the rather narrow window to the Universe in which all astronomical research was conducted prior to the middle of the last century. At wavelengths shorter than 0.3 μm, atmospheric absorption from oxygen, nitrogen and ozone becomes important.

One can improve the atmospheric opacity in certain wavelength ranges by making high-altitude observations from mountains, aircraft or balloons, to remove much of the overlying atmospheric mass, and by carefully selecting particular wavelength windows. However, for vast sections of the electromagnetic spectrum, the atmosphere remains opaque.

Let us consider the UV region. The classical visible window mentioned above terminates at about $0.3\,\mu m$ (300 nm or 3000 Å), which means that the window includes the near-UV region. Indeed, this can be used as a definition of that region.

However, the basic need for astronomical research in gamma-ray, X-ray and much of the UV and infrared spectrum is to get above the atmosphere, that is, into space. In short, we are only viewing a small fraction of the radiation from most extraterrestrial sources and that restricts our understanding of the physics of the Universe. Astronomy has to be done from space.

To avoid confusion, from this point on, we will tend to use the units of Å. Also, since the chapter is concerned with UV astronomy *from space platforms*, the emphasis is clearly geared towards wavelengths shorter than 3000 Å.

4.2.3. *Why bother with the EUV/UV range?*

So, why bother with the UV? If the UV wavelength range tells us nothing in addition to what we can learn from optical observation, then there is no reason to go into space to study that region of the electromagnetic spectrum. However, that is far from the case, and there are indeed many advantages of exploring the UV spectrum from astronomical objects.

The UV region is something of an atomic, ionic and molecular "window". Over 99% of the atoms in the Universe are hydrogen or helium and, of the remaining elements, the most common are oxygen, carbon, neon and nitrogen. For all of these elements either in atomic or ionic form, or indeed in simple molecular form such as H_2, N_2 and CO, the resonance spectral transitions occur at wavelengths under 2000 Å. Resonance transitions are transitions between energy states involving the ground state.

Most material in the Universe is in a plasma state. Spectral line identification, and the analyses of line shapes, shifts and intensities, can provide plasma diagnostic information, such as densities and flows, for a broad range of temperatures, and this effectively equips us with a set of tools for exploring the physics of astrophysical plasmas. This applies to exploration of our own star, as well as planetary atmospheres, comets, stars, interstellar gas and dust and extragalactic objects.

In addition, the back body curves of very hot stars, say, with temperatures in excess of 10,000 K, are such that they emit much of their radiation in the UV, that is, much of their radiation would be invisible from the ground.

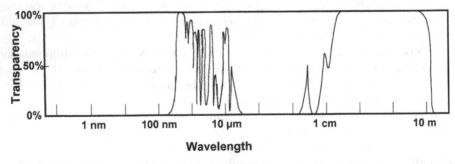

FIGURE 4.2. A sketch of the transparency of the Earth's atmosphere from sea level.

Finally, UV radiation and its variation from the solar atmosphere is of interest for its impact on the ionosphere as well as the effects on our skin. These facts provide a driver for understanding the radiation processes in this region from our own star.

There is no question, then, that the UV range is of fundamental interest to astronomical research across a broad range of disciplines. Indeed, for most astronomical phenomena, we are not able to make *in situ* measurements. We are totally dependent on remote sensing, and the UV region provides us with remote sensing tools, which can be used to identify and explore the characteristics of many astronomical targets.

4.3. Early space observations in the UV

4.3.1. *The opening of the field with the V-2 and the Aerobee rockets*

As we have seen, to develop UV astronomy fully, one needs to provide a space-borne platform. Thus, progress in UV astronomy has been dependent as much on the development of launch vehicles as on the development of detection and optical techniques.

The first opportunities to really break free of the restricted windows of the electromagnetic spectrum discussed above, came just after the Second World War. Groups in the USA began to utilize captured German V-2 rockets, which were capable of ascending to altitudes of 160 km. However, the post-war years were crowded with many attempts to develop sounding rockets, with a catalogue of failures, technical advances and some spectacular successes. The opportunities for astronomy were well recognized and some of the earliest flights were attempts to make basic astronomical, cosmic ray or upper atmospheric measurements. In retrospect it was a period of frenzied activity, which provided a foundation for the space activities we know today, and it is hard to believe that this is under 60 years ago.

A group led by Richard Tousey (Fig. 4.3), from the US Naval Research Laboratory (NRL), in Washington DC, was the first to make use of the V-2 opportunity for UV astronomy; their target was UV observation of the Sun.

FIGURE 4.3. Richard Tousey (courtesy US NRL).

An initial attempt on 28 June 1946 failed because the camera was never retrieved from the V-2 impact area at the White Sands Missile Range in New Mexico. A second attempt, on 10 October 1946 was successful. With this flight, their UV observations of the solar spectrum opened up the space age for astronomy.

The Tousey V-2 observation (Fig. 4.1) showed the solar spectrum, at differing altitudes during the flight, from 3400 Å, down to 2200 Å at the highest altitudes (see Tousey *et al.*, 1946). The rocket achieved an altitude of 180 km and landed 23 km downrange of the launch site. Note from Table 4.1 that the observed range extends from the near-UV to the middle-UV. There are a number of spectral features, which can be seen in Fig. 4.1, which will be discussed later.

It was no accident that UV solar astronomy opened up space age astronomy. The Sun is, of course, a very bright source and the basic optical and detection methods for the UV are not far removed from those of the visible. The first solar X-ray observations followed in 1949 when a payload of Geiger counters was launched aboard another V-2 rocket by Herbert Friedman, also of the US Naval Research Laboratory.

It is interesting to note that the early V-2 instruments were unpointed but nevertheless returned valuable new information on the solar UV spectrum. It was not until the 1950s that instrument pointing capabilities were developed for the Aerobee sounding rockets. The Aerobee rocket project (Van Allen *et al.*, 1948) began as early as May 1946 when a contract was given by the Applied Physics Laboratory to the Aerojet Engineering Corporation. The first research flights started in the Spring of 1948 with cosmic ray and magnetic field experiments flown from White Sands. Altitudes of about 115 km were achieved. The first solar UV Aerobee flight was in August 1948.

It is a difficult problem to point accurately a telescope or spectrometer aboard a sounding rocket. By definition, the rocket will have varying orientation, and may be spinning or tumbling. A servo-feedback-controlled gimbal mount system was developed by the University of Colorado, for the Air Force, that allowed such instruments to be pointed accurately from the Aerobee rocket. Once at the required altitude, the doors which protect the payload would be opened and Sun sensors would search for and lock onto the Sun. The Sun sensor fed positional information to the servo-motors on the gimbal mount supporting the instrument. During the flight, changes in orientation would be compensated for by the same procedure. This technology and experience was carried through into the later satellite programmes.

Numerous Aerobee flights were dedicated to solar research. A landmark instrument that flew on some six Aerobee rockets was the US NRL multispectrograph, developed in the late 1950s, which observed the far-UV and X-ray solar spectrum.

4.3.2. *The orbiting solar observatories and orbiting astronomical observatories satellites*

The first satellites dedicated to space astronomy were the orbiting solar observatories (OSOs). They were conceived in the late 1950s as a standard spin-stabilized platform for observing the Sun.

It is not the purpose of this chapter to discuss the details of every mission or series of spacecraft which involves UV astronomy, but the OSO and orbiting astronomical observatories (OAO) series are landmarks, not only opening up UV space-borne astronomy, but also introducing basic spacecraft techniques for astronomy which are familiar to us today. Thus, we devote this section to a description of the OSO and OAO missions.

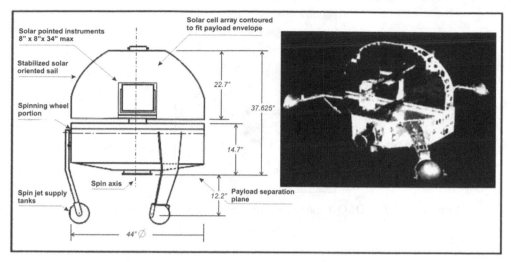

FIGURE 4.4. The OSO satellite concept (courtesy NASA).

Ball Brothers Research Corporation of Boulder, CO, USA, had developed stabilized pointing control devices for sounding rockets and balloon systems during the early days of the US space programme. The company was asked by John C. Lindsay of the Goddard Space Flight Center to develop an astronomical satellite design. By 1960 they produced a successful engineering prototype designed to study UV, X-ray and gamma radiation from the Sun. The first seven OSOs were all built to the same basic design as the prototype.

The OSO spacecraft concept is shown in Fig. 4.4. The spacecraft stood a little under 1 m in height. The top half was stabilized, with solar panels and solar pointed instruments facing the Sun. The bottom half spun on the central spacecraft axis. A photocell, amplifier, torque motor system turned the top section backwards with respect to the bottom half at the rate of spin, thus providing a constant pointing capability. Pointing accuracy was 2–3 arcmin.

Whilst the upper portion was designed to keep the pointed instruments (34 kg payload) directed at the Sun, the lower, spinning portion carried some 45 kg of instruments and rotated once every 2 s, allowing those instruments to scan the solar disk and atmosphere. The OSO had three protruding arms that extended after deployment, giving the system greater axial stability.

OSO-1 was launched on 7 March 1962 into a near-circular orbit of altitude 575 km, with 32.8° inclination. At 208 kg it was light compared to some of our modern solar missions, such as SOHO at 1.85 tons. OSO-1 carried a payload designed to measure the solar electromagnetic radiation in the UV, X-rays and gamma-rays, as well as dust particles in space. Thus, this was the dawn of space-borne UV astronomy. Data were received from the spacecraft until 6 August 1963 and the satellite reentered the Earth's atmosphere in October 1981.

The paper by Behring (1970), which reported on the operation and results of the EUV spectrometer on OSO-1, is considered to be something of a landmark and we highlight that here. The paper outlines the problems in space experiment operation which are all too familiar to us now, such as the concerns of launch vibration, power and telemetry limitations and the effects of the environment. The EUV spectrometer was designed to study the solar spectrum shorter than 400 Å. Behring reported the detection of lines down

FIGURE 4.5. The OAO-3 spacecraft, renamed Copernicus (courtesy NASA).

to 147 Å, with identification of emission lines from Fe, Si, N, Ca, and the bright He II Lyman line. In all, 6000 spectra were taken between 7 March and 22 May. Analysis of the variations of emission line intensities were used to investigate coronal temperatures, and several solar flares were detected, with the observation of high-temperature emission lines.

Following on from OSO-1, a series of missions were used to explore solar radiation and the upper atmosphere. The last of the series, OSO-8, was launched on 21 June 1975. The spacecraft had the same basic design concept of pointed instruments on a sail section, with a rotating base, though the details had been upgraded significantly from the original design.

The contribution of the OSO series to UV solar physics was significant. Not only had the solar UV – indeed the astronomical UV – window been opened from an orbiting platform, with the detection and identification of a range of UV emission lines, but the missions had provided a solid foundation in the development and operation of solar spectroscopic instrumentation in space, which was to feed into the next generation of missions.

The first successful launch of the OAO series was in 1968. OAO-2 surveyed the UV sky from about 1000 to 3000 Å, and was followed by the very successful OAO-3, in August 1972, which was renamed Copernicus, after launch. We discuss Copernicus in a little detail here because of its significance as a landmark mission in UV astronomy.

Copernicus was placed into a near-circular orbit of radius 7123 km, inclined at 35°, on 21 August 1972. The main experiment was an 80 cm aperture UV telescope spectrometer, the telescope tube occupying the central cylinder of the spacecraft (Fig. 4.5), but the spacecraft also contained a cosmic X-ray experiment. Solar panels were fixed at a 34° angle to the observing axis, and were kept within 30° of the Sun. This restriction resulted in parts of the sky being visible only at certain times of the year. The spacecraft was operational until February 1981.

The UV telescope consisted of a Cassegrain system feeding a spectrometer, with the entrance slit at the Cassegrain focus. Two movable carriages containing phototubes could be programmed to scan the spectrum from about 710 to 3280 Å, with a resolution of 5 Å at the best. The instrument was used to probe the interstellar medium and to study hot stars. In all, Copernicus obtained high-resolution far- (900–1560 Å) and near- (1650–3150 Å) UV spectra of 551 objects, primarily bright stars. The locations of these sources are shown in Fig. 4.6, which illustrates how Copernicus truly opened the field of stellar UV astronomy.

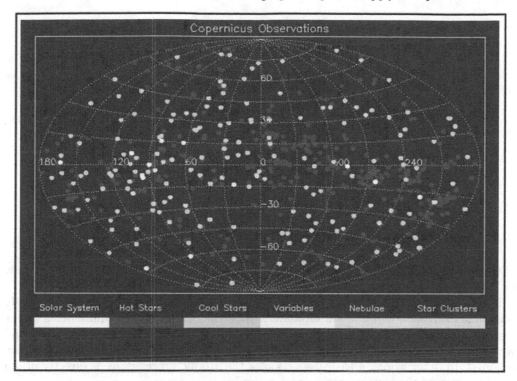

FIGURE 4.6. The Copernicus UV sources (courtesy NASA).

The OSO and OAO spacecraft paved the way for EUV/UV missions such as the Solar Maximum Mission (SMM), Spacelab 2, the SOHO, the Transition Region and Coronal Explorer (TRACE), Coronas-F for solar research, and missions such as the International UV Explorer (IUE), European X-ray Observatory Satellite (EXOSAT), Roentgen Satellite (ROSAT), Hubble Space Telescope (HST) and EUV Explorer (EUVE), for astrophysical research. It is the major results from these missions that we will describe later.

4.4. Basic processes leading to EUV/UV radiation from space plasmas

Before we embark on a description of instrument techniques and methods, we have to be aware of the basic requirements on our observations.

Suppose we have an astronomical source. How do we obtain any physical information from that source? We cannot travel to it to make *in situ* measurements, in most cases, so we are reliant on remote sensing information.

Figure 4.7 shows what you might call a generic UV astronomical source, which we have labelled as an "astrophysical plasma cloud", but it could be a piece of the solar atmosphere, a supernova remnant, a comet or hot star. Again, we note that most material in the Universe is in a plasma state, so it is this state that we consider in detail here.

Our basic concern is that the plasma cloud has parameters such as density, temperature, flows, abundance and composition, and it is these which we require in order to explore the physics of the source itself. However, as indicated by the arrows, all we can measure is the electromagnetic radiation from the source.

So, in this section we are asking the question: How is EUV/UV radiation produced in such a source? We know that the radiation emitted contains information on the nature

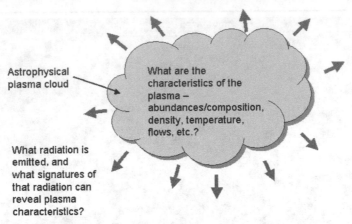

Astrophysical
plasma cloud

What are the
characteristics of the
plasma –
abundances/composition,
density, temperature,
flows, etc.?

What radiation is
emitted, and
what signatures of
that radiation can
reveal plasma
characteristics?

FIGURE 4.7. A cartoon of an astrophysical UV source.

of the plasma, and that must form the basis of our understanding of the source, but, how do we unlock that information?

In this section, then, we embark on an atomic physics tutorial. It is not a purely academic exercise because it will ultimately drive the observational requirements of our instrumentation. We must know how UV radiation is produced in order to understand what wavelengths are important, what methods are necessary and what resolutions are required. In other words, a basic inspection of the atomic processes feeds directly into a consideration of the instruments we wish to build.

Good reviews covering areas included in this section have been given by Dwivedi *et al.* (2003), Mason and Monsignori-Fossi (1994), Mariska (1992) and others.

So, where do we start? It seems that the best starting point is to consider the most basic composition of our plasma cloud. What is it made of? For this, we need to consider the ionization balance or degree of ionization of an element.

4.4.1. *Ionization balance*

Elements are ionized as a result of the loss of bound electrons through a variety of processes. However, recombination processes may occur by various means, so the ionization balance of an element is determined by equating the ionization and recombination rates. Consider an ion which has lost m electrons. The ionization processes, that will take the ion to the $m + 1$ ionization state, that is, the processes indicated in Fig. 4.8(a), include (i) collisional ionization, (ii) autoionization, (iii) charge transfer, or exchange and (iv) photoionization.

Collisional ionization occurs when free electrons with sufficient energy collide with the ion and transfer energy to release a bound electron. Autoionization involves the release of an electron following the decay of an ion from a doubly excited state. Charge transfer (charge exchange) involves the collisions between ions, resulting in the exchange of bound electrons. In this case, an electron is lost. Photoionization is ionization through the arrival of a photon, which releases an electron.

Photoionization is not of major importance for so-called optically thin plasmas but is of interest in some special astrophysical cases which are optically thick.

In equilibrium, then, the processes of ionization are balanced by recombination processes, that is, as indicated in Fig. 4.8(b). These recombination processes include (i) radiative recombination, (ii) dielectronic recombination and (iii) charge transfer (charge exchange).

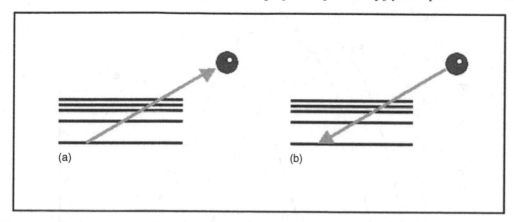

FIGURE 4.8. A schematic representation of the processes of ionization and recombination, the balance of which determines the ionization state of a plasma. In each case, a set of bound energy levels are drawn, with the ground state as the lowest level, and an electron is shown, either escaping from the ion (a) or being captured (b).

TABLE 4.3. The ionization and recombination processes.

Ionization processes

Collisional ionization	$X^{+m} + e \rightarrow X^{+m+1} + e + e$	An electron collides with the ion, resulting in the release of a bound electron
Autoionization	$(X^{+m})^{**} \rightarrow X^{+m+1} + e$	The doubly excited ion X^{+m} decays, releasing sufficient energy to free an electron
Charge transfer	$X^{m} + X^{+m+1} \rightarrow X^{+m+1} + X^{+m}$	A collision between two ions results in the exchange of charge
Photoionization	$X^{+m} + h\nu \rightarrow X^{+m+1} + e$	A collision with a photon of energy $h\nu$ results in the release of an electron. The reverse of radiative recombination

Recombination processes

Radiative recombination	$X^{+m+1} + e \rightarrow X^{+m} + h\nu$	A collision in which an electron is captured resulting in the release of the excess energy, $h\nu$, as a photon. The reverse of photoionization
Dielectronic recombination	$X^{+m+1} + e \rightarrow (X^{+m})^{**} \rightarrow X^{+m} + h\nu$	An electron collision with an ion results in a doubly excited state which then either decays through a radiative transition or may autoionize
Charge transfer	As above	As above

Radiative recombination is a collisional process where a free electron recombines with the ion, resulting in the emission of a photon. Dielectronic recombination is a process whereby an electron collision with an ion results in a doubly excited state which then either decays through radiative transitions, or may even autoionize. Charge transfer, or charge exchange, is described above, but in this case, an electron is captured.

We tabulate these processes, somewhat schematically in Table 4.3. We use the notation X^{+m} to denote the element X stripped of m electrons.

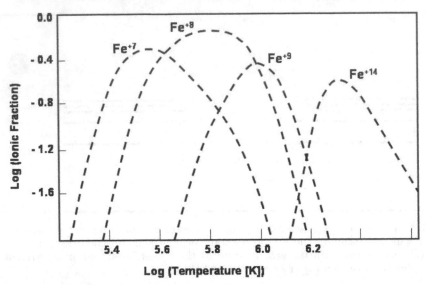

FIGURE 4.9. The ionization balance for a number of iron ions, showing the different ranges of temperature of formation of these ions (adapted from Arnaud and Rothenflug, 1985).

It should be noted that dielectronic recombination is an important recombination mechanism at high temperatures.

A full consideration of the ionization balance for any particular ion, as a function of temperature is clearly beyond the scope of this chapter. Good references include Arnaud and Rothenflug (1985), and Arnaud and Raymond (1992).

The principal feature of interest, when a full calculation of the ionization balance is made, is that any particular ion exists only across a narrow range in temperature. The most basic reasons for this are clear; as temperatures rise in a plasma, higher-energy collisional processes are more likely to result in more extreme ionization states.

An example of calculated ionization balances for range of ions is given in Fig. 4.9. This stresses that detection of radiation signatures from any particular ion, immediately provides information on the temperature of the plasma being observed. This is an important result, which feeds into any consideration of EUV/UV emission line observation.

Of course, one statement made at the start of this section was that we are considering a plasma in equilibrium, that is, ionization and recombination processes balance. It is quite possible to find situations were such a balance is not achieved, such as in rapid events. This makes the consideration of ionization states much more complex and time dependent. In practice, such rapid events are rare. It may apply, for example, to solar flares.

Now that we have an understanding of the ionic nature of a plasma, let us consider the generation of UV radiation from such a plasma.

4.4.2. *The components of the EUV/UV spectrum*

Figure 4.10 shows a portion of the solar spectrum, taken by the Solar UV Measurement of Emitted Radiation (SUMER) instrument on board the SOHO spacecraft, for the EUV/UV range 800–1500 Å. One is immediately struck by the complex nature of such a spectrum, which has three basic components. First, we see a forest of emission lines. These sit on a slowly varying continuum. An additional striking feature is the occasional precipitous edge, such as that at 912 Å. These three components are now described in some detail.

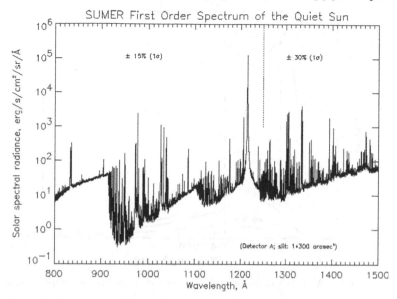

FIGURE 4.10. A solar spectrum from 800 to 1500 Å taken using the SUMER instrument aboard the SOHO spacecraft (courtesy SUMER SOHO Team, see e.g. Wilhelm *et al.*, 1997).

4.4.3. *The generation of EUV/UV radiation: continua*

Free electrons in a plasma can jump between energy states via the free–free transition process resulting from the deceleration of an electron in the Coulomb field of an ion. This is also known as bremsstrahlung (braking) radiation. In the initial state, the electron is free of any bound state, and that is also the case after the transition. The change in energy results in the emission of a photon and, because the electrons are free, the energy spectrum of the photons will form a continuum. The impact on the ion is negligible when compared to the electron, in such a collision.

When studying continuum spectra from astrophysical plasmas we must consider the bremsstrahlung radiation from a particle distribution, which, for most purposes, will be Maxwellian. That is, we are considering a plasma in thermodynamic equilibrium. In a Maxwellian state, the velocity distribution can be written as:

$$N(v)\,dv = 4\pi N_e \left(\frac{m_e}{2\pi\,kT}\right)^{3/2} \exp\left(\frac{-m_e v^2}{2kT}\right) v^2\,dV \qquad (4.1)$$

where N_e and m_e are the electron density and mass, k is Boltzmann's constant and T the temperature. A consideration of such a distribution allows one to calculate the rate of free–free emission by an ionized plasma as:

$$I_{ff} = 5.44 \times 10^{-40} Z^2 g \exp\left(\frac{-h\nu}{kT}\right) T^{-1/2} N_e N_i \quad (\text{J/m}^3/\text{s/ster/Hz}) \qquad (4.2)$$

The ion charge and number density are given by Z and N_i, and g is known as the Gaunt factor, which is itself a function of radiation frequency (ν) and temperature. Of course, in calculating the rate of free–free emission in a plasma source, we are heading towards the calculation of a measurable quantity.

However, let us take a more simple-minded view of a source in thermodynamic equilibrium. Planck's law describes the intensity distribution of radiation from a source in thermodynamic equilibrium. Such a radiation distribution is known as a black-body

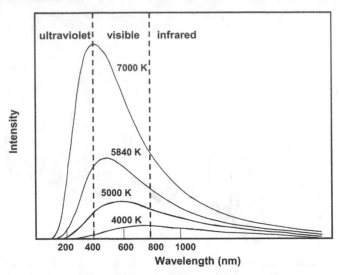

FIGURE 4.11. A sketch of black-body curves for a variety of temperatures.

radiation curve. It has the form (as a function of wavelength, λ):

$$I(\lambda)\,\mathrm{d}\lambda = \left(\frac{2hc}{\lambda^5}\right)\frac{\mathrm{d}\lambda}{\mathrm{e}^{\frac{hc}{kT\lambda}} - 1} \tag{4.3}$$

Note that this is dependent only on wavelength and temperature. Whilst the Planck function represents a useful tool for the comparison of plasmas in equilibrium, it should be realized that in reality, astronomical sources may have temporal and spatial variations, the superposition of free–bound and bound–bound transitions may complicate the spectrum, and the effects of optically thick and thin situations will influence what the observer sees, and this means that any comparisons will be approximate at the best.

Nevertheless, inspection of Equation (4.3), as plotted in Fig. 4. 11, can readily show the plasma temperature regimes that emit significantly in the EUV/UV wavelength bands. Black-body curves for temperatures below ∼5000 K show little emission in the EUV/UV region. Sources with temperatures in excess of this begin to show significant UV contributions. However, sources with temperatures in excess of 7000 K actually peak in the UV, and thus may be virtually invisible in some cases from the surface of the Earth.

The peak temperature of a black-body curve is given by:

$$\lambda_{\mathrm{peak}} = 2.9 \times 10^{-3}/T \tag{4.4}$$

The Sun's spectrum mimics a black-body curve of temperature 5800 K, which peaks in the visible at 5000 Å. However, as a spectral class G star, the Sun is relatively cool. The surface temperatures of F0, A0 and B0 stars are 7500, 11,000 and 20,000 K, respectively, so it is clear that analyses of EUV/UV spectral continua are important for stellar physics.

4.4.4. *The generation of EUV/UV radiation: lines*

In contrast to the free–free collisions, which produce the continuum radiation, transitions which involve electrons bound to energy levels within an atom or ion before and after the transition (i.e. bound–bound transitions), produce emission lines. The transition from one bound state to another lower-energy-bound state results in the emission of a photon at a particular energy. This is seen in the resulting spectrum as a spike in intensity at a

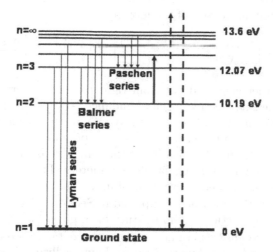

FIGURE 4.12. Transitions within the hydrogen atom.

particular wavelength. Figure 4.10 shows a forest of such features with varying intensities. Figure 4.12 shows the energy-level structure for neutral hydrogen. It clearly indicates that the energy levels (given with respect to the ground state) are well defined and thus, any bound–bound transition has a well-defined energy or wavelength.

A number of energy levels are shown; the ground state is level $n = 1$. Downward directed arrows represent transitions to lower-energy states, resulting in the emission of photons at specific energies. Transitions from the upper levels to the ground state are known as the Lyman series. Similarly, transitions from upper states which terminate on the $n = 2$ and $n = 3$ levels are known as the Balmer and Paschen series, respectively. Oppositely directed transitions, indicated by the upward directed arrow, are transitions which absorb radiation at specific wavelengths. The dashed lines are the ionization and recombination transitions from/to the ground state.

Let us consider an element, which we label arbitrarily as X. The element is said to be an ion if it has been stripped of any of its electrons. If it has lost m electrons, we use the notation X^{+m}. Our bound–bound transition can be written as:

$$X_j^{+m} \to X_i^{+m} + h\nu_{ji}$$

where the initial and final (lower energy) bound states of the ion are denoted by j and i. The right-hand term is the energy of the emitted photon, which can also be written as hc/λ_{ji}, where h is the Planck's constant and c is the speed of light.

The radiant power density of the spectral line, from this transition, is given by:

$$\varphi(\lambda_{ji}) = n_j A_{ji} h\nu_{ji} \quad (\text{J/cm}^3/\text{s}) \tag{4.5}$$

where the first two terms of the right-hand side are the number density of the level j and the Einstein spontaneous emission coefficient.

Our aim here is to explore the parameters that we can measure and clearly the radiant power density is related to the intensity we will ultimately measure. Note that we know the energy of the transition (wavelength) and the Einstein coefficient, so the term we wish to explore is the number density.

Let us take a rather surprising leap, which, in effect, is a useful trick.

We can write the number density of the level as $n_j(X^{+m})$ and express it in the form:

$$n_j(X^{+m}) = \left[\frac{n_j(X^{+m})}{n(X^{+m})}\right] \times \left[\frac{n(X^{+m})}{n(X)}\right] \times \left[\frac{n(X)}{n(H)}\right] \times \left[\frac{n(H)}{n_e}\right] \times n_e \qquad (4.6)$$

This form seems rather contrived but from right to left the terms of Equation (4.6) include, the electron density, the hydrogen abundance (relative to the electron density), the abundance of the element (X) relative to hydrogen, the ionization ratio of the ion (i.e. its abundance relative to the element), and the population of (number density in) level j, relative to the total number density of the ion. This breaks the consideration of the number density into a set of recognizable parameters. These are the parameters we either wish to establish or we believe we understand.

The ionization ratio is the function we explored in Section 4.4.1 and we have an understanding of its form as a function of temperature for a given ion.

We note that for low-density plasmas, collisional excitation processes are generally faster than ionization and recombination rates. Thus, collisional excitation dominates over the ionization–recombination processes in producing excited states.

So, in effect, a consideration of the electron density, and ionic and elemental abundances feed into any detailed calculation of energy-level populations and such populations are required to estimate the radiant power density for a particular plasma. In practice, the number density population of level j has to be calculated by solving statistical equilibrium equations for the ion, taking into account the collisional and radiative excitation and de-excitation processes.

In plasmas such as the solar atmosphere, conditions are such that collisional excitation from the ground state (g) can be considered to dominate over other excitation processes, and radiative decay is the principal process for de-population. Given equilibrium, we can equate the collisional and decay processes. Collisional excitation is dependent on the number density of the colliders and of the ions in the ground state, and on the electron collisional excitation rate coefficient, C_{gj} (for a transition from the ground state to level j). The decay process, being purely radiative, is simply dependent on the number density in level j and the Einstein coefficient for spontaneous emission, A_{jg} (for a transition from level j to the ground state).

Thus, in equilibrium, we may write:

$$n_g(X^{+m})n_e C_{gj} = n_j(X^{+m})A_{jg} \qquad (4.7)$$

Rewriting this, we can state that:

$$\frac{n_e C_{gj}}{A_{jg}} = \frac{n_j(X^{+m})}{n_g(X^{+m})}$$

If $n_e C_{gj} \ll A_{jg}$ then the population of level j is not significant compared to the ground state.

We can rewrite the radiant power density, given Equations (4.5) and (4.6), as:

$$\varphi(\lambda_{jg}) = [h\nu_{jg}n_e^2 C_{gj}] \times \left[\frac{n(X^{+m})}{n(X)}\right] \times \left[\frac{n(X)}{n(H)}\right] \times \left[\frac{n(H)}{n_e}\right] \qquad (4.8)$$

which means that the emission line intensity is proportional to the square of the electron density.

Of critical importance is the collisional excitation rate coefficient, which is dependent on the velocity distribution of the electron population. If we assume a Maxwellian (equilibrium) distribution, this can be written as:

$$C_{gj} = \left(\frac{8.63 \cdot 10^{-6}}{T_e^{1/2}}\right) \left(\frac{\gamma_{jg}(T_e)}{\omega_j}\right) \exp\left(\frac{-h\nu_{gj}}{kT_e}\right). \tag{4.9}$$

The parameters γ_{jg} and ω_j are the thermally averaged collision strength and the statistical weight for the level, j.

The radiant power density of an emission line, which is in units of $J/cm^3/s$, is clearly related to parameters that one can measure. Measurements are, of course, recorded as an intensity or energy per unit area per second at a detector/telescope, so it is sensible to convert the radiant power to something more convenient.

First, we derive the *irradiance* of the spectral line. This is the energy you will measure at a distance r from the source. For this, we must integrate the radiant power over the emitting volume, considering the radiation as isotropic. Thus, for the transition between levels j and i, the irradiance is given as

$$E(\lambda_{ji}) = \frac{1}{4\pi r^2} \int_v \varphi(\lambda_{ji}) \, dV \quad (J/cm^2/s/ster)$$

This is, of course, dependent on r.

If the source is spatially resolved, the emitting volume dV is the volume as projected into the spatial area under consideration, which may be a single pixel. Thus, we consider a solid angle, $d\Omega = dS/r^2$, where dS is the projected area of the volume in question. Thus, we talk of *radiance*, rather than irradiance, which is defined as irradiance per unit solid angle, or

$$L(\lambda_{ji}) = \frac{dE(\lambda_{ji})}{d\Omega} = \left(\frac{1}{4\pi r^2}\right) \varphi(\lambda_{ji}) \, dV \left(\frac{r^2}{dS}\right) = \left(\frac{1}{4\pi}\right) \varphi(\lambda_{ji}) \frac{dV}{dS} \tag{4.10}$$

Note that this does not depend on the distance to the source.

Emission line measurements are frequently given as irradiance or radiance. Thus, we have now linked our consideration of the emission line process and the measurement of line intensities.

It is clear that the atomic calculations associated with spectral line analysis are complex and, indeed, in some astrophysical plasma environments we may have the additional problem of non-Maxwellian, time-dependent electron distributions. It is beyond the scope of this review to extend this discussion further and the reader is referred to the reviews listed above.

We have been considering electron–ion collisions. It should be noted that proton collisions can become important, that is, comparable with electron collisional processes, only for transitions where $h\nu_{ji} \ll kT_e$. This is the case for transitions between fine structure levels at high temperatures.

4.4.5. *The generation of EUV/UV radiation: edges*

So far, we have discussed free–free and bound–bound transitions, resulting in a continuum and a line spectrum. Now, we consider free–bound collisions. In this case, a free electron of energy $h\nu'$ collides with an ion, and is captured. This is, in fact, the radiative recombination process mentioned above. Since the dominant process is for the electron to go to the ground-state level, the minimum energy that can be radiated by this process is the ground state ionization energy. Thus, this process results in continuum emission, with an edge or cut-off at this energy value. As can be seen from Fig. 4.12, this means that for a given ion, such as the hydrogen atom, one will see series of emission lines, such as the Lyman series, with progressively shorter wavelengths (increasing energies) up to a boundary, which is the edge – defined by the ionization energy.

4.4.6. An example spectrum

Figure 4.10 shows a spectrum of the Sun taken using the SUMER spectrometer on board the SOHO spacecraft (Wilhelm *et al.*, 1997). The spectrum displays a plethora of emission lines, dominated by the hydrogen Lyman–alpha line at 1216 Å, that is, the $n = 2$ to 1 transition (see Fig. 4.12). The emission lines are mainly from ions of oxygen, nitrogen, carbon, iron and magnesium, as well as hydrogen. It is clear that the emission line spectrum is superimposed on a continuum, and an edge can clearly be seen at 912 Å. This is the hydrogen Lyman edge.

4.5. The EUV/UV toolbox

The beauty of the processes mentioned in some detail in Section 4.4 is that they provide a set of plasma diagnostic tools for the astrophysicist. The remote sensing of plasma emissions provides information on plasma parameters that, in most astrophysical situations, we cannot determine by any other means. As we shall see in this section, we can use the emission characteristics of the EUV/UV region to derive information on astrophysical temperatures, densities, abundances, flows, etc. and, thus, the most basic characterization of space plasmas is possible, and this is essential for the study of fundamental processes in our Universe. There are a number of approaches open to us, and we run through these in turn now.

4.5.1. Plasma diagnostics: emission measure

A most important observational feature which we must remember at all times, for both imaging and spectroscopy, is that for optically thin plasmas the intensity at any particular wavelength for a specific direction, or line of sight, is formed from an integration along the line of sight. We are rarely viewing a single source, but a combination of contributions from plasmas along the line of sight. The source that we are interested in may dominate the emission, but it is prudent to remember that the intensity at any particular wavelength may be modified significantly by unrelated plasmas.

One technique which takes into account the integration along the line of sight is the *emission measure* analysis. In Section 4.4, we wrote the radiant power density in Equation (4.8), which we may rewrite as:

$$\varphi(\lambda_{jg}) = [h\nu_{jg} n_e^2 G(T, \lambda_{jg})] \times \left[\frac{n(X)}{n(H)}\right] \times \left[\frac{n(H)}{n_e}\right] \qquad (4.11)$$

We define the so-called contribution function as:

$$G(T, \lambda_{jg}) = \left[\frac{n(X^{+m})}{n(X)}\right] C_{gj}$$

This function is strongly peaked in temperature, because, as we saw earlier, the ionic abundance, $n(X^{+m})/n(X)$ peaks strongly with temperature.

If we reduce this to its most basic form (see Pottasch, 1963) for a single emission line emitted by a plasma which we take to be at a temperature corresponding to the maximum temperature of the contribution function for the emitting ion, and assuming that radiation is uniformly distributed over space, we may write the radiance as:

$$L(\lambda_{ji}) = \left(\frac{1}{4\pi}\right) \varphi(\lambda_{ji}) \frac{dV}{dS}$$

$$= \left(\frac{1}{4\pi}\right) \left[\frac{n(X)}{n(H)}\right] \times \left[\frac{n(H)}{n_e}\right] [h\nu_{ji} <EM> <G(T)>] \qquad (4.12)$$

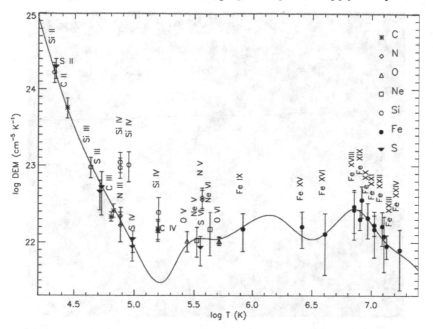

FIGURE 4.13. Emission measure as a function of temperature for the star AU Mic derived from FUSE, HST STIS and EUVE data (adapted from Del Zanna *et al.*, 2002).

where $<EM>$ is the emission measure, defined by:

$$<EM> = \int_s n_e^2 \, \mathrm{d}s \qquad (4.13)$$

Here, we assume that the emission is from a plasma near temperature $T_{\max}$, that is, the temperature of maximum abundance of the emitting ion (see Section 4.4.1). We assume that the temperature range, ΔT, is relatively narrow, which is in keeping with the narrow temperature ionization contributions. Thus, in this simplified view, we define:

$$<G(T)> = (1/\Delta T)\int_{\Delta T} G(T, \lambda_{jg}) \, \mathrm{d}T \qquad (4.14)$$

This approach allows us to estimate the average electron density from the emitting region, by crudely writing:

$$<n_e^2> = <EM>/s \qquad (4.15)$$

Note that this is an average, assuming a homogeneous, isothermal volume.

In addition, if we consider one line, we are in effect calculating the emission measure for the specific temperature of the ion emitting that line. Thus, if we consider the same calculation for a range of emission lines for the same line of sight, we can derive an emission measure as a function of temperature for the plasma under study – integrated along the line of sight. Put in its most basic interpretation, we are identifying the amount of material emitting at different temperatures along our integrated line of sight, a so-called differential emission measure.

Figure 4.13 shows a differential emission measure plot, as a function of temperature, calculated from Far-UV Spectroscopic Explorer (FUSE), HST space telescope imaging spectrograph (STIS) and EUVE data for the star AU Mic. This curve shows a dominant cool component, to temperatures of order 100,000 K and a hot component extending to temperatures over 10 million K.

FIGURE 4.14. NASA TRACE image of the solar atmosphere, showing the fine-scale structure at the 350 km/pixel level (courtesy TRACE team).

4.5.2. *Plasma diagnostics: Electron density*

Observations of the solar atmosphere demonstrate readily that plasma often resides in filamentary structures defined by, or confined to, magnetic fields. This is so for even the most fine-scale spatial resolutions, such as the 0.5 arcsec (350 km) pixels of the TRACE instrument shown in Fig. 4.14. Thus, the average density of the best spectrometers, which have larger-resolution values than the best imagers, are for volumes which include an integration over filamentary structures.

This means that the density measurement identified in the emission measure discussion, above, may be in error because of the need to include the effects of "filling factors", that is, how much of the area viewed by the pixel is actually emitting? However, it turns out that there is a method for the accurate determination of plasma density and that the density calculated from the emission measure method can be used in parallel, to determine the filling factor.

The density diagnostic we use is one based on emission line intensity ratios. It makes use of the fact that not all lines are populated and de-populated by the standard collisional excitation and radiative decay processes. Consider the presence of a metastable level – one from which the radiative decay rate is very small (i.e. decays from this level are rare). Such a decay is described as a forbidden transition. For such a metastable level, collisional decay can become important, or collisional excitation from the metastable level can lead to allowed transitions from upper levels.

For low electron density, radiative decay processes from the metastable level can dominate, due to the fact that collisions are infrequent. Thus, we see the forbidden line. The intensity in this case is proportional to the density. For high electron densities, collisional decay processes dominate and the line intensity varies as the square of the density. A thorough consideration of these two limiting conditions and the transition between them, as a function of density, provides us with a useful tool for density measurement.

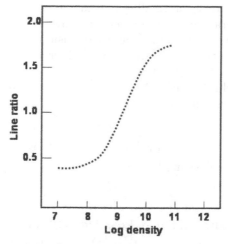

FIGURE 4.15. The Si X 347/356 Å intensity ratio as a density diagnostic (adapted from Mason *et al.*, 1997).

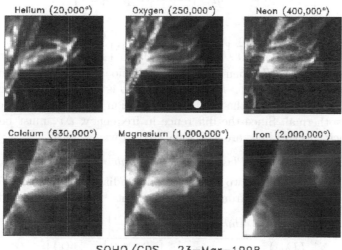

SOHO/CDS   23-Mar-1998
Loops of gas at different temperatures observed near the solar limb

FIGURE 4.16. Simultaneous images of an area of the solar limb from six selected emission lines, representing six different plasma temperatures. The ions are from the elements indicated and the formation temperatures are given. The images are 200,000 km × 200,000 km, directed at a complex active region magnetic loop system on the solar limb. The Earth is shown to scale in the upper-middle frame (courtesy SOHO Coronal Diagnostic Spectrometer (CDS) team).

The relevant density ranges for a particular ion/transition, depend on the atomic parameters. However, an example of such a density-sensitive ratio is shown in Fig. 4.15 for the Si X emission lines of 347 and 356 Å. The high-density and low-density limits are clearly identified, and a useful line ratio density diagnostic ratio can be seen in the range 10^8 to $5 \times 10^{10}/\text{cm}^3$.

4.5.3. *Plasma diagnostics: electron temperature*

In the discussion above, concerning ionization equilibrium, it was clear that the very existence of emission from lines from a particular ion can be regarded as a measure of the presence of a plasma at a particular temperature. This is often used as a quick indicator of plasma temperatures. The images shown in Fig. 4.16 show a region of the solar limb taken, simultaneously, in six differing emission lines from different ions formed at widely different temperatures. The temperatures are indicated. Whilst these images are indicative of the temperature ranges and the basic nature of plasmas at different temperatures in the solar atmosphere, we must consider a different approach to provide an accurate temperature measurement.

Let us consider the radiance ratio of two emission lines. We defined the radiance in Equation (4.10), and, thus, can write:

$$\frac{L(\lambda_{ji})}{L(\lambda_{kl})} = \frac{\left(\frac{1}{4\pi}\right)\varphi(\lambda_{ji})\frac{dV}{dS}}{\left(\frac{1}{4\pi}\right)\varphi(\lambda_{kl})\frac{dV}{dS}} = \frac{\varphi(\lambda_{ji})}{\varphi(\lambda_{kl})}$$

Thus, assuming that the lines are from the same ion,

$$\frac{L(\lambda_{ji})}{L(\lambda_{kl})} = \frac{(h\nu_{ji}C_{ij})}{(h\nu_{kl}C_{lk})}$$

Given the collisional excitation rate of Equation (4.9), we may write this as:

$$\frac{L(\lambda_{ji})}{L(\lambda_{kl})} = \frac{h\nu_{ji}\left[\frac{\gamma_{ji}(T_e)}{\omega_j}\right]\exp\left(\frac{-h\nu_{ji}}{kT_e}\right)}{h\nu_{kl}\left[\frac{\gamma_{kl}(T_e)}{\omega_l}\right]\exp\left(\frac{-h\nu_{kl}}{kT_e}\right)}$$

or

$$\frac{L(\lambda_{ji})}{L(\lambda_{kl})} = \left(\frac{h\nu_{ji}}{h\nu_{kl}}\right)\left\{\frac{[\gamma_{ji}(T_e)\omega_i]}{[\gamma_{kl}(T_e)\omega_l]}\right\}\exp\left[\frac{(h\nu_{kl}-h\nu_{ji})}{kT_e}\right] \tag{4.16}$$

Thus, if $(h\nu_{kl}-h\nu_{ji}) < kT_e$ then the radiance ratio is pretty much independent of temperature, but if $(h\nu_{kl}-h\nu_{ji}) > kT_e$ then the ratio is sensitive to changes in electron temperature. This assumes that the two lines in question are emitted by the same plasma and that it is isothermal. Since the difference in frequency, $\Delta\nu$, must be greater than (kT_e/h), two such spectral lines are well separated in wavelength.

4.5.4. *Plasma diagnostics: abundance*

Let us consider the ratio of two emission line radiances, with the form given in Equation (4.8), but in this case from differing elements, X and Y. We find,

$$\frac{L(\lambda_{ji})}{L(\lambda_{kl})} = \frac{(h\nu_{ji}n_e^2C_{ji}) \times \left[\frac{n(X^{\cdot\,m})}{n(X)}\right] \times \left[\frac{n(X)}{n(H)}\right]}{(h\nu_{kl}n_e^2C_{kl}) \times \left[\frac{n(Y^{\cdot\,mm})}{n(Y)}\right] \times \left[\frac{n(Y)}{n(H)}\right]} \tag{4.17}$$

If we select emission lines from the neutral ion in each case, such that $n(X^{+m})/n(X) = 1$ and $n(Y^{+mm})/n(Y) = 1$, we find,

$$\frac{L(\lambda_{ji})}{L(\lambda_{kl})} = \frac{(h\nu_{ji}n_e^2C_{ji}) \times \left[\frac{n(X)}{n(H)}\right]}{(h\nu_{kl}n_e^2C_{kl}) \times \left[\frac{n(Y)}{n(H)}\right]} \tag{4.18}$$

Thus, if we know the transition energies and the emission is from the same plasma, and if the collision rates are known, we have the relative abundance of the two ions, X and Y.

This technique can be extended across many lines and ions to derive abundances of a range of elements in a plasma under study.

4.5.5. *Plasma diagnostics: dynamics*

Information on the dynamics of plasmas can be obtained by the analysis of line profiles. The Doppler shifts of emission lines can be used to determine flow patterns in astrophysical plasmas. This includes broadening due to unresolved small-scale flow patterns or due

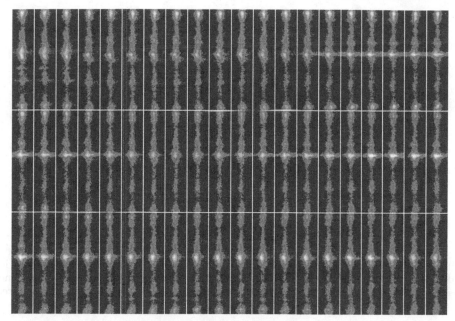

FIGURE 4.17. Explosive events detected in the solar atmosphere using the SUMER/SOHO instrument. Each image of the 20 × 3 array is an exposure of a slit using the 100,000 K Si IV line at 1393 Å. Time runs from top-left to bottom-right; each exposure is for 10 s and the total time interval is 10 min. The slit is 120 arcsec long (84,000 km on the Sun), from top to bottom of each image. The left–right axis is spectral, ranging from ±100 km/s. Significant Doppler shifts can be seen on several occasions (adapted from Innes *et al.*, 1997) (courtesy SOHO SUMER team).

to the line of sight oscillations of emitting ions during the passage of wave activity. Thus profile analyses are a powerful tool for plasma diagnostic work.

Of course, the most basic velocity event is a discrete jet of plasma, moving away from or towards the observer at velocity, V, relative to the surrounding plasma. If the emission line under examination is at wavelength, λ, then the shift in wavelength relative to the line center (rest wavelength) is given by,

$$\Delta\lambda = (V/c)\lambda \quad (\text{Å}) \tag{4.19}$$

where c is the speed of light. If the shift is towards the shorter-wavelength (blue) end of the spectrum, the jet is moving towards the observer and if the shift is towards the longer-wavelength (red) end of the spectrum, the jet is moving away from the observer.

An example of bi-directional jets detected through the observation of discrete line broadening is shown in Fig. 4.17. These are the so-called explosive events detected in the solar atmosphere.

An observed emission line profile is made up of three components. The line width can be written as:

$$\left(\frac{\Delta\lambda}{\lambda}\right) \propto \left[\left(\frac{kT}{M_i}\right) + \psi + \varepsilon\right]^{1/2} \tag{4.20}$$

where the three terms of Equation (4.20) are the thermal width, the non-thermal width and the instrumental width, from left to right.

The thermal width can be calculated for the spectral line under study because we know the temperature of maximum abundance of the ion and the mass of the ion (M_i).

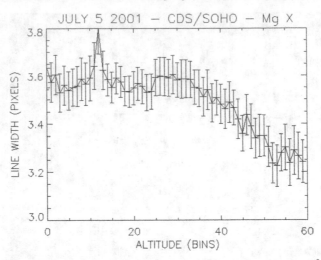

FIGURE 4.18. Emission line narrowing from the Mg X (million K) line at 625 Å. The line width is plotted against altitude (the edge of the Sun is in bin 11). The narrowing is possible evidence for the dissipation of Alfvén waves, that is, we are seeing particle motion transverse to the direction of the magnetic field lines (from Harrison *et al.*, 2003).

The instrumental width is the intrinsic line width of the instrument, that is, the width of the image of the spectrometer slit on the detector plane. This should be known from optical considerations of the instrument and pre-flight calibration. However, it will dominate the line widths for most astronomical spectrometers.

The non-thermal component may be due to Doppler shifts, either as discrete jets, or as broadening due to unresolved flows or wave activity.

A careful consideration of the instrumental and thermal components of an emission line can reveal the degree of the non-thermal component and this can be used to study the plasma dynamics. An example from a study of wave dissipation in the corona is shown in Fig. 4.18.

4.5.6. *Plasma diagnostics: the need for atomic data and calibration*

It is truly remarkable how much information can be gleaned from a careful analysis of emission lines and continua through the remote sensing of astrophysical plasma EUV/UV emission. Studies of abundances, temperatures, densities, flow patterns, etc. provide information which allow thorough studies of plasma conditions for numerous sources and this is critical information for understanding the physics of astrophysical plasmas.

However, all of this is dependent on two things. First, we have made use of a variety of atomic parameters, such as collision excitation rates, Einstein coefficients, contribution functions, and our analyses are dependent on an accurate knowledge of these parameters. Second, the accurate measurement of intensities is clearly subject to accurate instrument calibration, as a function of wavelength.

Thus, the EUV/UV astronomer is as reliant on accurate atomic modelling, and instrumental calibration as on the specific observation sequences. It is all very well making superb observations of spectral emission lines from an extraterrestrial source, but you need to convert the counts per second to physical units and then you need the atomic tools to understand the processes that resulted in the emission.

Whilst the detailed calculation of the required atomic parameters, and the methods of instrument calibration are beyond the scope of this review, we highlight the fact that these issues are critical to the efficient interpretation of EUV/UV spectral observations.

4.5.7. *Plasma diagnostics: observational scenarios*

We have explored the nature of the UV spectrum, and the plasma diagnostic methods. Before considering the instrumentation used to make EUV/UV space observations and results from these instruments, it is worth considering some basic observational points.

There are three basic types of EUV/UV observation of use to the astronomer:

• Direct EUV/UV radiation, generated by plasma processes within the body under study (e.g. emission line spectrum from the Sun);

• Modification of a reflected spectrum, such as that of the Sun, producing additional absorption features (e.g. sulphur dioxide absorption within the reflected solar spectrum from Venus);

• Modification of a spectrum (e.g. solar or stellar) due to line of sight absorption through an atmosphere or plasma region (e.g. absorption features superimposed on a stellar spectrum when viewed through the Earth's upper atmosphere).

Thus, in addition to the richness of the EUV/UV emission spectrum, one can make use of additional features depending on the target of observation and the circumstances of the observation.

4.6. EUV/UV instrumentation in space

4.6.1. *Some general points*

In broad terms, EUV/UV instrumentation can be divided into several types, including imagery, spectrometry and photometry. The fundamental approaches are not too dissimilar from those of visible optical systems. However, for the EUV/UV we find reduced transmissions and reflectivities for the optical components, relative to the visible, and this dictates the specific design details.

Each instrument consists of an optical system that brings light to a focus, using a lens or mirror system. Near-UV can be studied using instruments similar to visible devices. However, EUV/UV light at shorter wavelengths cannot be focused by lenses and novel techniques must be adopted.

At the focus, we place a detector or a slit, the former to produce images and the latter as the entrance to a spectrometer. Somewhere in the system one will either find a filter made out of a material such as aluminium, to block out the visible light whilst allowing the passage of the required radiation, or the detector system itself must be blind to visible light.

In this section we will explore some of the most basic points on methods and technology for focusing, dispersing and detecting EUV/UV radiation. This will be done with reasonable technical detail, for the general scientific reader. It is beyond the scope of this review to then outline the technical details of all space instruments operating in the EUV/UV, so we have taken three space instruments operating in the EUV/UV to illustrate the techniques. Finally, a brief description of all major EUV/UV missions to date, operated after the OSO and OAO series is presented.

Three areas are not covered in this chapter. However, although they are beyond the scope of the review, they do merit a brief mention at this stage.

First, as mentioned above, we stress that for any EUV/UV instrument, photometric calibration is extremely important, in particular for spectroscopic observations. Even if an instrument operates perfectly, without the knowledge of accurate conversion from counts per second to physical units, such as photons per second per unit area, the real physical insights of the instrument are limited. This should be obvious from the descriptions of the emission processes, above. We need to know the emissivity or the radiance of an emission line in physical units to explore the nature of the emitting plasma.

Second, contamination control is a major issue. Contamination of optical surfaces can have serious consequences in this wavelength range. Control of materials and the environment during the instrument build is an essential component of any mission. Likewise, we must control contaminants in space. For example, materials may outgas when they encounter the vacuum of space. Thus, it is wise to provide for long outgassing periods prior to opening doors to look at the Sun, for example. You do not want to bake contaminants on to optical surfaces.

Third, a major, yet often overlooked aspect to mission success is the operations scenario. Space missions are necessarily expensive and complex. Thus, for modern missions, detailed operations planning is essential, prior to the launch and during the mission. For some modern "hands on" missions, where experimenters have access to their instruments via operations facilities, the key to efficient exploitation has been in the detailed design of operations scenarios coupled with regular, even daily, planning meetings to assign observations.

Let us now consider some general principles of EUV/UV instrumentation.

The first question to consider is how you can focus EUV/UV light. In the early rocket experiments, magnesium fluoride lenses were used. Magnesium fluoride is transparent down to 1100 Å. The development of mirror coatings with high reflectivity in the UV was crucial for progress beyond those first simple experiments. Oxidized aluminium, on the surface of visible light mirrors, is not an efficient reflector in the UV and it was determined that coating pure aluminium with a layer of magnesium fluoride stopped oxidization and thus preserved reflectivity. However, for wavelengths below 1100 Å other reflecting materials had to be found. For these wavelengths, gold was used in the OSO, Skylab and SOHO missions and silicon carbide has been used (e.g. SOHO).

At short wavelengths we cannot use standard optical telescope systems with lenses and, as suggested above, make use of mirror systems. One of the major instrument developments was due to the German physicist Hans Wolter. He showed that reflection off two mirrors, at grazing incidence, namely a paraboloid and a hyperboloid, would provide a focus. This approach can be used for short-UV and -X-ray wavelengths, and has been used in missions such as SOHO, XMM and ROSAT.

Figure 4.19 shows that there are three basic Wolter telescope configurations, known as types I, II and III. All types involve two reflections in grazing incidence. The diagrams can be considered to be cross-sections of full-revolution concentric mirrors, though it is possible to make use of just a section of the full-revolution mirror. An example of the use of a Wolter II will be discussed in some detail later.

For longer wavelengths, and, with certain optical coatings, for shorter-EUV wavelengths, a normal incidence mirror system such as a Cassegrain can be considered. For most telescopes of this kind, a Ritchey–Chretien design is used. This is the case for the Hubble Space Telescope (HST) and for the IUE. The standard Cassegrain consists of a paraboloid primary mirror feeding a secondary mirror, which is a hyperboloid, and light is reflected back through the centre of the primary mirror to a focus. In the Ritchey–Chretien design, the primary is also a hyperboloid shape. This improves imaging over a wider field when compared to the basic Cassegrain.

Normal incidence reflections at the short-EUV wavelengths have low efficiency but we can provide, in effect, more reflections through the fabrication of multilayer optical coatings. For example, the EUV imaging telescope (EIT) instrument on board the SOHO spacecraft makes use of alternate layers of molybdenum and silicon, of varying layer

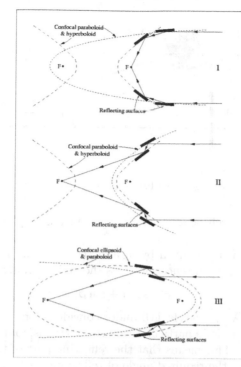

FIGURE 4.19. The Wolter grazing incidence mirror telescopes – types I, II and III (courtesy James Manners, Royal Observatory Edinburgh Ph.D. thesis).

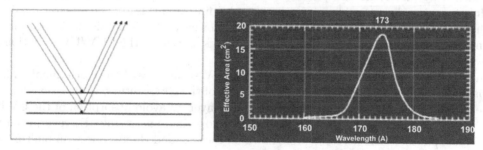

FIGURE 4.20. A schematic representation of the multilayer coating and an example of an enhanced reflectivity at a selected wavelength, for the TRACE instrument.

thickness, to enhance the reflectivity for selected wavelength bands on the mirrors. This is shown schematically in Fig. 4.20.

This multilayer technique can be applied to any optical surface, not just the telescope mirrors. An example of the use of multilayer optics is discussed in detail later. However, to provide some feeling for the performance of a typical multilayer optical application, an Mo/Si multilayer coating can be tuned to produce reflectivities of order 20–25% in a narrow band, 40 Å wide, centred on, say, 225 Å. We use this for illustration; the use of different materials, and multiple reflections, etc. can be used to select required bands.

There are technical limitations, for example to the longer-wavelength regions (say, longer than 300 Å), but there is much ongoing research devoted to multilayer technology and development.

In the case of a spectrometer, light is brought to a focus by the telescope at a slit. It is then incident on a dispersing element, and at the short-EUV/UV wavelengths, this is a

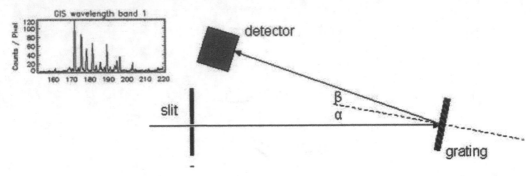

FIGURE 4.21. The basic spectrometer layout.

grating. This grating may itself be a concave shape, such as a spherical or toroidal form, to allow a spatial focus.

If the grating ruling spacing is given by d, and the incoming ray hits the grating at incident angle α, with an angle of reflection β, as shown in Fig. 4.21, then we may write:

$$n\lambda = d(\sin\alpha + \sin\beta)$$

where n is the order and λ the wavelength under consideration.

To increase the reflectivity for a particular value of β, that is, for a specified wavelength, the grating can be blazed. This means that the reflecting part of the grooves is tilted in such a way as to tune it to the required angle of reflection.

EUV/UV photons are detected by photographic film, by the fluorescence of certain chemicals, by photocells and photoelectric devices. Early space instruments, on rocket flights, for example, were able to use photographic film, and this applied on later missions such as Skylab (which was a manned mission). However, photoelectric devices are used in most space applications, for obvious reasons. In some cases, the EUV/UV radiation is converted to visible light before detection.

In the 1950s and 1960s, Kodak increased the UV sensitivity of their Schumann emulsions, which were originally developed at the turn of the 20th century. In these emulsions, silver halide grains are sticking out of the gelatin layer, to avoid absorption of the UV by the gelatin. This made UV imaging and spectroscopy more accessible. However, even in the 1950s, photoelectric detectors were flown on rockets, and channel electron multipliers came into use in the first satellites.

Detector arrays, which can be used for 2D detection of images or spectra were difficult to produce for the UV, so most early instruments making use of photoelectric detectors were pin-hole devices. Interlaced exposures and spatial scans were necessary to build up spectral maps. The modern solution to this is the charge-coupled device (CCD).

A CCD is a highly sensitive photon detector which consists of an array of a large number of light-sensitive pixels. In each pixel, light may be recorded as charge on arrival at a photodiode. The image is, thus, recorded usually as a 2D image of charge wells which may be read out after the exposure by transferring the charge to a read-out port.

Figure 4.22 shows some examples of CCDs of differing pixel array numbers and configurations, and a schematic of a typical light sensitivity curve. It is clear that such a device is not immediately useful for UV, let alone, EUV, observations, given the poor sensitivity below 4000 Å. To make use of these devices at the required wavelengths one will often use an intensifed-CCD approach or a back-thinned, back-illuminated approach.

The intensified-CCD approach is illustrated by the simple cartoon shown in Fig. 4.23. One makes use of a microchannel plate (MCP), which is a wafer of hexagonally packed

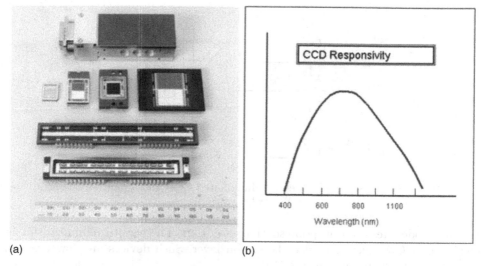

(a)   (b)

FIGURE 4.22. (a) Examples of CCD chips (courtesy Nick Waltham, RAL). (b) A schematic demonstration of the typical sensitivity of a CCD.

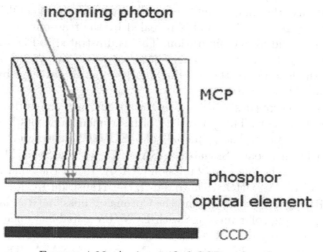

FIGURE 4.23. An intensified CCD approach.

lead glass tubes within which arriving photons produce electron showers, via the photo-electric effect. The electron shower may be incident on a phosphor, which converts the charge to a visible light signal ultimately detected by the CCD. Between the phosphor and CCD there may be an optical component such as a lens or fibre-optic taper. A voltage is applied across the MCP to enhance the electron shower, and this voltage can be used to control the exposure by setting to zero, or even reversing, at appropriate times.

Another way to make use of CCDs for EUV/UV wavelengths is to use them in a back-illuminated mode. As illustrated in Fig. 4.24, the back of the CCD consists of a silicon layer in which arriving UV photons can, through the photoelectric effect, release electrons, that is, in the chip itself. Figure 4.24(a) shows just one CCD pixel, the arrow indicating that it is front illuminated. The area shown is the side view of the pixel and the hatched area is the photodiode. If the CCD was reversed and back illuminated, the thickness of the silicon means that the charge cloud will spread significantly before detection at the

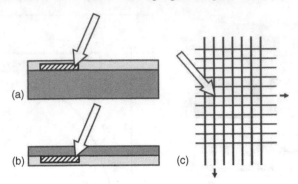

FIGURE 4.24. Front (a) and back (b) illuminated CCD and a multi-anode array (c) detector concept.

pixel photodiodes, resulting in poor spatial resolution. The way to minimize this is to back-thin the CCD, that is, reduce the silicon layer. Such devices are now used quite widely. This is illustrated in Fig. 4.24(b).

Another approach is to convert the arriving photons to charge, again using MCP devices, and allow the charge to fall on an anode system that is structured in such a way as to reveal the location of the charge cloud on the detector face. In the multi-anode or delay-line systems, the charge-cloud location is revealed by the read-out on two orthogonally arranged anodes in a grid-like configuration. This is illustrated in Fig. 4.24(c). A similar alternative approach is the wedge-and-strip detector where the anode face is divided in such a way that the location of the arriving charge cloud is revealed by the amount of charge read-out at different ports.

Beyond the basic concepts described above, the specific instrument design depends very much on the object of study. Thus, a variety of instrument concepts have been flown in space on numerous missions. The principal solar EUV/UV instruments, after the OSO series, have been flown aboard Skylab (NASA, 1973–4), the Solar Maximum Mission (NASA SMM, 1980–9), the Spacelab 2 mission (NASA, 1985), the Solar and Heliospheric Observatory (ESA/NASA SOHO, 1995 to date), the Transition Region and Coronal Explorer (NASA TRACE, 1998 to date), and the Coronas-F mission (Russian Space Agency, 2001 to date). Upcoming solar missions with EUV/UV capability include Solar Terrestrial Relations Observatory (STEREO) (NASA, due for launch in 2006), Solar-B (ISAS, due for launch in 2006), the Solar Dynamics Observatory (NASA SDO, due for launch in 2008) and Solar Orbiter (ESA, due for launch after 2012). The principal astrophysics missions include the International UV Explorer (NASA/ESA IUE, 1978–96), ROSAT (Germany/NASA/UK, 1990–9), HST (NASA/ESA HST, 1990 to date), the EUV Explorer (NASA EUVE, 1992–2001), FUSE (NASA FUSE, 1999 to date) and the Galactic Evolution Explorer (NASA GALEX, 2003 to date). Many examples of the results from these missions are given in the following sections. However, to explore the techniques used in EUV/UV instrumentation, we describe three instruments in some detail here and present brief details on the remaining missions. The detailed descriptions are given of Coronal Diagnostic Spectrometer (CDS) instrument on SOHO, the FUSE and TRACE instruments.

4.6.2. *Three detailed instrument descriptions: SOHO/CDS, FUSE and TRACE*

The CDS (Harrison *et al.*, 1995) is an EUV spectrometer operating in the 170–800 Å region. Carried aboard the ESA/NASA SOHO, it was designed to allow detailed analyses of the EUV spectrum of the solar atmosphere. Trace elements which exist in the solar

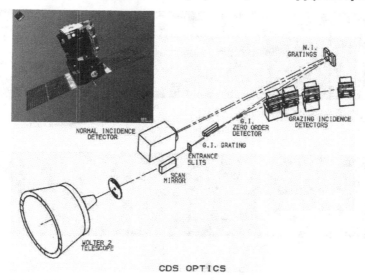

CDS OPTICS

FIGURE 4.25. Optical layout of the CDS instrument aboard the SOHO spacecraft, and, inset top-left, an image of SOHO (Harrison *et al.*, 1995).

atmosphere are revealed by the presence of emission lines in the solar EUV spectrum, and various ionization stages of elements such as iron, silicon, magnesium, oxygen, nitrogen etc.... can be identified in the solar spectrum. The range of identified ions exist across a broad range of temperatures in the solar atmosphere, from ten thousand to a few million K. Analyses of the intensities, relative intensities and line shapes of these lines can be used as a toolbox for diagnosing the solar atmospheric plasmas, and that has been done very effectively using CDS instrument aboard SOHO.

The CDS instrument, as shown in Fig. 4.25, consists of a full-revolution Wolter II telescope feeding two spectrometers. The telescope forms a focus at a slit, which acts as the entrance to the spectrometers. However, the full telescope aperture is not used; two doors at the front of the instrument allow illumination of two sections of the telescope. These two portions ultimately feed into each of the spectrometers, simultaneously.

The solar image is presented at the location of the slit. The slit itself can be moved in one direction (vertically, i.e. solar N–S) and a flat scan-mirror can be moved in the other direction (horizontal, i.e. solar E–W) to select which portion of the Sun is fed into the spectrometers. For that portion of the Sun's image, the light falls upon a grating and is dispersed onto a detector system. Even though they pass through the same focus, the two beams from the telescope separate and enter the two difference spectrometers. One beam is dispersed using a toroidal-shaped grating in normal incidence, which produces an image of the slit, across a wide wavelength range on a 2D detector system. Indeed, two gratings are placed at this location, with rulings of 2400 and 3600 lines per mm, resulting in wavelength bands of 308–380 and 517–633 Å on the detector face. These two bands are shown in Fig. 4.26 for a typical, quiet Sun observation. The toroidal normal incidence reflection allows a focus, that is, it is a stigmatic system and the image of the slit is focused (spatially). That is, we have long, thin images of a portion of the Sun from a whole series of selected emission lines at one time.

At the heart of the 2D detector is an intensified CCD with 1024×1024 $25\,\mu\text{m}$ pixels.

Figure 4.26 shows one exposure. We can identify many emission lines from this spectrum, from many different ions. Of particular note are the lines of He I 584 Å, O V 629 Å, Mg IX 368 Å, though many others can be identified, from ions of He, O,

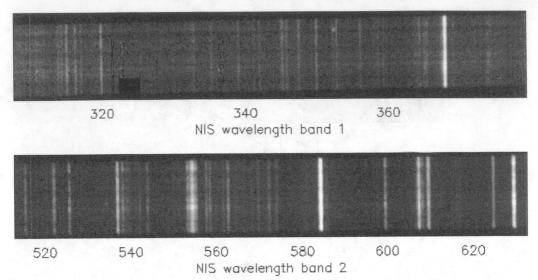

FIGURE 4.26. The two CDS/SOHO spectral bands in normal incidence. The image of a
slit-shaped area on the solar disc is dispersed in wavelength by two gratings
(Harrison *et al.*, 1997).

N, Mg, Si, Fe, with characteristic temperatures of formation from 20,000 to 2,000,000 K.
This is a solar observation, demonstrating the wide range of temperatures which occur
in the solar atmosphere.

If exposures, such as that of Fig. 4.26, are interlaced with incremental movements
of the flat mirror, one can build up images, simultaneously in the different wavelengths –
a procedure known as rastering. This is well illustrated by Fig. 4.16, which shows a series
of such CDS rastered images of an active region on the solar limb.

The CDS wavelength ranges selected were chosen because of the selection of emission
lines available for plasma diagnostic analyses. However, the normal incidence grating approach has a low reflectivity for wavelengths shorter than about 300 Å. Having said that,
we are aware of a rich set of emission lines from highly ionized iron at shorter wavelengths,
in particular in the range 170–220 Å, which provides a superb plasma diagnostic capability
for the higher temperatures of the solar atmosphere. For this reason, a second spectrometer is incorporated into the CDS instrument. Thus, the second light-path from the telescope falls upon a toroidal grating in grazing incidence, which disperses radiation around
four detectors on the so-called Rowland circle, providing four addition wavelength bands.
However, this is not a stigmatic system due to the required grazing incidence design.
Thus, a different slit is selected, a pinhole rather than a long, thin slit, and the detectors
are 1D systems. Again, an example of data from one exposure is given in Fig. 4.27.

Again, many emission lines are identified, from a wide range of ions and these are used
to characterize solar atmospheric plasmas.

In the grazing incidence system, the detectors consist of a structured anode system
behind an MCP stack. In this case, the anode is divided into a pattern of conducting
material such that the division of charge between anode components from any arriving
charge cloud can be read off and the ratio between the fraction of the cloud at different
read-out locations can be used to determine the location on the anode plate, that is, to
determine the wavelength.

The CDS instrument has been operational aboard the SOHO spacecraft since 1995 and
its plasma diagnostic capabilities have resulted in over 550 papers in 8 years, covering

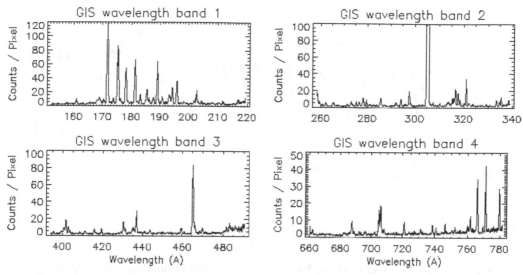

FIGURE 4.27. A CDS GIS spectrum, showing the four ranges (Harrison *et al.*, 1997).

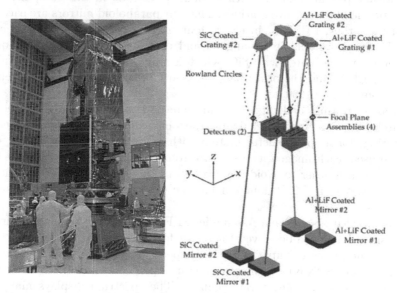

FIGURE 4.28. The FUSE spacecraft and optical layout (Moos *et al.*, 2000) (courtesy FUSE team).

many different aspects of solar physics. Some more results from this instrument are discussed in the review of solar EUV/UV astronomy below.

Another space-borne spectrometer is the FUSE instrument, which was launched on 24 June 1999. FUSE is dedicated to high-resolution UV astronomy in the range 905–1187 Å, with a spectral resolving power ($\lambda/\Delta\lambda$) of 20,000. An overview of the mission and instrument is given by Moos *et al.* (2000). The basic layout of the instrument is illustrated in Fig. 4.28, along with an image of the spacecraft shortly before launch. The optical layout consists of a four-component telescope-mirror system with differing optical coatings used to maximize the reflectivity across selected wavelength regions. The sensitivity of FUSE is a factor of 10,000 better than Copernicus.

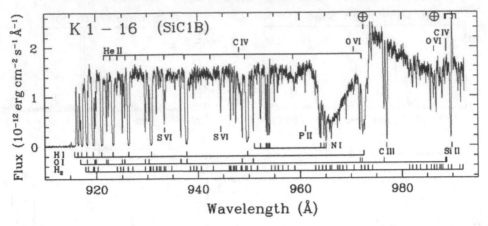

FIGURE 4.29. A FUSE spectrum of a planetary nebula source (see text) (Moos *et al.*, 2000).

The primary optical elements are four off-axis paraboloid mirrors. Although the basic layout has some similarities to the Cassegrain and Ritchey–Chretien approaches, with normal incidence primary mirrors with secondary mirrors, in this case, the secondary reflection is provided by concave gratings. The four paraboloid mirrors are each matched to one grating – thus, FUSE is, in effect, four telescope/spectrometer systems in one structure. The four systems are co-aligned and each has a band-pass of about 200 Å. Two of the mirrors are coated with SiC, which is suited to the reflection of wavelengths shorter than 1020 Å, and the other pair are coated with Al and a LiF "overcoat" for wavelengths longer than 1050 Å.

The detectors, gratings and slits (at focal plane in figure) are located in the so-called Rowland circle. This is a circle which defines the focus, and the locations of a spectrometer slit and grating, for a spectrometer system with no inherent magnification. Only two detectors are used, each imaging two spectra from two gratings. The gratings are tilted with respect to each other to avoid any overlap of the spectral images on the detector faces. The detectors are MCP double-delay-line detectors with effective pixel sizes of $6 \times 9 \, \mu$m at best.

The items labelled Focal Plane Assemblies in Fig. 4.28 consist of a flat mirror and a selection of slits. These can be moved to correct for focus and alignment.

Figure 4.29 shows one example of a FUSE spectrum. This spectrum is taken in the 912–992 Å range of CSPN K1-16, the central star of planetary nebula (Moos *et al.*, 2000). The spectrum was taken using a SiC channel. The spectrum displays many features. Along the bottom the transitions from intervening gas are highlighted (H I, O I, H_2), and above the spectrum, transitions thought to be from the central star are indicated (He II, C IV, O VI).

As with the SOHO CDS instrument, the FUSE spectrometer is a general purpose astronomical tool which is used by a large community on a range of targets. More results will be shown later, but in a review of this kind, we can only sample the output from such an instrument.

Having presented two spectrometers as examples of modern EUV/UV instrumentation, we now introduce an EUV/UV imager. For this, we describe the TRACE instrument (Handy *et al.*, 1998).

The TRACE instrument is shown in Fig. 4.30. It is a solar instrument flying on a dedicated NASA Small Explorer (SMEX) mission, launched in April 1998. The basic design is that of a 30 cm aperture classic Cassegrain system, with a primary mirror at the

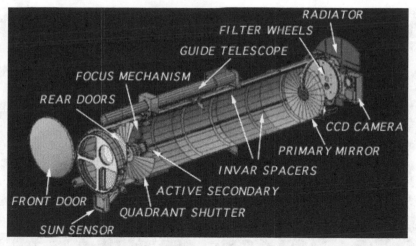

FIGURE 4.30. The TRACE (Handy *et al.*, 1998) (courtesy TRACE team).

back end of the telescope tube, with a smaller secondary (which can be actively adjusted to produce high-precision image stability) feeding light back through an aperture in the center of the primary mirror, to a 2D detector system. The telescope aperture is divided into four quadrants and behind this is a quadrant-shutter which allows selection of only one quadrant at a time. This is because each quadrant corresponds to an optical path which is sensitive to a particular narrow EUV/UV wavelength band.

When discussing the optical system of the CDS instrument, above, it was pointed out that the normal incidence reflection of shorter wavelengths was difficult. However, the relatively new technology (to space applications) of multilayer optics, as described above, has been utilized well for the TRACE instrument. Different multilayer spacings have been used to enhance specific narrow wavelength bands. Three of the four aperture sections are matched to sectors on the primary mirror which are multilayer coated. These are tuned to the three EUV wavelengths of 171, 195 and 284 Å – that is, centred on the bright lines of Fe IX (800,000 K), Fe XII (1,600,000 K) and Fe XV (2,000,000 K).

The fourth TRACE quadrant is devoted to the longer-UV wavelengths. Thus, the mirror quadrant is not multilayered, but coated in aluminium and a protective magnesium fluoride layer. We note that for the EUV quadrants of the telescope, visible light is blocked by thin metallic filters. For the UV quadrant, the UV is isolated by a combination of broad- and narrow-band interference filters.

The detector system is a 1024×1024 back-thinned back-illuminated CCD of $21\,\mu m$ pixels. This represents 0.5 arcsec per pixel (350 km on the solar surface) over a field of view 8.5×8.5 arcmin. A rotary shutter just in front of the CCD controls the exposure time.

Thus, TRACE is a solar EUV/UV imager for which we can select from a set of preset bands for observation with 0.5 arcsec spatial resolution elements. An example of an image from the instrument is given in Fig. 4.14, utilizing the 171 Å region. This shows a complex set of fine-scale magnetic loops on the solar limb.

4.6.3. *A tour of astronomical EUV/UV missions*

We have used the CDS, FUSE and TRACE instruments to illustrate some basic modern approaches to EUV/UV space observation and instrumentation. We now take a brief tour of other EUV/UV space instruments/missions in the post OAO/OSO era. Unfortunately there is not enough room in this review to provide a detailed presentation of each

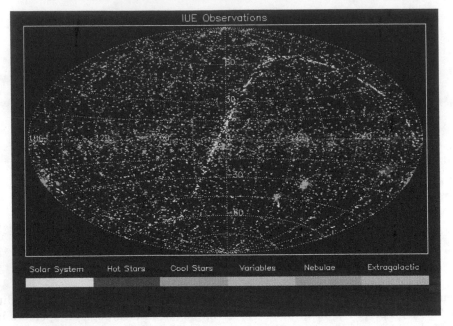

FIGURE 4.31. The IUE sky survey (courtesy NASA IUE team).

instrument and mission, so brief details are given. First, we consider the astronomical (non-solar) missions:

1. *IUE*: The NASA/ESA IUE spacecraft was launched on 26 January 1978, and operations were terminated after a record duration on 30 September 1996. The mission purpose was astronomical spectroscopy in the UV, with two wavelength ranges of 1100–2000 and 2000–3000 Å, and resolutions of 6–7 or 0.1–0.3 Å. Some 104,000 astronomical spectra were obtained. This is illustrated by the IUE sky survey map shown in Fig. 4.31, which provides a superb, thorough view of the distribution and nature of UV astronomical sources. This can be compared to the Copernicus map of Fig. 4.6. The instrument consisted of a 45 cm aperture Ritchey–Chretien telescope with a beryllium paraboloid primary mirror feeding two spectrographs with Vidicon cameras utilizing a UV convertor. For more detail on the IUE mission and instrumentation, the reader is referred to Boggess *et al.* (1978).

2. *HST*: The HST was launched from the Shuttle Discovery on 24 April 1990. It is a space-based astronomical observatory, with multiple instruments. Several UV instruments have been flown or have been planned to be flown aboard the HST. These include the following:

(i) The faint object spectrograph (FOS) provided low- and high-resolution (5 and 0.9 Å at 1200 Å) spectra across a broad UV-visible wavelength range, of targets down to 22nd magnitude. It operated from 1990 and was replaced during the 1997 servicing mission.

(ii) The Goddard high resolution spectrograph (GHRS) provided high-resolution (0.012-Å at 1200 Å) spectroscopy in the UV range 1050–3200 Å, in 50 Å windows. As with the FOS, it was operated from 1990 to the 1997 servicing mission.

(iii) The STIS provides UV spectroscopy and imaging in the range 1150–10,000 Å with a spectral resolving power $(\lambda/\Delta\lambda)$ 60,000. It was installed on HST during the second servicing mission in 1997.

(iv) The cosmic origins spectrograph (COS) is a much lower-resolution instrument ($\lambda/\Delta\lambda$ about 20,000) but with higher throughput. It is designed to obtain spectra from distant sources and was scheduled to be installed in 2004, though this depends on the Shuttle schedules after the return to service following the Columbia disaster.

There is no certainty that the instrument will be installed on the HST at this point. The HST optical layout consists of a Ritchey–Chretien design which, of course, feeds the scientific instruments placed at the focal plane.

3. *ROSAT*: The German, NASA UK ROSAT spacecraft was launched on 1 June 1990. Operations ceased in 1999. Named after the discoverer of X-rays, its main purpose was X-ray astronomy. However, it carried an instrument called the wide field camera, operated in the range 60–300 Å, which extends into the region we define as EUV. Developed in the UK, the Wide Field Camera consisted of a set of three nested, full-revolution Wolter I telescopes. The camera system made use of an MCP system and a set of interchangeable filters. For a description of the early operations of the Wide Field Camera the reader is referred to Pounds *et al.* (1991).

4. *EUVE*: The NASA EUVE was launched on 7 June 1992 and the mission ceased operation on 31 January 2001. It operated in the 70–760 Å region, and consisted of three grazing incidence scanning telescopes and an EUV spectrometer. For more details, see Bowyer and Malina (1991).

5. *FUSE*: The NASA FUSE mission was launched on 24 June 1999. It is a NASA Principal Investigator mission dedicated to high-resolution UV astronomy in the range 900–1100 λ. The spectral resolving power ($\lambda/\Delta\lambda$) is 20,000. A four-component telescope mirror system, with differing optical coatings is used to maximize the reflectivity across the wavelength range. Its sensitivity is 10,000 better than Copernicus. The FUSE instrument system is described in detail above.

6. *GALEX*: The NASA GALEX mission was launched aboard a Pegasus XI rocket on 25 March 2003. Its mission is a 28-month UV imaging and spectroscopic survey, using a 50 cm Ritchey–Chretien telescope with field of view 1.2°. The basic goal of this instrument is to measure the UV properties of local galaxies, to provide information on galaxy origins, star formation, extinction, metalicity, etc. With regard to imaging, GALEX will provide two surveys in the far-UV (1350–1800 Å) and near-UV (1800–3000 Å) with 3–5 arcsec resolution, and better than 1 arcsec astrometry, and a cosmic UV background map. On the spectroscopic side, GALEX is providing three overlapping slitless-grism spectroscopic surveys in the range 1350–3000 Å, with resolving power about 100, covering 100,000 galaxies with red-shifts in the range 0–2, with measurements of extinction and star formation rates.

This is not an exhaustive list of EUV/UV astronomy space missions. UV observations have been made, for example, aboard the Shuttle. This includes the Hopkins UV telescope (HUT) and orbiting retrievable far- and extreme-UV spectrometers (ORFEUS), which were flown in the 1990s and provided brief observations of the FUSE wavelength range. Also Ultraviolet Spectrometers have been flown aboard Voyager and Apollo.

4.6.4. *A tour of solar EUV/UV missions*

With regard to solar missions, beyond the OSO series, we note the following:

1. *Skylab*: The manned Skylab mission carried the Apollo telescope mount, which included a series of solar instruments which were active between 14 May 1973 and 3 February 1974. Three of the instruments operated in the UV range, these included:

(a) The UV Spectrometer Spectroheliometer provided spectra or spectroheliograms of any selected region of the spectrum from 280–1350 Å with spatial resolution 5 arcsec and spectral resolution 1.6 Å.

(b) The EUV spectroheliograph was a slitless spectrograph, which produced over-lapping wavelength dispersed images of the solar disc in the range 171–630 Å, with wavelength resolution of 0.13 Å and spatial resolution between 2 and 10 arcsec.

(c) The UV spectrograph covered the wavelength range 970–1970 Å with spectral resolution 0.06 Å, and the 1940–3940 Å range with resolution 0.12 Å. This instrument had a slit width of 60 arcsec and resolution 2 arcsec.

Further information on the Skylab mission and results can be found in the review by Eddy (1979).

2. *SMM*: The NASA SMM mission was launched on 14 February 1980 and ceased operation in 1989. It was a multiwavelength mission to study solar flares and included the UV Spectrometer and Polarimeter (UVSP) instrument which operated in the 1170–3500 Å range, with a spectral resolution element of 0.02 Å and spatial resolution of down to 2 arcsec and a field of view 256 arcsec^2 (the Sun is half a degree across). This instrument allowed observation of the lower-temperature flare, active region and quiet Sun plasmas (under 200,000 K). The basic design was a Gregorian telescope with Ebert spectrometer; details can be found at Woodgate and Tandberg-Hanssen (1980).

3. *SOHO*: The ESA/NASA SOHO spacecraft has been the workhorse of solar physics space efforts since its launch in 1995. It carries 12 instruments designed to study the solar interior, atmosphere and heliosphere, and several of them operate in the EUV/UV range. This includes:

(a) The SUMER instrument, which allows spectral studies of the 500–1600 Å range with resolution of order 1 arcsec and resolving power 20,000–40,000. SUMER is designed mainly to study the fine-scale structure, evolution and activity of the low-solar atmosphere. It consists of a primary normal incidence parabolic mirror feeding a secondary (also parabolic) and a spherical concave grating spectrometer with delay-line detectors (Wilhelm *et al.*, 1995).

(b) The CDS which is discussed above, covers the 170–800 Å region with spatial resolution of order 2–3 arcsec and spectral resolving power 1000–10,000. CDS extends the capabilities of SUMER by concentrating on the higher temperature plasmas (see Harrison *et al.*, 1995).

(c) The UV Coronagraph and Spectrometer (UVCS) covers selected UV lines observed off the solar disc (high corona) with spatial resolutions and spectral reso-lutions down to 12 arcsec and 0.08 Å. This instrument extends the plasma measure-ments to higher solar altitudes. The instrument consists of three spherical telescope mirrors feeding three spectrometer channels utilizing delay-line detectors (Kohl *et al.*, 1995).

(d) The EIT is an EUV imager which produces full disc 1024 × 1024 pixel images in four selected EUV lines. It consists of a Ritchey–Chretien telescope with multilayered optics and back-thinned CCD detectors (see Delaboudiniere *et al.*, 1995).

(e) The Solar Wind Anisotropies (SWAN) instrument, which maps the full sky in Lyman alpha to monitor the hydrogen distribution of the heliosphere. The instrument consists of two components each containing toroidal mirrors mounted on a periscope system, feeding a multi-anode detector system. The two instruments are mounted in locations which allow scans to be built up of the full sky (Bertaux *et al.*, 1995).

(f) In addition, the solar EUV monitor (SEM) is mounted on one of the SOHO particle instruments. It monitors the full-Sun intensity in the EUV. Although instru-ments of this kind have not been discussed in this review, it is effectively a photodiode system behind an aluminium filter and transmission grating. The instrument mea-sures the integrated Sun flux in the He II 304 Å line and in the (integrated) band 170–700 Å. The instrument is included as part of the Charge, Element and Isotope

Analysis System (CELIAS) instrument on SOHO and is described by Hovestadt *et al.* (1995).

4. *TRACE*: The NASA TRACE mission was launched in 1998 and is a high-resolution EUV imager, working closely with the SOHO mission. Details of the mission and instrument are given above.

5. *Coronas-F*: Coronas-F is a Russian-led solar mission launched on 31 July 2001. It is a polar orbiting platform aimed at studying solar activity, in particular flares, and the solar interior (Oraevsky and Sobelman, 2002). The 15-instrument payload includes the following instruments with an EUV or UV capability:

(a) The SPIRIT, full Sun XUV spectroscopy imaging instrument package includes the SRT, solar X-ray telescope, and the RES, X-ray spectroheliograph. The former includes three imaging components providing full-Sun images in a range of emission lines centred on 171, 195, 284 and 303 Å, with varying resolutions, down to 1.4 arcsec. The RES instrument images the Sun in X-ray channels covering Fe and Mg X-ray lines but includes an EUV band producing images in the range 180–200 and 285–335 Å, resolution elements of 6 arcsec and 0.03 Å/pixel.

(b) The SUFR – solar UV radiometer – monitors integrated solar variations in several spectral bands in the range 10–1300 Å.

(c) The VUSS – Solar UV Spectrophotometer – monitors solar integrated UV flux around the hydrogen Lyman–alpha line at 1200.

As with the non-solar missions, other solar space-borne observations have been made using the Shuttle, for example, the Coronal Helium Abundance Spacelab Experiment (CHASE), which was flown in 1985 and was very much a forerunner of the CDS instrument on SOHO. In addition, there are still active rocket flight programmes of UV instruments, such as the NASA SERTS (solar EUV rocket telescope spectrometer) instrument.

This section has given a flavour for the techniques and missions used to explore the UV universe. We have had to be somewhat selective and non-uniform in describing the instrumentation and technology and it should be noted that this is a field which is constantly developing. The next generation of instruments and missions are under development at this time and this will be discussed later.

4.7. EUV/UV observations and results

4.7.1. *The Sun*

The Sun was used to open the field of the UV astronomy, some 57 years ago, and it is a vigorously developing topic to this day. Indeed, with the advent of SOHO and the accompanying TRACE mission, one could argue that our observations of the EUV/UV Sun have revolutionized our understanding of the solar atmosphere over the last 8 years. In addition, of course, the Sun is an important target of study because (a) it is the only star that we can study in detail, (b) it is important to study and monitor the solar activity impacts on the Earth, and (c) the solar plasmas provide us with experimental capabilities to study fundamental astrophysical processes.

Although the Sun's spectrum mimics that of a black-body curve at a temperature a little under 6000 K (Fig. 4.11), this is not the full story. The visible Sun, the so-called photosphere, has a temperature of order 6000 K, but features superimposed on the black-body spectrum in the EUV/UV region are due to emission lines from ions with temperatures commonly up to a few million K. This is well illustrated in Fig. 4.10. Putting it another way, as we move through the UV regions we find the following: The 1600–3000 Å region is characterized by spectral emissions and absorption lines from the photosphere and a layer we call the chromosphere; the 150–1600 Å region of the spectrum is dominated by emission lines from the outer atmosphere, known as the corona, and from the transition region between the corona and chromosphere, and from hot solar flare plasmas.

These solar atmospheric layers are identified in the following, simple-minded way. The 6000 K photosphere is considered to be the solar surface. Initially, the plasma temperature drops with altitude, as one would expect, given the laws of thermodynamics. There is a temperature minimum, in the low atmosphere, with temperatures falling to 4000–5000 K. This lowest-atmospheric layer is the chromosphere. Above it is the corona, where plasma temperatures achieve an incredible 2–3 million K, and the heating of that layer of the solar atmosphere is still a major outstanding solar question. Between the superhot corona and the chromosphere is a steep temperature gradient. This thin layer is known as the transition region, which one typically characterizes as occupying the temperature range of a few tens of thousand K to under one million K.

The wide temperature range of the solar atmosphere produces a plethora of emission lines on top of the UV continuum spectrum and it is these lines which can be used to explore plasma diagnostics of the solar atmosphere.

Early results showed that the Sun has a hot atmosphere and that there are "holes" in that atmosphere. A modern example of that is shown in Fig. 4.1. This is an EIT/SOHO image of the corona, taken in emission from Fe IX at 171 Å. This emission is produced by an ion at a temperature of almost 1 million K so we know immediately that the corona is hot. The beauty of this kind of solar observation is that by selecting emission from the hot corona, one is not blinded by emission from the much cooler solar disc. Several features are visible in such images. First, we can see complex, bright loop systems. These are the so-called active regions, which include the sunspots and flare activity. The magnetic fields are produced by the complex motions of plasmas within the Sun. The loops are stressed magnetic arcade systems twisted by the motions of the Sun, which are trapping hot plasmas. These active regions can be the sites of magnetic rearrangement, if the fields are stressed enough and, through a process known as magnetic reconnection the fields can adopt a lower-energy state and release stored energy dramatically, into the local plasma, heating it to tens of millions of K and sending streams of accelerated particles into space. This is a solar flare.

The darker region at the top of the Sun is a coronal hole. This is an area of lower emission where the field lines are open to space. We note here that processes within the solar atmosphere also accelerate plasmas to form the so-called solar wind which appears as an almost continuous stream of plasma extending into space. A higher-speed solar wind originates from the polar holes.

The rest of the Sun appears to be somewhat diffuse, with occasional bright points (which we call bright points!).

The recent observations of the Sun have really stressed the dynamic nature of the solar atmosphere. Movies show striking evolution on all time scales. In effect, we are witnessing the writhing of magnetic fields rooted in a fluid body with the added complications of differential rotation (the poles rotate slower than the equator) and the turbulent convective flows. The result is a complex plasma-magnetic environment with many dynamic features, some of which have only recently been identified.

On the smallest scale, UV and EUV instruments have been used to home in on the so-called super-granulation of the solar surface, seen in cooler temperature emission lines. This super-granulation is an imprint of the convection cells which bring energy to the solar surface. The super-granulation can be seen clearly in the background full-Sun image of Fig. 4.32, which is taken in the 304 Å emission line of He II by the SOHO EIT instrument.

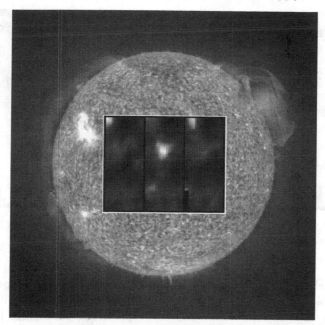

FIGURE 4.32. A SOHO/EIT full-Sun image (background) taken in the 60,000 K He II line at 304 Å, showing the super-granulation. Overlayed on this, is an observation of blinker activity. The same area of the solar disc is shown, at three times, in emission from O V (629 Å at 260,000 K), and a bright flash is seen in the middle frame (courtesy SOHO/CDS and EIT teams).

This ion has a temperature of formation of about 60,000 K. Overlayed on the full-Sun image, Fig. 4.32 shows a region of the Sun, some 30,000 km × 74,000 km in size at three times, within minutes of one another, taken by the SOHO CDS instrument. We can see a close-up of the super-granulation; 2–3 network cells can be seen, with the pattern persisting between images. However, there is a bright flash in the middle frame. This is an Earth-sized brightening, commonly known as a blinker, of which several thousand may be occurring on the solar disc at any moment in time (Harrison *et al.*, 1999 and 2003). In addition, also studying the small-scale events of the network, Fig. 4.17 showed some so-called explosive events, which are small jets of plasma of size typically 1500 km, with speeds of up to 150 km/s. These blinkers and explosive events are basic phenomena currently under active study because of their potential role in fundamental processes in the Sun's (a stellar) atmosphere. Are they signatures of magnetic reconnection on small scales? Since these small-scale activities occur throughout the solar disc, could the net effect relate to the fundamental processes of solar wind acceleration and/or coronal plasma heating?

Still on the subject of basic processes, the emission line narrowing as a function of altitude shown in Fig. 4.18 has been interpreted as a signal of Alfvén wave dissipation. Alfvén waves may be generated by the convective motion of the solar interior and are in effect transverse waves on which the particles ride; the net effect to an observer is to see Doppler broadening. A narrowing of lines with altitude can be a signal of the dissipation of such waves. This is an area of current debate.

On more intermediate spatial scales, Fig. 4.33 illustrates a dramatic spiraling plasma jet. This was identified, also, by selecting the 250,000 K O V emission line at 629 Å (Pike and Harrison, 1997). The left-hand frame shows the raw image, taken on the solar limb.

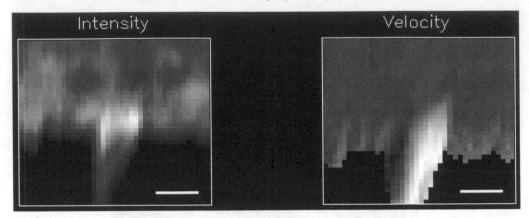

FIGURE 4.33. A spiraling, jet-like ejection (Pike and Harrison, 1997) (courtesy SOHO CDS team).

A ghostly vertical jet is detected, extending above the solar southern polar region. The right-hand Doppler image reveals that the jet is spiraling. The white and black regions are moving away and towards the observer, respectively. The white line denotes the diameter of the Earth to the same scale. Such tornado-like events appear to be driven by the convergence of magnetic structures.

Finally, we show a solar flare event detected using the TRACE instrument (see Fig. 4.14). It shows a complex loop system using the 171 Å 800,000 K Fe IX line. The striking point about such images is the fine-scale structure of the loops. The pixel size is 350 km and the images reveal an almost hair-like structure. In addition, movies of events such as this are breathtaking, showing truly rapid, dynamic activity.

We have only scratched the surface of solar EUV/UV research – there are many facets which we have not covered but choices had to be made due to space limitations. However, this quick tour should be sufficient to illustrate both the complex and fascinating nature of the Sun's atmosphere and the vigorous research efforts which are under way to understand our own star.

One aspect which we have not touched on in any great detail here is the fact that the EUV/UV emission from the Sun is of great importance because such radiation influences the Earth's upper atmosphere. These influences include the maintenance of the Earth's ionosphere and UV radiation at wavelengths below 2000 Å which dissociates molecular oxygen, indirectly influencing the production of ozone. For reasons such as this, scientists are attempting to monitor the total EUV/UV dose arriving at the Earth, and to understand the principal sources of the wavelengths of particular interest. This links us rather nicely to the EUV/UV studies of planetary atmospheres.

4.7.2. *Planetary atmospheres*

Studies of planetary atmospheres in the UV – including that of the Earth – are important for the identification and measurement of atmospheric gases. The resonance absorption and emission spectral features are mainly in the UV region for most common atmospheric gases. For example, oxygen absorbs strongly at wavelengths below 2000 Å; hydrogen can be detected by its scattering of incoming H I Lyman alpha radiation at 1216 Å.

The far-UV camera/spectrograph deployed on the Moon by the crew of Apollo 16 took the image of the Earth shown in Fig. 4.34. The sunward side of the Earth reflects much of the UV light. Perhaps of greater interest is the side facing away from the Sun, where

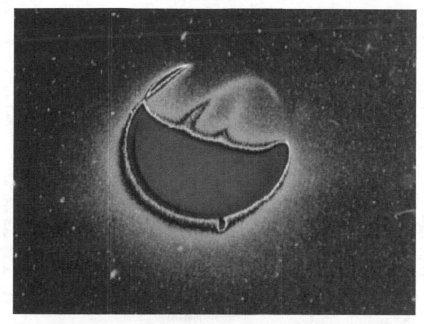

FIGURE 4.34. An UV image of the Earth taken by the Apollo 16 far-UV camera/spectrograph from the Moon (courtesy NASA).

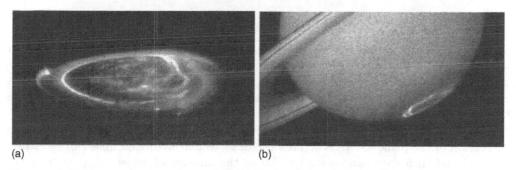

(a)   (b)

FIGURE 4.35. Aurora in the polar regions of Jupiter (a) and Saturn (b) observed from HST (courtesy NASA).

UV emitting bands are also apparent. In this case, these are the result of aurora caused by the influx of charged particles as a result of the arrival of coronal mass ejection from the Sun.

Figure 4.34 highlights the fact that even though we can fly through and sample our own atmosphere, we can produce, and make use of, global maps using remote sensing from space. The auroral features are the result of emission following excitation by incoming charged particles. This can be seen at Earth and has also been identified using the HST STIS instrument in the polar regions of Jupiter and Saturn (Fig. 4.35).

The solar spectrum reflected off a planetary atmosphere may be modified in many ways. For example, middle-UV absorption features characteristic of sulphur dioxide are found in reflected UV radiation from Venus. Similarly, the Earth's atmosphere shows absorption characteristic of ozone. We know from UV observations that the atmospheres of Mars and Venus consist mainly of carbon dioxide. Atomic oxygen is found in the spectra from Venus and, of course, Earth, but not Jupiter. Jupiter's spectrum displays

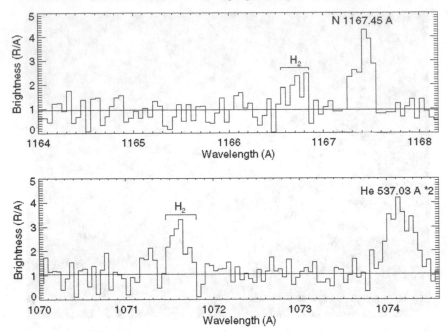

FIGURE 4.36. FUSE observation of H_2 lines in the Martian spectrum (courtesy NASA/JHU FUSE team).

signatures from atomic and molecular hydrogen. The Earth and Titan (Saturn's largest moon) are unique in displaying signatures of atomic and molecular nitrogen.

Although the atmospheres of the Moon, Mercury and Ganymede (Jupiter's largest moon) appear to be non-existent, upper limits on atmospheric densities have been obtained from space-based UV observations.

Figure 4.36 shows FUSE spectra of the 1070–1075 and 1164–1168 Å regions from an observation of the planet Mars. The presence of molecular hydrogen lines can be clearly identified and such data can be used to assess the amount of water escaping from the Martian atmosphere (Krasnopolsky and Feldman, 2001).

4.7.3. *Comets*

Comets are minor solar system bodies composed mainly of volatile materials such as water and ice. They usually occupy highly eccentric orbits and may approach very close to the Sun at perihelion. Indeed, many comets, known as sun grazers, do not survive the extreme environmental conditions at perihelion and simply vaporize completely. As comets approach the Sun, a significant amount of mass vaporizes to form a halo, or coma. This halo is visible in the UV, and in particular the dominant emission is from hydrogen in the form of the Lyman–alpha line at 1216 Å.

Observations of Halley's comet in 1986, for example, showed that the UV halo (obtained in Lyman alpha) was considerably larger than the visible halo and extended to a volume which was much larger than the Sun. Such observations reveal not only the composition of the halo but also provide valuable information on the vaporization rate.

In addition to the Lyman–alpha emission line, another common spectral feature is the OH molecular emission near 3100 Å. These signatures indicate that water is the dominant volatile component of comets. Other spectral features detected in the UV are from CH,

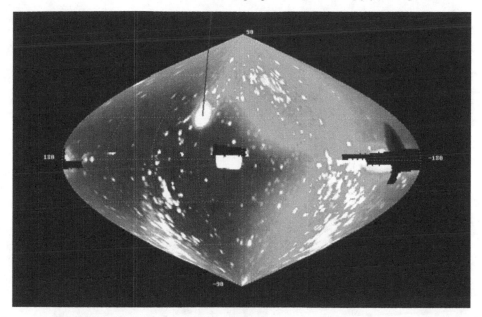

FIGURE 4.37. SWAN/SOHO full-sky image showing Lyman–alpha emission (Combi *et al.*, 2000) (courtesy SOHO SWAN team).

CN, C_2 and NH. Observations of comets West (seen in 1975) and Halley (seen in 1986) also revealed signatures of CO.

Figure 4.37 shows a full-sky image of hydrogen Lyman–alpha distribution, taken by the SWAN instrument aboard the SOHO spacecraft on 3 April 1997 (Combi *et al.*, 2000). Comet Hale–Bopp is visible as an oval shape just above-left of the centre of the image. Whereas the nucleus of Hale–Bopp is believed to be 40 km across, the SWAN image reveals a halo of size 100 million km. In addition, analysis of the brightness and size of the emitting area reveal that 600 ton of ice evaporates each second from the nucleus.

4.7.4. Stars

Those stars which we categorize as hot, with surface temperatures of over 10,000 K, which includes stellar classifications of B and A, emit much of their radiation in the UV. This was predicted years before observations were made from space.

Of course, the principal star under study using EUV/UV techniques is the Sun, which we would categorize as a cool star, that is, in the surface temperature range under 10,000 K. This includes stars classified as G, K and M – those which we think of a yellow, orange and red. Such stars emit most of their radiation in the visible part of the spectrum and can be observed well from the ground. However, we know that the Sun has a hot corona and EUV/UV spectral features which reveal the structure and evolution of its atmosphere. Thus, we may expect the same from other stars.

The pioneering survey work of Copernicus was discussed above, but the area of EUV/UV stellar astronomy was really opened up by the long-lived IUE mission. Spectra were obtained and quite naturally compared to our Sun. Some general results have been obtained. For example, cool supergiants do not show evidence for hot coronae, and this is assumed to result from the presence of a strong stellar wind. Even a null result is important in identifying the physics of astronomical bodies! However, some stars are much more active than the Sun. One well-studied star in that category is the giant

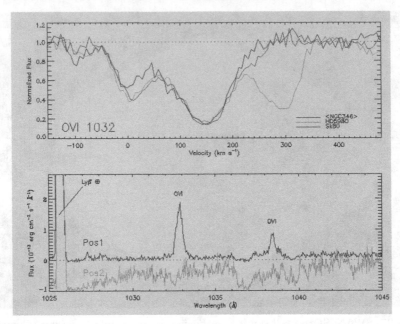

FIGURE 4.38. FUSE observation of a supernova remnant (Danforth *et al.*, 2003) (courtesy FUSE team).

Capella. Also, the cool M-class stars (surface temperatures about 3000 K) show surprising activity and this can include flare activity with individual flare energies far exceeding those we witness on the Sun.

In the shorter-wavelength EUV region, most stars are not observable, even from space-based platforms. At wavelengths under 912 Å the radiation is absorbed by interstellar atomic hydrogen. This is the hydrogen edge mentioned above. However, the spatial variability of interstellar atomic hydrogen does mean that EUVE has been successful in detecting and measuring EUV spectra from a number of hot stars in this short-wavelength range.

As an example of the state and development of UV stellar survey work, the early survey of Copernicus, shown in Fig. 4.6, should be compared with the IUE survey of Fig. 4.31. The latter shows the huge number of sources detected by IUE. A good, specific example of a detailed (non-solar) spectrum from a star, from the FUSE spacecraft, is shown in Fig. 4.29. As with the Sun, such spectra provide the tools for detailed plasma diagnostic analyses.

Figure 4.38 shows a specific EUV/UV observation made using the FUSE spacecraft, in this case, providing details of the end of star's life.

The plots show data obtained from an observation in the vicinity of a star formation region of the Small Magellanic Cloud galaxy. The star-formation region is a diffuse cloud, which glows because of embedded hot stars – those formed within the cloud. However, a combination of X-ray and UV observations have identified a supernova remnant within the cloud.

In general, an expanding blast from a supernova heats the gas shell to millions of K – seen as X-ray emission. In this case, there is no visible outer shell. Chandra observations reveal an extended region of X-ray emission, within the cloud. The FUSE spacecraft has been used to study the spectra of various stellar sources in the region of the cloud. Data

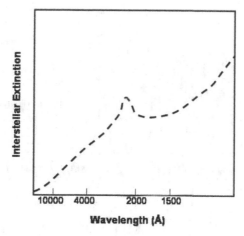

FIGURE 4.39. A schematic of average interstellar extinction as a function of wavelength.

from the O VI 1032 Å region of the stellar spectra are shown in Fig. 4.38. The top panel reveals a normalized flux, specifically for the 1032 Å line, for three stellar sources, and this shows that one of the stars is indeed behind an absorbing region, not detected in visible light. In addition, two spectra are shown in the lower panel: one apparently inside the supernova remnant, (emitting the O VI lines) and one outside. For more details, see Danforth *et al.* (2003).

4.7.5. *Interstellar and intergalactic gas and dust*

Interstellar space contains traces of both gas and dust – distributed rather irregularly.

Hydrogen is the most common element in the Universe. In interstellar space, it exists in the ground state and can absorb incident UV light at the Lyman line wavelengths and a continuous wavelength range beyond the Lyman edge at 912 Å. Thus, interstellar space is expected to be rather opaque below this limit.

Interstellar dust particles consist mainly of silica, magnesium and iron silicates, amorphous carbon or water ice. Continuous absorption by interstellar dust rises steadily with decreasing wavelength throughout the visible region, so distant stars appear reddened. The absorption continues to rise in the UV, with a significant "bump" at 2200 Å. This form is shown schematically in Fig. 4.39.

The 2200 Å peak in the extinction curve, as shown in Fig. 4.39, matches theoretical predictions from graphite particles. However, Fig. 4.39 shows an average extinction profile, and it has become clear that the UV extinction curve does differ in magnitude and shape in different directions within the galaxy and within the Magellanic Clouds. This indicates that there must be several (at least three) independent components of the interstellar dust, in which relative abundances vary in different ways in different locations.

As an example of interstellar observations, Fig. 4.40 shows a FUSE observation of the distant quasar HE2347-4342 (Kriss *et al.*, 2001). It shows dips, mainly caused by helium absorption in the intergalactic medium. The lower panel shows a comparison projected spectrum from the 10 m Keck telescope observations – scaled from observations of hydrogen absorption from the same source. The absorption traces look very different and a number of bands are highlighted in the top panel where the helium data reveal greater absorption.

Hydrogen is more easily ionized, that is, it becomes more invisible more easily. Therefore, the helium data more accurately reveals the intergalactic material. This is an

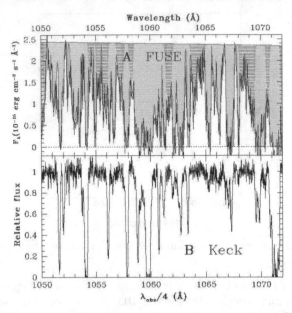

FIGURE 4.40. A FUSE spectrum of a distant quasar, showing helium absorption in the intergalactic medium. The lower panel shows a comparison to projected hydrogen data using the Keck telescope (Kriss *et al.*, 2001).

important result; since there is more material in the interstellar/intergalactic medium than in all stars put together, measurements of such material is critical for understanding the Universe.

Reflection spectra from interstellar dust have also revealed much information. For example, in some reflection nebulae the dust particles are very efficient scatterers of far-UV radiation. This suggests that, except for the 2200 Å extinction peak, which is tuned to absorption from graphite particles, it seems that most interstellar extinction is due to scattering rather than pure absorption.

Most galactic H II regions appear as refection nebulae in the far-UV. This is due to the relatively high dust densities in these regions, combined with the presence of UV-bright stars which both illuminate the dust and excite UV and visible emission lines. However, the dust in the galactic H II regions is probably not typical of dust in the general interstellar medium and in other reflection nebulae. This is due to the fact that in regions such as M42, the Great Nebula in Orion, the intense radiation field can significantly modify the size distributions and compositions of dust particles. This is indicated by a comparison of extinction curves and reflection spectra for differing regions.

4.7.6. *Extragalactic objects*

Galaxies are giant clusters of stars and associated interstellar material and, as such, one would expect to see EUV/UV radiation, mainly from the hot stellar sources. In addition, UV observations are important for the study of quasars and active galactic nuclei. The energy output for such sources peaks in the UV – many argue that this is expected from an accretion disc around a super-massive black hole. The Lyman-alpha absorption lines from such sources are red-shifted into the visible – multiple lines are seen, probably absorbed by a series of clouds between Earth and the quasar.

Figure 4.41 shows three different galaxies taken in UV light by NASA's Ultraviolet Imaging Telescope (UIT) on the Astro-1 mission in 1990. Such images reveal several

FIGURE 4.41. Three spiral galaxies imaged in UV light from the UIT instrument aboard the Shuttle (courtesy NASA).

issues. First, the appearance of the galaxies is due mainly to clouds of gas containing newly formed stars many times more massive than the Sun, which glow strongly in the UV. In contrast, visible light pictures of such galaxies show mostly the yellow and red light of older stars. The massive, bright stars detected in the UV are relatively short-lived and do not move far from the spiral arms in which they are formed. Thus, they are good tracers of the spiral arm structure. Clearly, by comparing UV and visible images of galaxies, astronomers can glean information about the structure and evolution of galaxies.

This concludes our brief tour of some space observations in the EUV/UV, in a number of different areas of astronomy. Of course, the tour has been somewhat selective and biased; we cannot hope to review the details of all aspects of EUV/UV space astronomy. However, it has served to illustrate the advances we have made in EUV/UV astronomy from space and outlines the basic information we are obtaining at this time.

4.8. The future

We now turn to the future. What are the possibilities for major advances in the area of EUV/UV space astronomy? What missions are in the pipeline and what should we expect from these?

In the solar area, we have the ongoing SOHO and TRACE missions, which are healthy, and we anticipate operations for some years to come. In addition, there is a strong programme of future international solar missions with EUV/UV capabilities. In the non-solar area, the FUSE, GALEX and HST missions are continuing. However, in contrast to the solar case, there is little planned for the future and the anticipated HST UV developments look to be in question in the shake-up after the Columbia disaster.

The principal new non-solar UV instrument was the COS instrument, mentioned above, which was to be installed on the HST. Clearly, it is still hoped that this will take place – especially as this is the only major UV development planned for non-solar astronomy. HST already carries the STIS instrument, which provides UV spectroscopy and imaging in the range 1150–10,000 Å with spectral resolving power $(\lambda/\Delta\lambda)$ 60,000. This was installed in 1997. The new COS instrument has lower resolution $(\lambda/\Delta\lambda \sim 20,000)$ but has a higher throughput. Designed to obtain spectra from distant sources. The COS instrument is designed for studies of (i) the formation and evolution of galaxies, (ii) the origins of the large-scale structure of the Universe, (iii) the origins of stellar and planetary systems and (iv) the cold interstellar medium. Its sensitivity is over an order of magnitude better

FIGURE 4.42. The STEREO mission: two spacecraft out of the Sun–Earth line designed to study the Sun in 3D and Earth-directed mass ejection (courtesy NASA).

than previous HST UV instruments. COS was scheduled to be installed aboard HST in 2004.

Of course, we do hope that the existing, on-going FUSE, GALEX and HST instruments continue well into the future. We also note that the next-generation HST, the James Webb space telescope (JWST), is not a follow-on from HST; there will be no UV capability of that spacecraft. In effect, in space astronomy (non-solar) there is something of a "red-shift" in our observational capabilities.

Other opportunities may arise, for example, through the NASA SMEX opportunities, and this can happen over relatively short time scales.

In great contrast to the non-solar case, the solar community has the ongoing SOHO and TRACE missions and something of an international strategy of space missions. If one considers where we are now in terms of general solar space observations, we have not yet left the cradle. All solar remote sensing has been done from near-Earth space (Earth orbit or the L1 point), that is, from one viewpoint in the Sun–Earth line and on the ecliptic plane. The next few years see us leaving the cradle, with missions travelling out of the Sun–Earth line, with spacecraft making widely separated observations of the Sun, with missions travelling out of the ecliptic and with solar encounter missions. These missions include STEREO, Solar-B, SDO, Solar Orbiter and Solar Probe. All have an EUV/UV capability and all build on the experiences of missions such as SOHO. We describe them in turn here.

STEREO is a NASA mission due for launch in 2006. It consist of two identical space-craft, in 1 AU solar orbits, one leading and one lagging the Earth – and separating from each other by 22.5° per year. From these vantage points, the aim is to provide 3D images of the solar atmosphere (with a combination of well-spaced 2D images) and to monitor the space between the Sun and Earth (to detect Earth-directed solar ejecta). Remember that the Sun is the only star we can view from widely separated platforms, so this is a unique mission in terms of stellar physics. The payload of each spacecraft includes an EUV imager, in addition to coronagraphs and wide-angle heliospheric imagers, which

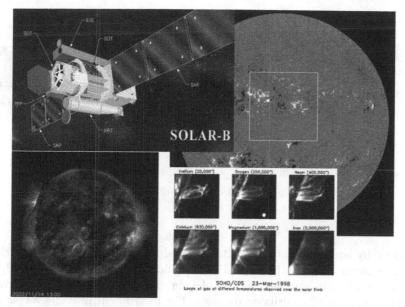

FIGURE 4.43. The Solar-B mission: a representation of the products from Solar-B using examples from SOHO, including EUV spectroscopy, EUV full-Sun imaging and magnetograms (courtesy ISAS and SOHO teams).

together provide coverage from the solar atmosphere to the Earth. In addition, the spacecraft include particle, field and radio instruments.

The EUV imager, known as EUVI, has twice the spatial resolution of the SOHO/EIT instrument, and it has improved cadence. In addition, there are two such devices on two platforms, of course. This is a powerful combination. The imagers use similar wavelength bands to SOHO and TRACE, namely 171 Å Fe IX (800,000 K), 195 Å Fe XII (1,000,000 K), 284 Å Fe XV (2,000,000 K), and 304 Å He II (60,000 K). They utilize multilayered optics, similar to TRACE, and 2k × 2k back-thinned CCD camera systems.

One aspect of EUV/UV observation not covered by STEREO is spectroscopy.

The Japanese Solar-B mission, is the follow-on from the highly successful Yohkoh mission. Unlike Yohkoh, it includes an EUV/UV capability. The mission is a collaboration, led by the Japanese but including the USA and UK. Launch is planned for 2006. Solar-B is not going beyond Earth orbit. It is one of a new breed of high-resolution missions whose telemetry and mass requirements dictate that low-Earth orbits are necessary. It will occupy a polar, Sun-synchronous orbit of altitude 600 km. The basic mission goal is to map the solar atmosphere across all layers from the photosphere to the corona.

Solar-B carries a 50 cm optical telescope, and X-ray telescope and an EUV imaging spectrometer. Figure 4.43 shows the basic spacecraft configuration and a representation of the data from these instruments using images from SOHO. The optical telescope will provide high-resolution images of the photosphere and the X-ray imager will provide images of the corona. The spectrometer, known as EIS – the EUV Imaging Spectrometer, is very much a follow-on from the CDS instrument described above. The instrument layout is shown in Fig. 4.44. The concept is sufficiently different from past instruments to warrant some description.

The EIS instrument incorporates an off-axis 15 cm paraboloid mirror, used in normal incidence. The secondary reflection is in fact a shaped (toroidal) grating. This provides a system with only two reflections, thus, maximizing reflectivity. As with CDS, the grating

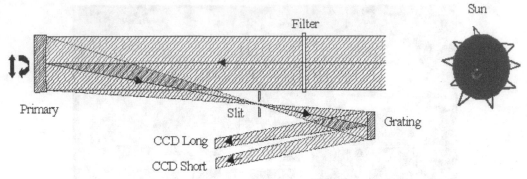

FIGURE 4.44. The Solar-B EIS instrument layout (courtesy MSSL).

assembly actually houses two gratings tilted to produce two spectra on a 2D back-thinned CCD detector, with wavelength ranges 170–210 and 250–290 Å. As with CDS, this covers a wide range of emission lines used for plasma diagnostic analysis. These wavelength ranges are obtained using multilayered optics. The instrument provides a resolution element (pixel) of 1 arcsec, and rasters (builds up images) by motion of the primary mirror. Despite the large size of the instrument, which is 3 m in length, its carbon fibre structure allows the mass to be kept to 60 kg. This compares to the 100 kg of the SOHO/CDS instrument.

The NASA SDO is planned for launch in 2008. As with Solar-B, SDO is a low-Earth orbit high-resolution mission. It is, in fact, the flagship mission of the NASA Living with a Star programme. The principal aims of SDO are, to investigate fine-scale structure and evolution of the solar atmosphere – to understand the links from the solar interior to corona, and to develop local-area helioseismology. The instruments include a high-resolution visible imager (for helioseismology and magnetic mapping), an EUV/UV imager and a full-Sun EUV monitor. The imager, named the Atmospheric Imaging Assembly (AIA), is a multi-wavelength imager, much like TRACE, but with simultaneous imaging. For both the EIT/SOHO and TRACE solar EUV/UV instruments, images are recorded and transmitted in succession.

Figure 4.45 shows a schematic of the observations from SDO. The high-resolution photospheric imaging will provide both magnetic mapping and, through local area helioseismology, a method for investigating sub-surface structure and evolution. This is indicated by the background image of Fig. 4.45, from the SOHO MDI instrument, which shows a photospheric magnetic map, and the image, top-right, of sub-surface structure during the early development of a sunspot. The inclusion of high-resolution EUV/UV imaging is indicated by the TRACE image (bottom-right) – and this capability, linked to the photospheric data provides a method for tracing the photospheric structure and evolution into the atmosphere. As with STEREO, there is no spectroscopic capability on SDO.

The NASA Solar Probe mission has long been discussed and planned but is, at this time, not a formally approved mission. A few years ago, it was an approved mission and the subject of a payload announcement of opportunity. It was cancelled but some studies have continued. For this reason, we include it here.

Solar Probe is a Jupiter gravity assist mission which slices through the corona at 4 solar radii from Sun centre. In effect it is a one-shot solar flyby mission. The duration from launch, through Jupiter flyby to solar encounter takes 3.8 years. The launch date is, at this time, unknown.

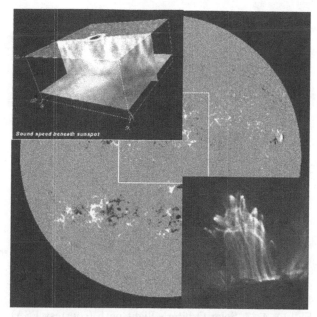

FIGURE 4.45. A schematic of the observations from SDO (courtesy SOHO MDI and TRACE teams).

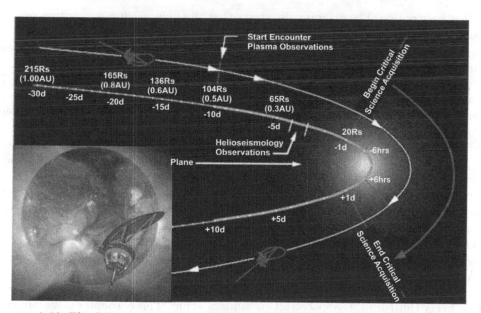

FIGURE 4.46. The Solar Probe mission. Bottom left: The spacecraft, showing the extensive heat-shield. Background: The planned flight path (courtesy NASA).

Probe would include some basic imaging, and particle and field instruments. All instruments would be extremely lightweight, and the planned EUV imager would view part of the solar disc through a narrow straw-like baffle which views the Sun through the extensive heat shield at the front of the spacecraft. This basic layout is shown in Fig. 4.46; the instrument package is in the shadow of the shield. Clearly the instrument details are not fixed at this time. However, the low-mass, low-telemetry requirements preclude any

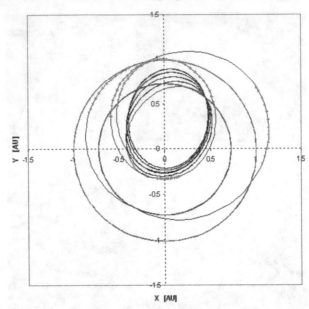

FIGURE 4.47. Solar Orbiter's flight path (courtesy ESA).

spectroscopic option. The diagram also shows that the flyby science-acquisition period is about 2 days.

In order to explore more fully the methods we use to develop missions and instruments, we now explore one case in some detail. We examine the next ESA solar mission, the Solar Orbiter, and concentrate in particular on the proposed EUV spectrometer. The factors that come into the consideration of instrumental parameters are unique to each mission, but taking one mission as an example is a useful way of illustrating how decisions are made and how instruments develop.

Solar Orbiter is a high-resolution frequent solar-encounter and high-latitude mission. The most novel aspect of the mission is its unique orbit. Thus, we describe this in some detail. Figure 4.47 shows the orbit configuration. At the centre of the plot is the Sun. The orbits of Venus and Earth are shown as circles of radius 1 and 0.67 AU. The eccentric orbits, which include dotted sections denote the path of Solar Orbiter. The dotted sections are where the Solar Electric Propulsion (SEP) system is firing; this system is required to provide an efficient means to get to the inner solar system. After several Earth and Venus flybys, the spacecraft occupies a 149-day solar orbit of perihelion 0.2 AU and aphelion 0.8 AU. Thus, the mission has close solar passes every 149 days, of duration about 30 days each. As can be seen from the plot, the orbit has a "focus", in the top-right sector, where, every third orbit, the spacecraft encounters Venus. This encounter is used to crank the orbit out of the ecliptic plane – initially by 12°. This can be seen quite clearly in Fig. 4.48.

Thus, we have a solar close-encounter mission – with repeated encounters – which climbs out of the ecliptic, to solar latitudes approaching 40°. The cruise phase (until the 149-day solar orbit is achieved) lasts some 1.9 years.

This mission provides a totally new solar aspect, opening up a number of new possibilities. This includes:

• First exploration of the innermost solar system – Orbiter will travel closer to the Sun than any other spacecraft to date;

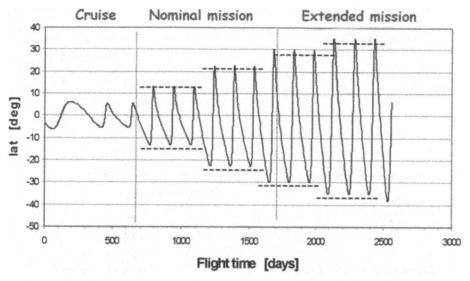

FIGURE 4.48. The latitude of the Solar Orbiter orbit (courtesy ESA).

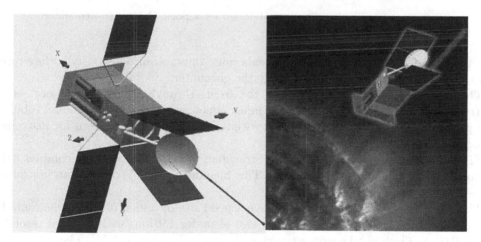

FIGURE 4.49. The Solar Orbiter spacecraft (courtesy ESA).

- Studies of the Sun from close-up;
- Co-rotation studies with the Sun, for part of the orbit – allowing extended studies of some features;
- First out of ecliptic studies of the Sun's polar regions.

Solar Orbiter is currently entering an industrial study phase, having completed payload studies with the research community. The mission is likely to be launched in the period 2012–14.

The spacecraft configuration is shown in Fig. 4.49. It shows two large solar panels (only in left frame) which are dedicated to the SEP system. Once the final burn of the SEP is done, when the 149-day orbit is achieved, these panels are ejected. Two other solar panels provide power to the spacecraft. However, to cope with the varying heat from the Sun, they swing back from the Sun, to present a grazing-incidence angle, at closest approach. At the same time, during the encounters, the high-gain antenna is folded back into the shadow of the spacecraft. Thus, during the close encounter periods, Orbiter is

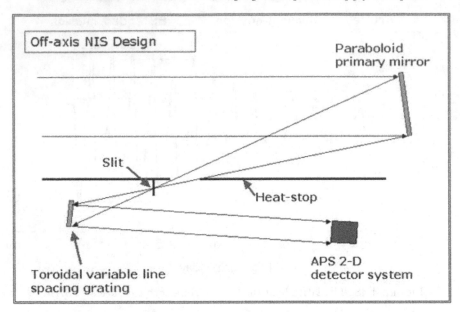

FIGURE 4.50. The baseline optical design for a Solar Orbiter spectrometer.

not in contact with Earth; the instruments must operate autonomously – they return data from the encounter after completing the encounter.

The spacecraft has a heat shield at the front, through which the remote sensing instruments view the Sun. The thermal input is not as extreme as for Solar Probe, but the perihelion heat load and the orbital variations are of major concern for instrument design. This is discussed later.

The spacecraft has a well-established strawman payload containing a complex set of remote sensing and *in situ* instruments. This includes two EUV devices, an imager and a spectrometer.

The imager is currently planned to have a pixel size of under 70 km on the Sun. The spectrometer is planned to have a pixel size of under 150 km, with spectral resolution about 0.01 Å. These pixel sizes are tuned to the perihelion, not to 1 AU. They compare to the best current EUV capabilities of 350 km for an EUV imager (TRACE) and 1500 km for an EUV spectrometer (CDS).

In addition, there is a possibility of a UV capability included in a coronagraph, to extend the work of the UVCS instrument on SOHO, observing not only the hydrogen Lyman-alpha line, but emission from ions of oxygen, iron and magnesium.

Figure 4.50 shows the current favoured design for the Solar Orbiter spectrometer. It is an off-axis normal incidence telescope, with a paraboloid primary mirror. This is similar to the Solar-B EIS. The secondary is a toroidal Variable Line Spaced (VLS) grating. As with the EIS instrument, the shaped grating removes the need for an extra reflection. The new concept of a VLS grating allows the spectrometer arm to move closer to the axis of the instrument, relative to a uniform grating system. This reduces the size, and thus the mass, of the instrument. The VLS grating also has better off-axis performance.

The orbit of Solar Orbiter provides a solar "constant" varying between 2 and 34 kW/m^2, at aphelion and perihelion, respectively. Recognizing these thermal extremes, the light stop between the primary mirror and the slit-grating acts as a heat stop. This is designed to reflect most of the incoming radiation out of the front of the instrument, that is, not only does it only allow the required portion of the solar image into the spectrometer, but it only allows a small fraction of the solar heat into the spectrometer. This means that the thermal problems are confined mainly to the primary mirror. It must have a dedicated radiator, possibly with heat-switches to smooth the orbital variations. This concept has to be refined as the design is optimized.

The design of the spectrometer is guided by requirements appropriate for an encounter mission of this kind, and building on the SOHO experience. In addition to a pixel size of under 150 km (under 1 arcsec at 0.2 AU), and a spectral resolution element of order 0.01 Å, the field of view must be some tens of arcmin square to view a reasonable area of the Sun from close-up. In addition, the spectral bands for line selection must include a wide temperature range and prime bands have already been identified as 170–220, 580–630 and >912 Å. This would provide coverage in the chromosphere, transition and corona.

In addition to high-resolution studies due to proximity, of particular interest, are the high-latitude observations. Polar regions are difficult to observe effectively from the ecliptic plane. For example, radial flows near the poles, including flows within the so-called plume and inter-plume regions, are very difficult to view using Doppler shift analysis. Of course, as with SOHO, Solar-B, etc. a spectrometer aboard Solar Orbiter is a general purpose plasma diagnostic tool box for many applications in solar physics.

Another design driver for the spectrometer is mass. The total payload mass for Solar Orbiter is about 140 kg. We must be looking at an instrument of order under 30 kg. This compares to the SOHO CDS spectrometer at 100 kg and the Solar-B EIS at 60 kg. The strategy is to adopt a low-mass carbon fibre structure. In addition, the detector option, mentioned below, is a low-mass option. The instrument will have no pointing system, which will also save mass; the instruments will be mounted together and co-pointed.

As mentioned above, during the solar encounter periods, the spacecraft is not in contact with the Earth. These encounters are of order 30 days. This means that data must be stored effectively for later transmission and that instruments must operate autonomously. To cope with the latter, pre-planned sequences must be designed and practiced during other portions of the orbits, and these must be co-ordinated ahead of time with other instruments and programmed into the spacecraft just prior to the encounter. In addition, it does mean that instruments must be able to cope with some problems, such as latch-up, without ground contact.

With regard to the data storage, the estimated on-board data memory, and the ground station support during the orbits for data return, have been used to calculate initial telemetry rates for the instruments. For the spectrometer this is currently given as 17 kbits/s. This may increase with further studies, improvements in memory capacity and ground station support. However, in any case, it is clear that significant data selection (i.e. which lines are required) and compression will be required. This is the subject of an active study.

There is one area of major concern that has not been covered and that is the particle environment. The cosmic ray environment will be similar to SOHO, or perhaps less intense, due to the influence of solar magnetic fields. However, the solar wind flux will be 25 times that at SOHO and, in addition, the 15 min lifetime of neutrons means that whereas SOHO does not see solar flare neutrons, Solar Orbiter will. In addition, particle storms from flares and mass ejection shocks will be much more intense than for SOHO and may be more frequent.

The major concern that this is generating is the impact on detector systems. In particular, the adverse effects from hadrons in the CCD lattice, effecting charge transfer processes means that another option is being sought. We need a detector that is radiation tolerant.

FIGURE 4.51. An early image from the 4k × 3k 5 μm pixel APS prototype (courtesy Nick Waltham, RAL).

Another consideration which is related to this is the fact that the smaller the detector pixels, the smaller the instrument. The current baseline is for a 4k × 4k pixel array of 5 μm pixels, and such a small pixel also precludes a CCD option.

Finally, the drive for low mass suggests that on-chip electronics would be favoured.

One option which appears to cater for these requirements is the Active Pixel Sensor (APS) detector. Such devices are being suggested for several Orbiter instruments as well as other missions.

An APS is a CMOS device. It is a silicon image sensor with pixels, as is a CCD. Its wavelength coverage is the same as a CCD. The difference is the fact that the charge is sensed inside the pixel. A smaller CMOS processing geometry leads to smaller pixels, which can lead to smaller instruments. CMOS allows on-chip readout circuitry, providing a lower mass and lower power than CCDs. There is no large-scale charge transfer across an APS device, which means that it is more radiation tolerant than a CCD.

These devices are being developed specifically with Solar Orbiter in mind. A 512 × 512 pixel array prototype with 25 μm pixels has been produced and back-thinned for EUV sensitivity. Subsequently, a 4k × 3k 5 μm array has been manufactured. An early image is shown in Figure 4.51.

The back-thinning process is of critical importance. This has been tested, in visible light, for the 512 × 512 APS device. This showed that the thinning process works, but etching down to the so-called epitaxial layer, at a thickness of only 3 μm was difficult. For the 4k × 3k APS this layer is 8 μm, therefore making the process more manageable. It is, of course, anticipated that a 4k × 4k device of 5 μm pixels will be demonstrated, for EUV sensitivity, by the time of the announcement of opportunity for the payload, which is expected no earlier than 2006.

Clearly, a consideration of a number of issues, including coping with thermal and particle extremes, instrument autonomy, high resolutions, low mass and telemetry are the major drivers for the instrumentation on Solar Orbiter, quite apart from the detailed scientific requirements. For other missions there are different drivers, though issues such as telemetry and mass frequently recur. However, it is clear that for any mission there are trade-offs and for many missions there are requirements for novel technology.

4.9. Closing words

The EUV/UV region has provided a rich harvest for the investigation of space plasmas in particular. Although the field was opened in 1946, we are still witnessing startling results and major advances at this time. The aim of this review was to not only provide an historical view of missions, instruments and results, but also outline the emission processes of the EUV/UV region, the scientific methods which the EUV/UV provides and the techniques required to exploit them. We also glanced into future opportunities and explored, in particular, the case of Solar Orbiter to illustrate how missions and instruments develop. The review should serve to indicate that we anticipate significant advances in coming years in terms of missions, instruments and observational results. This is an exciting field in space physics and there is much more to come . . . so, watch this space!

Acknowledgements

The author wishes to acknowledge the generosity of NASA and ESA, and other groups named in the text, for permission to reproduce images. In addition, he would like to thank the organizers of the XV IAC Winter School.

REFERENCES

ARNAUD, M. & RAYMOND, J.C. 1992 Iron ionization and recombination rates and ionization equilibrium. *Astrophys. J.*, **398**, 394–406.

ARNAUD, M. & ROTHENFLUG, R. 1985 An updated evaluation of recombination and ionization rates. *Astron. Astrophys. Suppl. Ser.*, **60**, 425–58.

BEHRING, W.E. 1970 A spectrometer for observations of the solar extreme ultraviolet from the OSO-1 satellite. *Appl. Optics*, **9–5**, 1006–12.

BERTAUX, J.L., KYROLA, E., QUENERAIS, E. & 22 CO-AUTHORS 1995 SWAN: A study of solar wind anisotropies on SOHO with Lyman alpha sky mapping. *Solar Phys.*, **162**, 403–39.

BOGGESS, A., CARR, F.A., EVANS, D.C. & 30 CO-AUTHORS 1978 The IUE spacecraft and instrumentation. *Nature*, **275**, 372.

BOWYER, S. & MALINA, R.F. 1991 The Extreme UV Explorer mission. *Adv. Space Res.*, **11**, 205–16.

COMBI, M.R., REINARD, A.A., BERTAUX, J.L., QUEMERIS, E. & MAKINEN, T. 2000 SOHO/SWAN observations of the structure and evolution of the hydrogen Lyman–alpha coma of comet Hale–Bopp. *Icarus*, **144**, 191–202.

DANFORTH, C., SANKRIT, R., BLAIR, W., HOWK, C. & CHU, Y.H. 2003 Far-UV and H-alpha spectroscopy of SNR 0057-7226 in the Small Magellanic Cloud H II region N66. *Astrophys. J.*, **586**, 1179–90.

DELABOUDINIERE, G., LANDINI, M. & MASON, H.E. 2002 Spectroscopic diagnostics of stellar transition regions and coronae in the XUV: AU Mic in quiescence. *Astron. Astrophys.*, **385**, 968.

DEL ZANNA, G., LANDINI, M., MASON, H.E. 2002 Spectroscopic Diagnostics of Stellar Transition Regions and Coronae in the EUV: AU Mic in Quiescence, *Astron. Astrophys.* **385**, 968–985.

DWIVEDI, B.N., MOHAN, A. & WILHELM, K. 2003 Vacuum UV emission line diagnostics for solar plasmas. In: *Dynamic Sun*, B.N. Dwivedi (Ed.), pp. 353–73. Cambridge University Press, Cambridge, UK (ISBN 0 521 81057 4).

EDDY, J.A. 1979 A New Sun: The Solar Results from Skylab. (Ed.), NASA, Washington DC.

HANDY, B.N., BRUNER, M.E., TARBELL, T.D., TITLE, A.M., WOLFSON, C.J., LaFORGE, M.J. & OLIVER, J.J. 1998 UV observations with the Transition Region and Coronal Explorer. *Solar Phys.*, **183**, 29–43.

HARRISON, R.A., SAWYER, E.C., CARTER, M.K. & 36 CO-AUTHORS 1995 The coronal diagnostic spectrometer for the solar and heliospheric observatory. *Solar Phys.*, **162**, 233–90.

HARRISON, R.A., FLUDRA, A., PIKE, C.D., PAYNE, J., THOMPSON, W.T., POLAND, A.I., BREEVELD, E.R., BREEVELD, A.A., CULHANE, J.L., KJELDSETH-MOE, O., HUBER, M.C.E., & ASCHENBACH, B. 1997 High resolution observations of the extreme UV Sun. *Solar Phys.*, **170**, 123–41.

HARRISON, R.A., LANG, J., BROOKS, D.H. & INNES, D.E. 1999 A study of EUV blinker activity. *Astron. Astrophys.*, **351**, 1115.

HARRISON, R.A., HOOD, A. & PIKE, C.D. 2002 Off-limb EUV line profiles and the search for wave activity in the low corona. *Astron. Astrophys.*, **392**, 319–27.

HARRISON, R.A., HARRA, L.K., BRKOVIC, A. & PARNELL, C.E. 2003 A study of the unification of quiet-Sun transient event phenomena. *Astron. Astrophys.*, **409**, 755–64.

HOVESTADT, D., HILCHENBACH, M., BURGI, A. & 30 CO-AUTHORS 1995 CELIAS – charge, element and isotope analysis system for SOHO. *Solar Phys.*, **162**, 441–81.

INNES, D.E., INHESTER, B., AXFORD, W.I. & WILHELM, K. 1997 Bi-directional plasma jets produced by magnetic reconnection on the sun. *Nature*, **386**, 811–13.

KOHL, J.L., ESSER, R., GARDNER, L.D. & 37 CO-AUTHORS 1995 The UV coronagraph spectrometer for the solar and heliospheric observatory. *Solar Phys.*, **162**, 313–56.

KRASNOPOLSKY, V. & FELDMAN, P. 2001 Detection of molecular hydrogen in the atmosphere of Mars. *Science*, **294**, 1914–17.

KRISS, G.A., SHULL, J.M., OEGERLE, W. & 16 CO-AUTHORS 2001 Resolving the structure of ionized helium in the intergalactic medium with the Far UV Spectroscopic Explorer. *Science*, **293**, 1112–16.

MARISKA, J.T. 1992 *The Solar Transition Region*. Cambridge University Press, Cambridge, UK (ISBN 0 521 38261 0).

MASON, H.E. & MONSIGNORI-FOSSI, B.C. 1994 Spectroscopic diagnostics in the VUV for solar and stellar plasmas. *Astron. Astrophys. Rev.*, **6**, 123–79.

MASON, H.E., YOUNG, P.R., PIKE, C.D., HARRISON, R.A., FLUDRA, A., BROMAGE, B.J.I. & DEL ZANNA, G. 1997 Application of spectroscopic diagnostics to early observations with the SOHO CDS. *Solar Phys.*, **170**, 143–61.

MOOS, H.W., CASH, W.C., COWIE, L.L. & 54 CO-AUTHORS 2000 Overview of the Far UV Spectroscopic Explorer mission. *Astrophys. J.*, **538**, L1–6.

ORAEVSKY, V.N. & SOBELMAN, I.I. 2002 Comprehensive studies of solar activity on the Coronas-F satellite. *Astron. Lett.*, **28**, 401-10.

PIKE, C.D. & HARRISON, R.A. 1997 EUV observations of a macrospicule: Evidence for solar wind acceleration? *Solar Phys.*, **175**, 457–65.

POTTASCH, S.R. 1963 The lower solar corona: Interpretation of the ultraviolet spectrum, *Astrophys. J.*, **137**, 945–966.

POUNDS, K.A. & WELLS, A.A. 1991 The UK Wide Field Camera on ROSAT – First Results, *Adv. Space Res.* Vol. 11, **11**, 125–134.

TOUSEY, R., BAUM, W.A., JOHNSON, F.S, OBERLY, J.J., ROCKWOOD, C.C. & STRAIN, C.V. 1946 Solar UV spectrum to 88 km. *Phys. Rev.*, **70**, 781–2.

VAN ALLEN, J.A., FRASER, L.W. & FLOYD, J.F.R. 1948 The aerobee sounding rocket – a new vehicle for research in the upper atmosphere. *Science*, **108**, 746–7.

WILHELM, K., CURDT, W., MARSCH, E. & 13 CO-AUTHORS 1995 SUMER – solar UV measurements of emitted radiation. *Solar Phys.*, **162**, 189–231.

WILHELM, K., LEMAIRE, P., CURDT, W. & 16 CO-AUTHORS 1997 First Results of the SUMER Telescope and Spectrometer on SOHO. *Solar Phys.*, **170**, 75–104.

WOODGATE, B.E., & TANDBERG-HANSSEN, E.A. 1980 The UV spectrometer and polarimeter for the Solar Maximum mission. *Solar Phys.*, **65**, 73–90.

5. The Luminosity Oscillations Imager, a space instrument: from design to science

By THIERRY APPOURCHAUX

European Space Agency, Advanced Concept and Science Payloads Office,
Noordwijk, The Netherlands
Present address: Institut d'Astrophysique Spatiale, Université Paris XI - C.N.R.S,
Bâtiment 121, 91405 Orsay Cedex, France

The Luminosity Oscillations Imager (LOI) is a part of the VIRGO instrument aboard the Solar and Heliospheric Observatory (SOHO) launched on 2 December 1995. The main scientific objectives of the instrument were to detect solar g and p modes in intensity. The instrument is very simple. It consists of a telescope making an image of the Sun onto a silicon detector. This detector resolves the solar disk into 12 spatial elements allowing the detection of degrees lower than seven. The guiding is provided by two piezoelectric actuators that keep the Sun centred on the detector to better than 0.1''. The LOI serves here as an example for understanding the logical steps required for building a space instrument. The steps encompasses the initial scientific objectives, the conceptual design, the detailed design, the testing, the operations and the fulfilment of the initial scientific objectives. Each step is described in details for the LOI. The in-flight and ground-based performances, and the scientific achievements of the LOI are mentioned. When the loop is looped, it can be assessed whether a Next Generation LOI could be useful. This short course can serve as a guide when one wishes to propose a space instrument for a new space mission.

5.1. Introduction

The building of a space instrument is not a simple affair. In order to justify the building of a space instrument, one should demonstrate that there is a definite need implying that the scientific objectives cannot be achieved from the ground. It is sometimes easier to justify going to space when light from specific spectral region Extreme Ultraviolet, Ultraviolet (EUV, UV) does not reach the ground. In the case of an optical instrument, other arguments are required for going to space. For instance, the absence of atmosphere can considerably improve image quality for there is no seeing in space. With the advent of adaptive optics, a large range of space mission may not be required anymore. Nevertheless, the stability and homogeneity of the images obtained from space cannot be matched by those of the ground. In recent years with the flourishing of helio- and astero-seismology, other needs related to the measurement of light-intensity fluctuations became a sound justification for going to space.

In this lecture, I would like to guide the reader in all the steps, (or in as many as I can), that are necessary for achieving Science with a space instrument. To that end, I will take as an example the instrument I built for the Solar and Heliospheric Observatory (SOHO): the Luminosity Oscillations Imager (LOI).

5.2. From science to instrument building

The design and building phase of any instrument can be summarized by Fig. 5.1. There are two distinct phases: the design and the building of the instrument. The first phase includes the following steps:

(*a*) Scientific objectives
(*b*) Scientific requirements

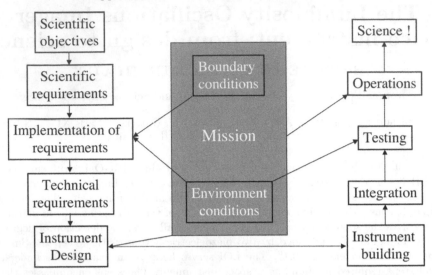

FIGURE 5.1. The burden of the instrumentalist: from concept to building. The flow chart on the left-hand side summarizes the conceptual phase of the design of the instrument. The flow chart of the right-hand side summarizes the building phase of the instrument. The difference between a ground-based and a space-based instrument is exemplified by the grey box in the middle. The boundary conditions are different.

    (*c*) Implementation of requirements
    (*d*) Technical requirements
    (*e*) Instrument design.

Step (a) defines what kind of science one wishes to achieve. Step (b) defines what kind of observables is needed in order to fulfil Step (a). The observables can be images in selected wavelength, spectrum analysis in any wavelength, polarimetry, particle distribution, and it can be any information carried out by photons or any other particles (protons, neutrons, electrons, neutrino). Step (c) defines the concept of the instrument that will return the observables. It is not yet a detailed instrument design but more a conceptual approach on the instrument: spectrograph, narrow band spectrum analyzer, polarimeter, particle counter, imager. Step (d) transforms the concept into an instrument and imposes technical requirements on the instrument such as temperature stability, resolution, field-of-view. At this step this is where one usually performs a tradeoff analysis between different technical solutions. Finally, Step (e) sets the detailed design of the instrument into an instrument to be assembled. All the technical requirements are then translated into building and interface requirements. The second phase or building phase includes the following steps:

    (*a*) Instrument building
    (*b*) Integration
    (*c*) Testing
    (*d*) Operations
    (*e*) Achievements of the objectives: science.

Step (a) sets the building requirements and the interface requirements between the different parts to be assembled. These can be mechanical, electrical or thermal interfaces. Step (b) is the integration of the building block of the instrument that should fit together given the interface requirements. Step (c) is the test of the whole instrument or part of the instrument. The testing of the parts can be done before integration. The testing also involves calibration of the instrument and environmental testing if required. Finally, Step

(d) is the successful operation of the instrument that will provide in the end the Science in Step (e). This last step can then be compared with the initial scientific objectives set before building the instrument. The achievements (or lack of) will then be used for building the Next Generation of instruments: more demanding and more performing.

5.3. What is different about a space instrument?

The two phases described above are not necessarily different for a ground- or space-based instrument. The main difference lies in the boundary and environment conditions that are very different for a space mission compared to a ground-based observatory. The boundary conditions are the following:

(a) mass (kg),

(b) volume (cm^3),

(c) power (W),

(d) telemetry (kbps),

(e) long lifetime (years),

(f) money (euro).

The mass of the instrument is limited by the number of instruments making the payload of the mission and by the capability of the launcher used for putting the spacecraft into space. The volume is limited for it has to fit somehow inside the fairing of the launcher. The power is also limited by the number of instruments and by the size of the solar array. The telemetry rate, the amount of data to be transmitted to the ground, is also limited by the number of instruments, by the power of the transponder, by the distance of the spacecraft to the ground and by the size of the receiving antenna. Last but not least, although it is commonly admitted that space instrumentation is more expensive than ground-based instrumentation, it is nevertheless required to build the instrument within a limited budget. The higher cost is related to the fact that repair is not an option in space† and that high-reliability parts should be used; this will ensure a long lifetime. All these boundary conditions are taken into account during the implementation of the requirements and at all latter steps till the instrument design.

The environment conditions are the following:

(a) launch loads,

(b) thermal interfaces and vacuum,

(c) electromagnetic interfaces,

(d) energetic particles,

(e) solar irradiations (or from other bodies).

During the launch, random vibrations created by a combination of acoustic noise and structural deformations are going to impact onto the physical integrity of the instrument. Random loads can be up to 20 g. If care is not taken about internal resonance in the instrument, it is likely that much higher load of about 100–200 g are likely to break critical part of the instrument. When in space, the instrument has to establish a thermal equilibrium that is achieved by radiative and conductive exchange; there is no more luxury for convection cooling in vacuum. Electromagnetic emission and compatibility should ensure that electric noise will be set to a minimum. This point can be very critical to instruments measuring electric or magnetic fields in space. Energetic particles (protons, electrons, heavy ions) will affect the functioning of analog an digital electronics by creating performance degradation or even transient disruption (single event upset or latch

---

†The Hubble Space Telescope was made repairable and upgradable but it should not be formally ranked as a space instrument.

ups). Finally, solar irradiations (or from other bodies) will radiatively heat the instrument but will also affect the optical properties of glass materials and thermal coatings. These environment conditions set the scene for the testing phase that will be significantly different from that of a ground-based instrument. The following tests have to be performed:

- Vibrations (Step a)
- Thermal vacuum (Step b)
- Electromagnetic Interference (EMC)/Electromagnetic Compatibility (EMI) (Step c)
- Irradiations (Step d and e)
- Thermal balance (Step e).

When the first four steps are passed with success that is, the instrument performs nominally after these stringent tests, the instrument is then so-called space qualified. It means that the instrument should nominally operate during the mission. The last test is usually not performed at the instrument level but when the overall spacecraft is built. It consists in simulating the illumination of the Sun in a realistic thermal environment in vacuum.

The building of a space instrument requires more attention than any other type of instrument. The amount of product assurance (i.e. paper work) is such that each single change needs to be recorded, documented and approved. This additional work (or burden) is an insurance that everything that needs to be known about the instrument can be latter used in case of malfunction during the mission. This is the only way that the instrument can be *repaired* in flight.

5.4. The Luminosity Oscillations Imager

5.4.1. *Scientific objectives*

In the mid-80s, helioseismology was emerging as a powerful tool to infer the internal structure of the Sun. The observations of many eigenmodes of the Sun was starting to enable inferences about the internal structure of the Sun. The Tenerife conference held in Puerto de la Cruz in September 1988 was a major milestone for helioseismic research European Space Agency (ESA SP-286). This conference provides a proper snapshot of the knowledge derived from helioseismology at the time. This conference was held right after the selection of the payload for the SOHO mission. At the time only the p modes had been detected. The p modes were observed in solar radial velocity with amplitude of about a few cm/s and periods of about 5 min. The p modes propagate from the visible surface down to $0.2\,R_\odot$ enabling to infer the structure and dynamics of the Sun from the surface to below the convection zone ($r \approx 0.4\,R_\odot$). The undetected g modes are confined under the convection zone. From that property, it was clear that the detection of g modes would have considerably improved our knowledge of the structure and dynamics of the solar core. The detection of g modes had been claimed by Delache and Scherrer (1983) but confirmations were still seek by van der Raay (1988) (and references in ESA SP-286). The g-mode detection was one of the main scientific objectives of the helioseismic instruments aboard SOHO. There were three investigations aboard SOHO aiming at detecting the elusive g modes: GOLF (Global Oscillations at Low Frequencies) (Damé, 1988), SOI/MDI (Solar Oscillations Investigation/Michelson Doppler Imager) (Scherrer *et al.*, 1988), and VIRGO (Variability of solar IRradiance and Gravity Oscillations) (Fröhlich *et al.*, 1988a). A sub-unit of VIRGO was specially dedicated to the detection of g modes: the LOI.

5.4.2. *Scientific requirements*

The main scientific requirements of the LOI were the following:

- detect low-degree oscillations modes (p and g modes at $l < 6$),
- detection of the modes in intensity fluctuations,

- determine the frequencies, amplitudes for eigenmodes in the frequency range of $1 \mu\text{Hz}-10\,\text{mHz}$,
  - solar noise limited,
  - uninterrupted observations.

The degree of a mode (l) is one of the quantum number of a spherical harmonics that represents the spatial distribution of the eigenfunctions. At the time, it was thought that only the low-degree g modes would have a significant amplitude (Berthomieu and Provost, 1990). In addition, it was believed that the modes needed to be detected not only in solar radial velocities (GOLF, MDI) but also in intensity; the difference in noise generated by convection would help to detect the faint g-mode signals. For instance, the amplitudes to be detected were about a few part-per-million (ppm) for the p modes, and predicted to be about a few tenth of ppm for g modes. Given the mode frequencies, a rather stable instrumentation was required for time period ranging from a few hours to a few minutes. The aim was also to be limited not by the instrumental noise but by the solar noise. At the time, the solar noise had been measured in intensity by ACRIM aboard the Solar Maximum Mission by Woodard (1984) and by the Inter Planetary Helioseismology with Iradiance Observation (IPHIR) instrument aboard Phobos by Fröhlich et al. (1988b) (See Fig. 5.8).

5.4.3. *Implementation of requirements*

The scientific requirements given above imposed several constraints on the instrument design that is even more severely constrained by the various boundary and environment conditions of the SOHO mission. The implementation of the requirements leads to the following design:

- low-resolution detector,
- image of the Sun (telescope),
- internal pointing,
- observe solar continuum avoiding variable Fraünhofer lines,
- stable detector,
- stable acquisition system.

The detector should have tens of pixels in order to detect the low-degree modes, requiring therefore an imaging capability for the instrument. In order to compensate for likely jittering of the image on the detector due to spacecraft motion and misalignment, an internal pointing system should keep the Sun centered on the detector. The latter should have a quantum efficiency as insensitive as possible to temperature changes. In addition, the acquisition system (analog to digital conversion) should not degrade the performance of the overall instrument. The instrument to be built following these requirements will be subject to several external conditions. The main boundary conditions are the following:

- mass less than 2.5 kg;
- power less than 2.5 W;
- volume: $30 \times 9 \times 9 \,\text{cm}^3$;
- telemetry less than 10 bits/s;
- pointing:
  - absolute better than 5 arcmin,
  - short term: relative better than 1 arcsec over 15 min,
  - medium term: relative better than 10 arcsec over 6 months.

Some of these conditions are somewhat easy to achieve (telemetry). One should keep in mind that the instrument will be subject of various environmental conditions as given

by the Experiment Interface Document Part A (spacecraft requirements also known as
the SOHO EID-A). They read as follows:

- nominal mission lifetime: 2 years,
- orbit: halo orbit around L1 Lagrangian point of the Sun-Earth system,
- dose rate for nominal mission: less than 10 krads behind 2-mm Aluminium shield
for a 2-year mission,
- stringent cleanliness requirements,
- launch loads of 12 g's rms all axes,
- temperature range: 0–40°C.

The lifetime of the mission is set by the cost of operating a spacecraft and by the achievements of the scientific objectives. The orbit chosen for the SOHO here ensures a continuous viewing of the Sun that is so essential for helioseismic investigations. The dose rate
is the energy deposited by electrons, protons and heavy ions particles that are likely to
damage optics and electronics alike. The cleanliness requirements are imposed by the UV
instruments aboard SOHO that are extremely sensitive to any polymerization of hydrocarbon contaminants by the solar UV and blue light (Bonnet *et al.*, 1978). Loads during
the launch plays also a major rôle in the design of the instrument. The temperature
variations are mainly set by the orbit chosen and by the thermal control performed by
the spacecraft.

5.4.4. *Technical requirements and design*

5.4.4.1. *General concept*

The design of the LOI instrument was an original idea of Andersen *et al.* (1988). The
instrument concept was to make an image of the Sun onto a low-resolution detector, and
to detect intensity fluctuations of the surface of the Sun. The instrument as designed
and as built was described by Appourchaux *et al.* (1995a). A cutout of the instrument is
shown in Fig. 5.3 and the conceptual design in Fig. 5.2. The LOI consists of a telescope
imaging the Sun through a narrow passband filter. The image is projected onto a photodiode array detector. The detector is made out of 12 scientific pixels and of four guiding
pixels. The shape of the scientific pixels is trimmed to detect modes for $l < 7$. The great
advantage of this configuration is that we can isolate and identify individual l, m modes
for up to $l = 6$ (Appourchaux and Andersen, 1990). The error signals given by the guiding pixels are used to keep the Sun centred on the detector using piezoelectric actuators.
The current of each scientific pixel is amplified and digitized using voltage-to-frequency
converters (VFCs). The 12 scientific pixels are read simultaneously while the four guiding
pixels are read sequentially. The sampling time of the instrument is 60 s. The integration
time of the scientific pixels is about 99% of the sampling time. Thanks to a sun that never
sets on SOHO, the temperature variations of the spacecraft will be minimal. That is why
the instrument is not thermally stabilized. It is sometimes better to have open loops than
closed loops. I will hereafter describe each main component of the LOI, these are the:

- Telescope
- Entrance filter
- Detector
- Guiding system
- Acquisition system.

For each the technical requirements and the design is detailed. The interface specifications for each will be quickly explained. These latter specifications are usually very
technical and drafted by the engineering team. Nevertheless, these interactions between
that team and the instrument scientist is key in developing a sound instrument.

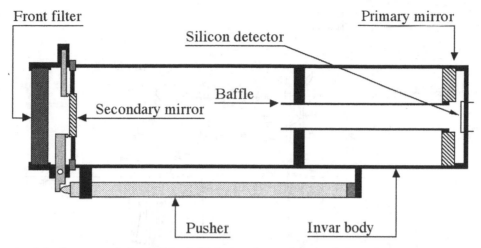

FIGURE 5.2. Conceptual view of the LOI design. It can be noted that the cylindrical telescope tube is closed by the entrance filter and by the detector. The only openings are the necessary gaps for the lever arm pushing on the secondary mirror mount.

5.4.4.2. *Telescope: requirements*

The telescope design was driven by the following set of constraints:
- image size,
- compactness,
- limited aberrations,
- stable focal length,
- pointing needs.

The size of the image was driven by the solid-state detector. We chose to have a large detector in order to minimize the size of the tracks (see Section 5.4.5). At the time, in the 90s, a detector based on a silicon wafer of 15 mm square was possible without pushing the technology. The mean solar image size was chosen to be 13-mm diameter. This will set the constraint for the focal length needed for the observations. It should be pointed out that the location of the SOHO spacecraft at L1 is such that the solar angular size is on average 1% larger. The focal length was chosen to be $f = 1300$ mm providing a mean solar angular size of $D = 12.22$ mm. These two numbers will drive the design of the telescope and of the size of the detector.

The compactness of the telescope was an important constraint. The overall instrument including electronics should be fitted in a box 35-cm long by 6-cm^2. Given the large focal length and the size of the telescope this imposed to have a two-mirror telescope with a primary and a secondary mirror.

Like in any optical design, it is crucial to minimize any sort of aberrations. From this point of view, a simple Cassegrain telescope with a parabolic primary mirror and an hyperbolic secondary mirror is not sufficient. The choice was to have a Ritchey–Chrétien with hyperbolic mirrors. This combination has the advantage of minimizing the coma on the edge of the field (Chrétien, 1958).

Given the sensitivity of the design to the intermirror distance, it is required that the material making the cylinder holding the mirror be insensitive to change in temperature.

Last but not least, the telescope must be able to correct for pointing offset as large as 7 arcmin with a stability better than 0.1 arcsec over 60 s. The pointing correction can be achieved either with the primary mirror or the secondary mirror. Given the requirement on the compactness of the instrument, it is easier to move the secondary

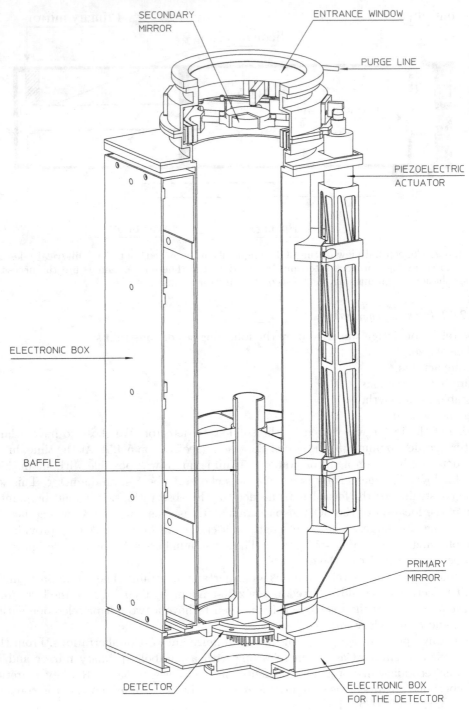

FIGURE 5.3. Cut-out of the LOI. The electronic box is located at the outside of the tube. The separation is needed for minimizing the contamination from the outgassing electronics parts and coatings.

mirror. The choice of the location of the rotation point was also driven by the available room. Although a rotation of the secondary mirror about the focus of the primary mirror would provide a symmetrical aberration (around the chief ray) and a lack of coma, we chose to rotate the secondary mirror around its vertex.

5.4.4.3. *Telescope: design*

The telescope is a Ritchey–Chrétien with a 1300-mm focal length. The effective diameter of the telescope is 54 mm, with a central obstruction of 26-mm diameter. The concave primary has a focal length of 289.63 mm and an aspherization of −1.02 (parabolic). The convex secondary has a focal length of 70.30 mm and an aspherization of −2.7 (hyperbolic). The focal length f of the mirror is then given by:

$$f = \frac{R}{2} \tag{5.1}$$

The shape of the mirrors is given from the following equation of a conic:

$$r^2 + 2Rz + (1+k)z^2 = 0 \tag{5.2}$$

where r is the radial distance to the mirror vertex, R is the curvature at the vertex, z is the height on the revolution axis (i.e. often the normal to the mirror) and k is the conic constant. It can easily be shown from Equation (5.2) that the surface is a sphere of radius R for $k=0$ and a paraboloid for $k=-1$. When $-1<k<0$, we have an ellipsoid. When $k<-1$, we have a hyperboloid. Equation (5.2) can be solved to yield z as a function of r:

$$z = \frac{r^2}{R} \frac{1}{1 + \sqrt{1 - \frac{s^\cdot (1+k)}{R^\cdot}}} \tag{5.3}$$

The focal length f of the telescope depends upon the focal length of the mirrors (f_p and f_s) and on the intermirror distance d:

$$\frac{1}{f} = \frac{1}{f_p} + \frac{1}{f_s} - \frac{d}{f_s f_p} \tag{5.4}$$

The focal length is positive for a concave mirror, and negative for a convex. The imposed focal length provide an intermirror distance of 235 mm.

There are several tolerances that will constrain the telescope design. For instance, we wish to impose that the focal length does not vary by more than ±0.5 due to temperature variations. In addition, the location of the focal plane is to be at 10 mm from the vertex of the primary mirror; this location must not vary by more than ±1 mm in order to minimize the change in the point spread function (PSF). It can be derived from Equation (5.4) that the sensitivity of the focal length to the intermirror distance is given as:

$$\frac{\Delta f}{f} = \Delta d \frac{f}{f_s f_p} \tag{5.5}$$

It can also be computed that the distance e of the focal plane from the vertex primary mirror is:

$$e = \frac{(f_p - d)f_s}{f_s + f_p - d} - d \tag{5.6}$$

From Equation (5.6), the sensitivity of e to the intermirror distance is given by:

$$\Delta e = \Delta d \left(\frac{-f_s^2}{(f_s + f_p - d)^2} - 1 \right) \tag{5.7}$$

The most stringent requirement from the two is the location of the focal plane that will impose a variation to the intermirror distance not larger than ±0.05 mm. Given the

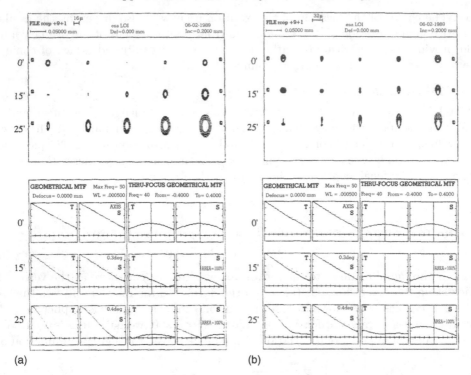

FIGURE 5.4. (a) Spot diagram and MTF of the telescope for an on-axis image, for different location in the field of view, and for various focal plane location. (b) Same diagrams for an off-axis image compensated by a tilted secondary mirror. Spot diagram and MTF can be obtained with a ray-tracing programme such as ZEMAX.

temperature variations encountered by the spacecraft of about 20 K, this will impose the maximum Coefficient of Thermal Expansion (CTE) that we can tolerate for the structure of the telescope, that is 10^{-5}/K. The chosen material for the structure is Invar with a CTE of 0.5×10^{-6}/K, a factor 18 better than required. Invar is a special type of stainless steel having a higher content of nickel (36%) providing a very low CTE.

This also force to choose mirror material with a low CTE in order to reduce the differential CTE. To minimize this effect, the mirrors are made of ceramic glass Zerodur with a CTE of 10^{-7}/K. The primary mirror is registered on three small Invar pads of 2-mm diameter; they are built in the Invar tube. At the back of the primary mirror, opposite of the pads, three 2-mm diameter pins apply each a 10-N force; this insures that the mirror is kept in place for accelerations up to 30 g. The secondary mirror is mounted with three similar pads, but instead of a known force, the ring supporting the pads is just lightly clamped and its holding screws are glued. The secondary mirror is mounted in a gimbal allowing to keep the Sun centred on the detector (see Section 5.4.6.1). The design is such that the Modulation Transfer Function (MTF) of the telescope is not degraded even for a 10' offset of the Sun (see Fig. 5.4).

Although, we had no requirement for scattered light, we studied the baffles of the telescope using the APART programme. The main source of scattered light will come from the mirror themselves for which the baffles are of limited use. The design of the internal baffle was such that no direct light from the Sun could reach the detector. The baffle was held in place with a spider in the shadow of the secondary mirror spider. In order to minimize scattered light from the telescope tube, the inside was machined to reduce the surface of the cylinder by removing material in the form of a cutting saw.

The tube (and the secondary gimbal) were painted with a low-outgassing paint known as Electrodag 501.

5.4.4.4. *Filter: requirements*

The design of the entrance filter was to have the following requirements:
- Narrow band (5 nm) at 500 nm
- Entrance window warmer than the structure
- Minimize radiation damage
- Controlled transmission
- Minimize out-of-band light contribution.

The spectral selection is performed by a 5-nm passband interference filter at 500 nm, used as an entrance window. The wavelength and passband was chosen as to minimize the amount of solar Fraünhofer lines likely to be sensitive to solar activity. The filter closes the Invar tube and isolate the mirrors and detector from the outside (see Fig. 5.2). The thermal design of the filter must insure that its temperature is hotter than the rest of the structure. This will reduce the risk of condensing contaminants coming from the space-craft, hence reducing the chance of a drastic drop in transmission due to polymerization. A temperature difference of about 10 K is sufficient to avoid condensation.

Since the filter faces directly the solar particle flux, the radiation damages to the filter should be minimized. The design of the filter should take into account both the thermal and the radiation aspects. The admissible degradation over the nominal mission lifetime should be lower than 90% transmission drop.

The transmission of the filter should be limited in order to avoid reaching the non-linear response of the detector. The transmission will depend upon the choice of material used for the detection, and upon the design of the detector.

Finally, since the solar flux is broadband, the flux outside the narrow passband should be reduced by 10^{-5} compared to the in-band transmission.

5.4.4.5. *Filter: design*

A transmitting filter with a narrow passband of 5 nm is achieved with multilayer coat-ings making constructive and destructive interferences. The choice of the wavelength is achieved in a similar way as in Fabry–Perot filter which transmission peaks are given by:

$$k\lambda = 2ne \cos i \tag{5.8}$$

where k is an integer, λ is the peak wavelength, n is the index of refraction of the gap, e is the thickness of the gap, i is the angle of incidence. In the case of a multilayer coating, a low index of refraction material of optical thickness (ne) of half-wavelength thickness is surrounded by sandwiches of quarter-wavelength thick material. The materials used are typically zinc sulphide, cryolite† or the likes. The idea is to deposit in alternance high and low index of refraction, respectively zinc sulphide and cryolite. The peak wavelength λ_0 is then simply given by:

$$\lambda_0 = 2ne \tag{5.9}$$

The design of interference filters is out of the scope of this course, but one can read with deeper interest MacLeod (1986). The additional transmission peaks obtained at λ_0/k (at shorter wavelengths) need to be suppressed by additional blocking either performed by additional multilayer coatings and/or coloured glass.

The overall design of the filter is also constrained by the need to have a warm filter. At the time of the design, there was very little literature on the thermal design of such

†Sodium aluminium fluoride.

interference filters. In addition, the filter sandwich was also designed as to minimize the radiation dose reaching the multilayer coatings. The thermal and radiation protection designs are constrained by each other. The thermal solution, studied using ESATAN, is described by Appourchaux and Cislaghi (1992). It requires the knowledge of the thermal characteristics of the glass substrates, the transmission, absorption and reflection properties of the glass and coatings, the thermal boundary conditions and finally the thermal characteristics of the interface. The thermal design of the filter insures that its temperature is hotter than the VIRGO structure by 10 K (Appourchaux and Cislaghi, 1992). The filter is made of three different substrates, all are 3-mm thick and from Schott. The first one is a BK7G18 glass, cerium doped and radiation resistant; it is used to reduce the radiation dose and proton flux received by the other substrates and coatings. The second substrate is a GG475 coloured filter which absorbs about 20% of the incoming solar radiation, it is also used to reduce the amount of blue light received by the coating. The coatings making the spectral selection are deposited on the GG475. The last substrate is a BG18 heat absorbing glass, it is used as a back-up in case the coatings reflecting the infrared get degraded. Finally a chrome coating is deposited at the back of the BG18 glass for getting a 7% transmission; this low transmission insures that the detector is used in a linear regime. The design ensures that the out-of-band contribution is less than $10^{\cdot}$.

5.4.4.6. *Detector: requirements*

The detector was to have the following requirements:
- low resolution sensitive in the visible (@ 500 nm);
- sensitive to low-degree modes ($l < 6$);
- quantum efficiency less than 500 ppm/K;
- pixel crosstalk less than 0.5%;
- response uniformity better than 1%;
- provide guiding signals;
- low noise.

The detector is based on silicon that is sensitive in the visible. The number of pixels was limited to 12 in order to reduce the burden of the read-out electronics. The sensitivity to low-degree modes is related to the geometry of the 12 pixels. The temperature dependence of the quantum efficiency of the silicon detector should be minimized for reducing the impact of the temperature variations on the intensity measurements. For instance, a peak-to-peak variation of 1 K would provide a 80 ppm rms variation of the measured intensity or when translated in terms of a white noise background in a power spectrum 0.4 ppm^2/μHz. The crosstalk between adjacent pixel should be kept to a minimum. The detector will also provide the guiding signals thereby ensuring that the solar image will properly centered onto the detector, and that guiding noise will be minimized. Finally, the detector was used as a current generator (photovoltaic mode) and was not polarized by a bias voltage. This is for getting a lower noise and for avoiding to control a few ppm the bias voltage.

5.4.4.7. *Detector: design*

The configuration of the 12 scientific pixels shown in Fig. 5.6 was optimized by Appourchaux and Andersen (1990). The sensitivity of that configuration to the modes is shown in Fig. 5.5. Other effects related to sensitivities to guiding offset can also be found in Appourchaux and Andersen (1990).

In order to optimize the temperature dependence of the pixels, several different processes were tested (see Section 5.4.9.4). After selection of the best process, the photodiodes were deep diffused p–n junction with an epitaxial substrate (low resistivity). In addition, four temperature diodes are built in the detector for having the temperature across the

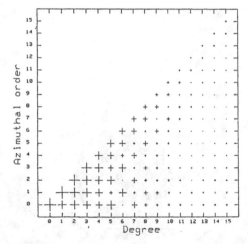

FIGURE 5.5. Response of the detector to spherical harmonics function as a function of the degree and of the azimuthal order. This configuration offered the best compromise between sensitivity to $m = l$ modes and $m = 0$ modes.

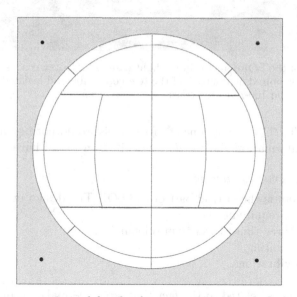

FIGURE 5.6. LOI detector optimized for the detection of low-degree modes. The guiding pixels are four quadrants of an annulus with an inner and outer radius of about 0.95 and 1.05 solar radius, respectively. Each guiding pixel is physically and electrically divided in two pixels for redundancy purposes. The four black spots at the corners are the temperature diodes. These diodes are below an aluminium layer shown here in grey.

substrate (see Fig. 5.6). These temperature diodes provides a signal in photovoltaic mode that is proportional to temperature. For improving the definition of the sensitive area of the scientific pixels, a layer of aluminium is deposited outside the pixels and over the temperature diodes. In addition, each pixel and each temperature diode is surrounded by an n^+ deep-diffused guard ring; this will minimize the crosstalk between the pixels. It will also insure that the signal of the pixels do not get contaminated by a pixel for which the bonding wire breaks; in this case the electrons flow to the ground and not the other pixels. Each guiding pixel is made of two independent pixels that are connected to

FIGURE 5.7. Photo of the LOI detector in its flight case. The Kovar case is soldered to a Kovar mounting plate that is bolted at the back of the telescope tube and isolated from the tube with an Ertapeek washer. The 100-μm wide tracks are easily seen.

the same output (Fig. 5.7). In case one of these pixels would not be functional, we would still have guiding information and the image of the Sun would have a slight offset.

5.4.4.8. *Guiding system: requirements*

The guiding system is the critical part of the LOI. The design of guiding system was to have the following requirements:

- pointing noise three times lower than solar noise,
- gross and fine pointing,
- survive launch vibrations,
- long lifetime.

The first requirement is related to the fact that since we make an image of the Sun on a low-resolution detector, the pointing jitter will introduce noise due to the solar limb darkening that can amount up to 1000 ppm/arcsec (Appourchaux and Andersen, 1990). At the times, measurements made by Inter Planetary Helioseismology with Iradiance Observation (IPHIR) aboard the Phobos spacecraft was the only reference of the level of the measured solar noise (Fröhlich *et al.*, 1988b). The typical solar noise was about 0.36 ppm^2/μHz @ 3 mHz (p-mode range), and 100 ppm^2/μHz @ 100 μHz (g-mode range) with ν^{-2} dependence to the power spectrum in that range. This would set the level of the power spectrum to 0.04 μarcsec^2/μHz @ 3 mHz (p-mode range), and 11 μarcsec^2/μHz @ 100 μHz (g-mode range). After integration over the whole spectral range, the rms pointing jitter should be lower than 0.12″, that can be roughly translated by a peak-to-peak value of 0.7″ (6 times larger). This requirement is to be achieved at the level of the detector, that is, after compensation by an eventual close loop system, and after integration by the electronics.

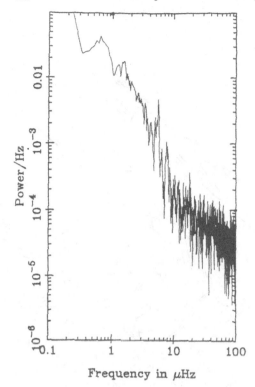

FIGURE 5.8. Power spectrum of the IPHIR instrument in the g-mode range. The power spectrum is only limited by the solar noise.

According to the SOHO EID-A, the variation of the absolute pointing between on-ground alignment and end of nominal mission is better than 5 arcmin (3σ value), while the relative pointing accuracy is better than an arcsec over 15 min, and better than 10 arcsec over 6 months. The latter is out of the requirements set by the solar noise. An active pointing system is therefore required to reduce the 10-arcsec figure to less than an arcsec. The guiding system must be able to cope with large misalignment (arcmin) as well as finer and faster jitter (arcsec).

As soon as the requirement for a mechanism is set, there is the obvious need that this mechanism be able to survive the vibrations encountered by the launch. And of course, the mechanism being needed for achieving the scientific objectives there is an additional requirement that the guiding system be able to function during the nominal mission and beyond.

5.4.4.9. *Guiding system: design*

The guiding system is based on the following principle: two piezoelectric actuators rotate two lever arms that push on a gimbal in which the secondary mirror is mounted. The rotation of this mirror is done around its vertex. Two spring arms opposite of the two lever arms provide a restoring force (about 20 N) and act as a locking mechanism (Fig. 5.9). In order to compensate for 1 arcmin off pointing on the detector, the secondary mirror needs to be rotated by 2.65 arcmin. This is due to the larger apparent size of the solar image as reimaged by the secondary mirror.

For coarse guiding, the total range needed is ± 7 arcmin. This is for allowing for the long-term drift of the spacecraft (± 5 arcmin), the temperature setting of the actuators

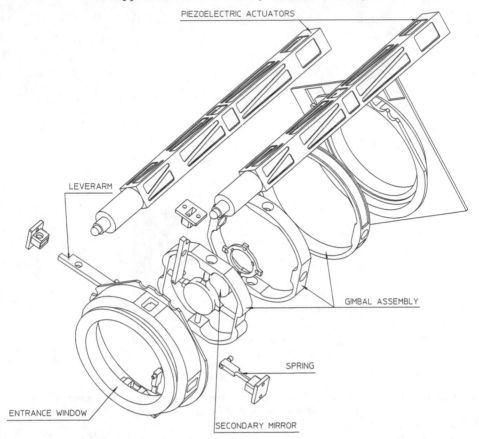

PIEZOELECTRIC ACTUATORS

LEVERARM

GIMBAL ASSEMBLY

SPRING

ENTRANCE WINDOW

SECONDARY MIRROR

FIGURE 5.9. Artistic blow-up of the guiding system. The assembly of the three rings making the gimbal is very critical. Friction between the rings modify the normal behaviour of the guider. The piezoelectric actuators are longitudinally kept in place by back screws, not shown here. For each axis, the fine adjustment is made with a screw located on the lever arm. The 20-N force is adjusted with a screw located on the spring arm.

(± 1 arcmin†) and alignment errors of the entrance filter (± 1 arcmin). The required extension d of the actuator is given by:

$$d = re\frac{A_s}{a_l}\frac{1}{C(K_s, K_l, K_p)} \qquad (5.10)$$

where r is the required range (± 7 arcmin), e is the distance of the lever arm, pushing on the secondary mirror mount, to the vertex of that mirror (=12 mm), A_s is the amplification giving the actual range at the secondary mirror (=2.65), a_l is the amplification of the lever arm (the piezoelectric actuator pushes on the other end of the lever arm with a reduction factor of 2.25), C is a reduction factor taking into account the stinesses of the actuator K_p, of the lever arm K_l and of the spring arm K_s. Unfortunately, the varying load seen by the actuator will reduce the effectiveness of the extension. The reduction factor is given by:

$$C(K_s, K_l, K_p) = \frac{K_l}{K_l + K_s}\frac{K_p}{K_p + \tilde{K}} \qquad (5.11)$$

†Corresponding to 0.5 μm/K.

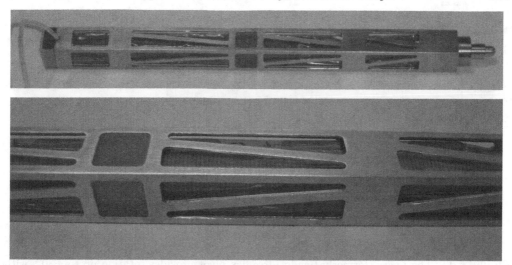

FIGURE 5.10. Photos of the piezoelectric actuators. Top, the actuator is about 230-mm long. For saving mass the case is ribbed. The actuator is held laterally at the back by a screw pushing into an hemispherical cut-out, in the middle by a bracket and at the top by a square cut-out in the mounting frame. The ball end of the actuator pushes onto a flat part of the lever arm. Bottom, enlarged view of the central part of the actuator. This is where one can see each disc making the actuator, and also where the stack is held by a potting material. This will considerably reduce the amplitude of the lateral motion of that part during vibrations.

where $\tilde{K}$ is the equivalent stiffness of the spring arm at the level of the piezoelectric actuator given by:

$$\tilde{K} = a_l^2 \frac{K_s K_l}{K_l + K_s} \qquad (5.12)$$

The guiding system had also to be designed bearing in mind that it must survive launch vibrations. The rôle of the spring arm is to provide a restoring force as well as to serve as a locking mechanism. The reduction of extension as given by Equation (5.11) was imposed to be less than 15%. This sets the requirements on the stiffness of the spring arm and lever arm system. Therefore according to Equation (5.10), the extension of the actuator should be at least 68 μm.

There are two additional design constraints that needs to be taken into account: the operating voltage and the failure scenario. These two constraints are related to the design of the piezoelectric actuator themselves. The actuator are made of about 240 elements of lead zirconium titanate (PZT) that are about 1-mm thick each. They are soldered back-to-back with 240 electrodes providing the electrical polarization for each element. The length of the 240-disc stack is about 230 mm. To avoid to break the stack during launch vibrations, the middle of the stack is strapped to the case over 10 mm; this design is qualified for the launch of SOHO (see Fig. 5.10). The failure mode of a single element will give a short circuit. Since all the elements are fed in parallel this would destroy the output of the voltage power supply. In order to avoid such a failure modes, it was decided to construct the actuator as a sum of 12 independent actuators by putting a resistor in series of the 20 elements. The failure of one element would reduce the extension by 1/12. Therefore, the extension of the actuator should be at least 74 μm. Last, the maximum operating voltage of the actuator is typically 700 V. At the time of design, lower voltage, say 100 V, would have implied thinner piezoelectric elements in the same ratio. These were considered still in the development stage for space application and the choice of having low-voltage was therefore not an option. We chose to operate for derating reason at a

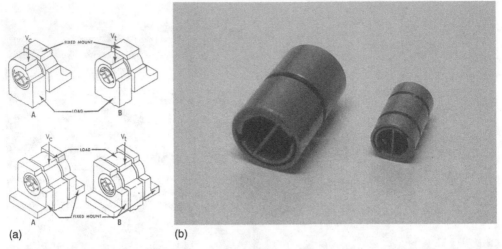

(a)   (b)

FIGURE 5.11. The Lucas (formerly Bendix) free-flex pivot provides rotation up to ±30°, without friction. The pivot is made of flat, crossed springs supporting rotating sleeves. There are two kinds of pivot: cantilevered (used in the gimbal, cartoon at left, top; and photo at left); double-ended (used for the lever arms, cartoon (a), and photo (b)).

maximum voltage of 600 V. This then induced a further extension reduction compared to the 700 V operation. The final requirement for the actuator extension was then 86 μm at 700 V.

The force applied by the spring arm is used as a locking mechanism. This is a passive lock and there is nothing to unlock. The complete guiding systems comprising the spring and lever arms, the two piezoelectric actuators, the gimbal rings (inner and outer) was subject to a finite elements model (f.e.m.) analysis performed with NASTRAN. This analysis permits to describe the mechanical behaviour of the system under various static and dynamical loads. The goal of that analysis was to assess whether the guiding system would survive launch (e.g. avoid gapping at the actuator level that could cause destructive shocks), and estimate the resonance frequencies of the guiding system that could lead to instabilities for the closed loop system. The f.e.m. analysis confirmed that the spring arm would effectively act as a lock, and that the first two resonance frequencies (250 and 380 Hz) would be high enough for not being excited by the servo loop system.

For the fine guiding, the error signals, provided by the guiding pixels, are fed back to the high-voltage power supply that drives the actuators. The guiding regulator is an integrator. The cut-off frequency of the complete guiding system is 10 Hz @ 3 dB.

Finally, the long lifetime was ensured by using Bendix free-flex pivots that provide rotation using a flexure mechanism. In case of a space application, ball bearings are to be avoided at all cost to the long continuous operation and likely wear occurring in such mechanism. In addition, ball bearing are considered dirty because of the use of a lubricating medium likely to outgas hydrocarbons. The free-flex pivots are shown in Fig. 5.11.

5.4.4.10. *Electronics: requirements*

The electronics was to have the following requirements:
- low noise,
- long integration time,
- high-voltage power supply,
- guiding servo loop,
- provide housekeeping signals.

The first requirement is obviously related to the tiny variation intensity measured on the Sun (ppm).

The second requirement is related to the sampling criterion of Shannon. When one wants to reconstruct an analog signal with a digital one, it is required that the sampling time be twice shorter than the highest frequency to be recovered (i.e. if one wants to access the highest frequency corresponding to a 2-min periodic signal) one must sample the signal to 1 min. Unfortunately, it is never mentioned anywhere how the sampling should be performed. As a matter of fact, a shorter integration time compared to the sampling time will considerably "fold" in high-frequency noise. It is then essential that the integration time during the sampling interval be optimized to be as long as feasible, (i.e. the integration cycle should be close to 100%.)

The electronics should also provide the guiding analog loop by taking the error signals from the detector and fed them back to the actuators via a high-voltage power supply driving them. The overall design of a control loop is well known but still great care should be taken for avoiding to have an unstable closed loop system.

Last several housekeeping signals such as temperatures, high voltages are needed to make sure that the instrument performs as expected.

5.4.4.11. *Electronics: design*

For low noise, the 12 scientific pixel currents are converted into voltages by 12 low-noise operational amplifiers (AD-OP97) in a zero-bias configuration. This latter will minimize the $1/\nu$ noise. The analog-to-digital conversion is performed by synchronous VFCs (AD652). VFC are intrinsically more stable than analog-to-digital converter because the reference is not a voltage but the frequency of a quartz. The VFCs are co-located on an aluminium-oxide substrate (hybrid technology) for improved thermal stability. The VFC outputs are accumulated into 24-bit counter application specific integrated circuit (ASICs). The integration time is 99% of the sampling time of 60s. The ASICs are read by the VIRGO data acquisition system (DAS) (Fröhlich *et al.*, 1995). The resolution of the data chain is better than 0.1 ppm.

The guiding pixels are amplified in the same way as the scientific pixels. The North–South and East–West guiding errors are obtained using differential amplifiers. The error signals are fed into servo regulators driving high-voltage output stages (bipolar discrete technology). The high-voltage amplifiers have a fixed gain. They are built with high voltage DC/DC converters with a fixed output voltage (600 V) and series regulator circuits. The negative lead of the actuators is kept at a constant voltage of 150-V. The positive lead is variable between 0 and 600 V. If needed, the servo loop can be switched off and a fixed voltage can be applied to the actuators. The fixed voltages are selectable by telecommands. The servo loop regulator is of the proportional-integral (PI) type (order 2 servo loop). This type of servo loop provides a zero steady-state pointing error. The closed-loop characteristics of the servo is a first-order low pass with a corner frequency of 10 Hz with a phase margin of about 65°.

5.4.5. *Integration*

At this stage of the game, each subsystem is defined with its own set of requirements. The subsystems have to be assembled together. This assembly requires the definition of interfaces for each subsystem. These interfaces can be mechanical, thermal or electrical. The interface definition can be somewhat complex and subject to extensive discussions or even negotiations. In the case of the LOI, the internal interfaces were defined by the LOI team, while the external interfaces (with the PI instrument) were defined with the VIRGO team (see Fig. 5.17). Then at a higher level, the VIRGO team defined the interfaces to the SOHO spacecraft in agreement with the SOHO project (see Fig. 5.18).

The main objective when defining the internal LOI interfaces was to ensure that the requirements imposed on each subsystem would not be degraded by the whole assembly. For instance, the primary mirror was mounted on three small pads machined inside the invar body. The primary mirror was held in place on the three pads by three opposite spring-loaded pads pushing at the back of the mirror with a predetermined force. Mounting tolerances ensured that the primary mirror would not move by more than $100\,\mu m$ axially on the pads. The secondary mirror had similar pads but instead of force applied by spring, a ring holding the pads was lightly clamp and its screws glued. This simple minded approach was simple and effective as shown by the telescope test. A similar approach was used for mounting the entrance filter. The case of the actuators is clamped for avoiding lateral motion, while a screw at the back of the actuator serves as for reference and for adjustment. The detector is directly mounted into a Kovar plate which is bolted on the Invar structure behind the primary mirror. The ground lead of the detector is connected to the Kovar plate, and this plate is isolated from the Invar using an Ertapeek washer.

The contamination programme put in place for avoiding contamination of the optical instruments, especially the instruments working in the UV and EUV, placed additional design constraints on the geometry of the instrument. For instance, it was suggested in the Experiment Interface Document Part C (SOHO EID-C) that the optical part of the instruments be physically separated as much as feasible from the electronics. The rationale for that suggestion was related to the large quantity of potential hydrocarbons contaminants present in the various electronic coating and parts. This suggestion was implemented in the LOI by having the invar tube nearly hermetical to the outside closed by the entrance filter and by the detector; the only openings were the necessary gaps for the two lever arms (see Fig. 5.2 and 5.3). The two other suggestions of the SOHO EID-C was to have an entrance window warmer than the structure (already implemented) and a cover. The warmer window reduces the amount of material that could be condensing on the entrance window. The cover is used for sunlight protection during critical operations (thruster operations generate plumes of hydrazine surrounding the spacecraft), but also is used for letting the whole instrument getting cooked and outgassed during its journey to the L1 Lagrangian point. The overall cleanliness requirement of the SOHO programme imposed severe constraints on the daily integration activities in the clean room. A constant monitoring of the dust particle and of the hydrocarbon contamination must be maintained. In addition, all parts are to be packed in a clean material that was simply regular aluminium cleaned and wiped with chloroform. Last but not least, all electronic parts and paints are baked at high temperature and in vacuum for reducing the amount of outgassing material they may contain.

Finally, another simple requirement is that the integration of the instrument be as simple as possible. This requires the interaction of many actors in the team to arrive at a sensible integration design.

5.4.6. *Testing*

The testing of the instrument is related to the derivation of its intrinsic performances and to its calibration. It ensures that the scientific and technical requirements are achieved. But the testing is also related to the simulation of the environment of the mission (vibrations, thermal radiations, etc.). Either type of test can be achieved at instrument level or at sub-assembly level.

FIGURE 5.12. Interferometric tests of the telescope performed in a clean room having a Zygo interferometer. The interferometric test consist in making the interference between a reference beam and the beam as transmitted by the telescope in our case going through the instrument twice. The interferences resulting are fringes that are supposed to be perfectly straight if the telescope is perfect; any deviation to the straight line provides information about the optical quality of telescope. That optical quality is usually referred to the wavelength of test (here a helium-laser at 632.8 nm) that sizes the wavefront deformation. In the case of the LOI, that wave front deformation is not worse than $\lambda/30$ rms in double passage. The four spokes of spider holding the secondary mirror arc clearly.

5.4.6.1. *Telescope*

The telescope was subject at instrument level to interferometric tests. The tests were performed with the mirrors integrated in the overall structure and with the mounting scheme mentioned above. The tests were performed on axis (i.e. the secondary was not tilted). Fig. 5.12 shows the results of the test. The optical quality is better than $\lambda/30$ rms.

The focal length of the telescope was measured using a collimator with a focal length known to better than 0.005%. It allowed to measure the focal length of the telescope to better than 0.06%.

5.4.6.2. *Filter*

The main concern for the entrance filter was its ability to survive under solar particles irradiations. γ-ray irradiations of the various glass were performed. It resulted in the understanding of how the irradiations would increase the absorption of the glasses by creating colour centres (Appourchaux et al., 1994). The tests performed under γ-ray irradiations are rather easy to implement. But in order to be more realistic, it was also necessary to irradiate the filter with protons. The main difference is that in the former case the dose is deposited homogeneously throughout the thickness of the glass, while for the latter the dose deposited will depend upon the energy of the proton. For protons, the dose will increase as the proton loses energy inside the glass thickness. Tests performed under protons irradiations simulating the 6-year SOHO radiation dose showed that the filter would survive the mission as shown in Fig. 5.13 (see also Appourchaux, 1993).

In addition, the filter after a 1-year dose of UV irradiation showed no significant sign of degradation either in its absorptance (Marco et al., 1994) or spectral characteristics.

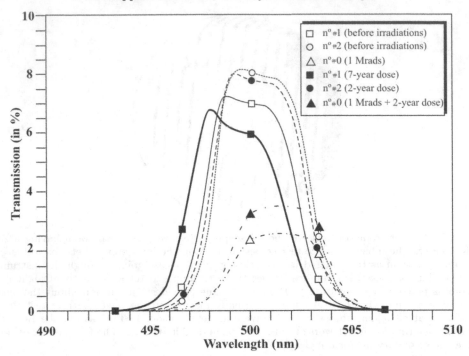

FIGURE 5.13. Two filters were irradiated with protons simulating a 2-y and 7-year dose. The degradation is not negligible for the 7-year dose providing a 15% transmission drop and significant passband shift. These two effects shows that both the glasses and the coatings were damaged by the irradiations. A third filter was irradiated with γ-rays and with a 2-year proton dose. The apparent increase in transmission is due to the annealing (or recovery effect) of the irradiations since the proton irradiation took place a few weeks after the γ-ray irradiation.

5.4.6.3. *Guider*

We encountered a major difficulty when making the guider functions properly. We found out that the assembly procedure for mounting the three rings making the gimbal system needed to set such that there was no friction between the rings. The slightest amount of friction would prevent the gimbal to rotate properly making the system extremely non-linear. In addition, the pivots are mounted in the rings with shear force making the gimbal almost impossible to disassemble if a problem is encountered. After setting the proper mounting procedure (using schemes) the guider can be fully assembled in the Invar tube with the actuators and the lever and spring arms. The overall alignment procedure is rather delicate as a predetermined force has to be applied by the spring arm to act as a locking mechanism: too little force may break the actuator during launch, too much force would prevent the actuator from rotating the gimbal ring.

Each actuator was also tested for its extension. Fig. 5.14 shows a typical response of the actuator extension. The response will differ depending whether the high voltage was increased or decreased, this is the hysteresis effect of the piezoelectric material. This effect has no impact on the servo loop of the guiding. The only impact is in the reproducibility of the voltage needed for aligning the telescope.

The servo loop of the guiding system was tested with an external optical stimuli that was *static* with respect to the telescope. The effectiveness of the servo loop was tested by injecting *directly* a fake electrical error signal. The test performed this way validate the electrical functioning of the servo loop, and allows to derive the characteristics of

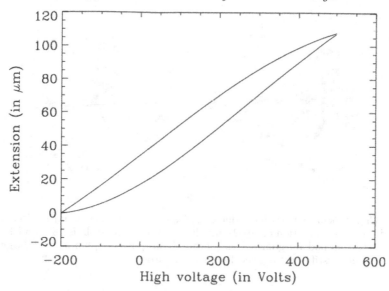

FIGURE 5.14. Extension of the piezoelectric actuator as a function of voltage. The hysteresis is an effect due to the property of PZT: the extension is different when the voltage is increased and then decreased, or vice versa. The hysteresis effect can be compensated by measuring the absolute extension of the actuator with capacitor sensors.

the filter required for providing the correction signal (integrator or multiple pole filter). The dynamic characteristics of the guiding system is then fully determined, and there is no need to provide a *dynamic* optical stimuli because to the response of the servo loop system is the same to the first order. Such a *dynamic* stimuli would only be needed if the response of the optical system would strongly be non-linear. In the case of the LOI, the non-linearities of the optical system are manifest for time scales of the order of months not seconds.

5.4.6.4. *Detector*

The choice of the detector substrate was the subject of an extensive study implying the testing of several type of junction. A photodiode is nothing less but a junction that is exposed to the light. Given, the bandgap of silicon that is about a few eV, photons falling on the junction will free a hole–electron pair that may travel the junction depending of the energy of the photons and on the strength of the electrical field across the junction, the latter depending on whether the junction is polarized or not. The modelling of the quantum efficiency of a junction is not a very complex affair whereas, the prediction of its dependence upon temperature is much more difficult. That is the main reason why several type of junction were fabricated and tested as can be attested by Appourchaux *et al.*, (1992). The different processes tested were the following:
- deep diffused p–n junction on $10\,\Omega\,cm$ material,
- ion implanted p–n junction on $10\,\Omega\,cm$ material.

Each grown on material with the following characteristics:
- $10\,\Omega\,cm$ material,
- $100\,\Omega\,cm$ material,
- epitaxial material (low resistivity).

The processes on epitaxial material had also n^+ deep-diffused guarded ring around the temperature diodes and the pixels; and also an aluminium layer deposited outside the

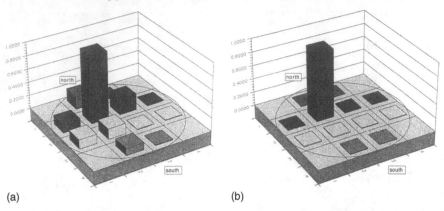

(a) (b)

FIGURE 5.15. Crosstalk measured on the detector: (a) for deep diffused p–n junction on $100\,\Omega\,\mathrm{cm}$; (b) for deep diffused p–n junction on $10\,\Omega\,\mathrm{cm}$. The maximum crosstalk is 18% and 1.3%, respectively. These detectors had no guarded ring around the pixels explaining the larger crosstalk measured, and hence justifying the need for such a guarded ring.

pixels above the sensitive area. In total, more than 35 detectors were tested. The spectral dependence of the temperature coefficient of the quantum efficiency was reported by Appourchaux *et al.* (1992). We also measured the crosstalk between neighbouring pixels (i.e. the amount of hole–electron pair flowing from one pixel to the other). Figure 5.15 shows a typical example of a crosstalk response that measured to be less than 0.03% when using guarded ring. The process chosen for the flight models was the *deep-diffused p--n junction on epitaxial material*. The main advantage is that this process is inherently more stable than that of the ion implanted. Additional tests included the calibration of the photoresponse of the photodiode (in A/W) using a calibrated radiometer. The tests on the Flight Models showed that the temperature dependence of the detector was less than $50\,\mathrm{ppm/K}$ @ $500\,\mathrm{nm}$.

5.4.6.5. *Electronics*

The stability of the VFC was measured using a stable voltage supply. The results showed that the output noise on the digital signal was $6.3 \times 10^{-5}\,\mathrm{ppm}^2/\mu\mathrm{Hz}$ @ $3\,\mathrm{mHz}$, and $10^{-2}\,\mathrm{ppm}^2/\mu\mathrm{Hz}$ @ $100\,\mu\mathrm{Hz}$. This stability is about at least a factor 100 less than the solar noise. This noise sources were the stability of the voltage standard used for calibration, and the temperature coefficient of the VFC which was about $1.5\,\mathrm{ppm/K}$.

As mentioned above, the guiding servo loop electronics was set during a system test of the whole guider. Additional tests included the measurement of the impedance of the actuators. The impedance of the actuator introduces phase gain that needs to be taken into account in the design of the servo loop.

5.4.6.6. *Instrument level test*

The tests performed at the level of the instrument are of several nature. The launch is simulated by vibrating the instrument on a hydropneumatic pot that can simulate either the sinusoidal loads or random loads. The electromagnetic emission (EMI) and susceptibility (EMC) are tested to ensure that the instrument does not radiate too much for the other instruments of the payload, and the other instruments will not induce noise inside the instrument. Finally, the thermal conditions encountered in space are simulated in a thermal chamber and in vacuum. All these stringent tests should not affect the performance of the instrument.

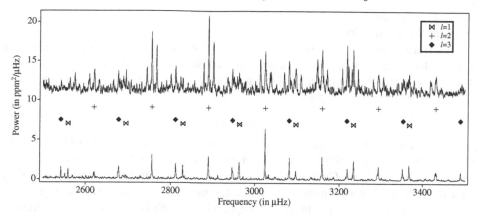

FIGURE 5.16. Power spectra in the p-mode range obtained for $l = 1$, $m = 0$ with the QM instrument (shifted by $+10\,\mathrm{ppm}^2/\mu\mathrm{Hz}$) compared to the FM instrument (unshifted). The atmospheric noise is the main source of noise for the QM. It demonstrates what we need to go to space. The daily interruption (day-night) can be seen as frequency aliases of the main peaks in the power spectra for the QM; two aliases are visually detectable. The uninterrupted view of the Sun does not introduce such aliases in the power spectra for the FM. Please note that the $l = 0$ modes are filtered out by the spatial mask.

These tests are performed at two level: qualification and acceptance. The qualification levels are usually higher and more stringent (larger load, larger temperature range) than the acceptance levels. The Qualification Model (QM) of a space instrument is almost like a Flight Model (FM). The former was subjected to qualification levels while the latter was subjected to acceptance levels.

The LOI programme followed the path: Structural Model, Electrical Model, QM, FM and Flight Spare. This is referred as the Model philosophy. For some faster programme, only an Electrical Model and a Flight Model (called Proto-Flight) are built. The choice of the model philosophy is a choice only based on trade-off between risk and speed. In the case of the LOI, there was no risk taken. The tests performed on the QM were very stringent. For instance, the QM had to be vibrated twice to qualification levels because the first time the screws of the guiding system were not properly locked by glue. Finally, the QM was used for checking the scientific and technical performances of the LOI. It was installed in Tenerife on 2 May 1994. The QM permitted to show that the instrument was well designed and gave an insurance that the space instrument would perform equally well. The LOI QM was the first instrument to detect the p modes in intensity from the ground; all modes above $l = 0$ were detected (Appourchaux *et al.*, 1995c). The results provided by the LOI QM are reported in Appourchaux *et al.* (1995c), Appourchaux *et al.* (1995b), Rabello-Soares *et al.* (1997a,b). The QM was nominally operated without any malfunctions until the summer 2000 when it was decided that it would not give any additional science return compared to its space counterpart. The spectrum obtained with the QM is compared to that of the FM on Fig. 5.16.

5.4.6.7. *Calibrations*

In addition to all the tests performed above, calibration are also required. Calibrations needed are the following:
- preamplifier-VFC chain (counts to V),
- pixel-preamplifier (V to A),
- quantum efficiency (A to W),
- temperature sensors (counts to degree),

- actuators (counts to V),
- guiding (V to arcmin).

These calibration are used for converting counts (so-called level-0 data) to engineering data (so-called level-1 data). The next step (level-2 data) involved *a posteriori* corrections related to the actual functioning of the instruments or to specific scientific massage.

5.4.7. *Operations*

After integrating the LOI in the VIRGO structure (Fig. 5.17), the VIRGO instrument is then integrated in the SOHO spacecraft (Fig. 5.18). The SOHO spacecraft is then put atop the rocket (Atlas-Centaur). The launch occurred on 2 December 1995. Simpler written than done. These sequence of events are usually extremely time consuming due to the great care necessary in these last steps. The launch is of course the riskiest part of the whole mission but when that hurdle is passed the mission and operation can start.

The operation of the LOI is rather simple: switch on and observe. There is an other mode that allows to disengage the servo loop system, and to apply a fixed voltage on the actuators. The story of the LOI cover that could be opened on the Christmas eve of 1995 but then that could not be opened 2 weeks later is well known by the SOHO community. The complete description of how and why the LOI cover was opened almost 3 months later (on 27 March 1996) is described in Appourchaux *et al.* (1997). Here, I would not like to expand on the marvellous power of the intellect over matter but this is clearly a simple example of how to solve remotely a problem without sending an astronaut. Of course, this *success* story is much less known (and rightly so) than the recovery of the spacecraft by the SOHO operation team in the summer 1998 (see ESA press release 33-98†). It shows that a well designed spacecraft (and instrument) can survive the most severe environment. From the whole payload complement, only one sub-unit of an instrument could not be operated after the so-called SOHO vacations of 1998. Here, I must point out that the LOI survived the severe cold and high temperature to which it was exposed during these SOHO vacations.

Finally, the LOI has been performing without a glitch for the past 8 years. It is likely that this will still be the case at the time this article goes to press. At the time of writing, it is anticipated that the SOHO spacecraft shall be operated until 2007, that is a full solar cycle compared with the initial 2-year mission.

5.4.8. *Performances and science*

The performance of the LOI were reported by Appourchaux *et al.* (1997). Since then very little has changed (see Fig. 5.19). For instance, the SOHO pointing has been remarkably stable providing short-term variations of the order of 0.1 arcsec p–p. This stability allowed to detect the solar p modes in the high voltages feeding the actuators but also in the guiding pixels themselves (see Fig. 5.20 Appourchaux and Toutain, 1998). The signal measured in the high voltage corresponds to an error signal of a few milliarcsec (corresponding to a few mV or a few Å of extension) at a p-mode frequency.

The temperature stability provided by the spacecraft has also been quite remarkable. The yearly temperature effect due to the varying SOHO-Sun distance is about 3 K p–p. Over periods of minutes to hours, the typical temperature variation is about 0.1 K p–p. This kind of variation is so small that the temperature dependence of the quantum

†available at sohowww.nascom.nasa.gov/gallery/ESAPR/press33.html.

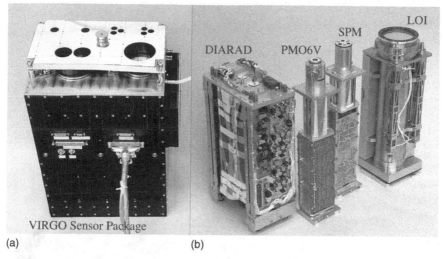

(a) (b)

FIGURE 5.17. The VIRGO instrument shown at (a) is made of several sub-units (b): two radiometers (PMO6, DIARAD), three colour sunphotometers (SPM) and of the LOI.

FIGURE 5.18. Photo of the Sun-facing part of the SOHO spacecraft where VIRGO is integrated. The largest opening is the LOI instrument.

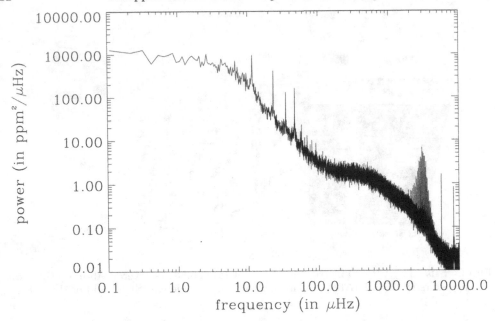

FIGURE 5.19. In-flight mean power spectrum of 26 times series of 108-day long each. The observation is about 7.7 years. The human interaction with the spacecraft manifests itself with the 24-hour frequency (11.57 μHz) and its first three aliases. The two large peaks at higher frequencies are due to the acquisition rate of the VIRGO instrument (3 min) and to the basic sampling of 1 min. The p modes are easily seen at around 3 mHz, with an amplitude not larger than 8 ppm^2/μHz.

efficiency of the detector does not contribute to the noise (about a ppm rms!). Until now, the effect of the temperature has not been detected.

The instrument is also degrading due to the exposure to vacuum and to solar light. Fig. 5.21 shows how the throughput has decreased over the past 8 years. It does not affect yet the noise performance of the instrument. Another minor problem has been detected in the VIRGO/SPM. The term coined is *attractor* as shown in Fig. 5.22 where it can be seen that the output of the instrument seems to be sometimes *attracted* to specific digital values. It does not occur very frequently for the LOI and contribute negligibly to the noise.

The major problem encountered is related to offpointing variations and/or temperature variations that are due to daily operations. These two effects introduce large offpointing inside the LOI of the order of tens of arcsec. Since the LOI operates in such a way that the axial symmetry is not kept (the secondary mirror being tilted), the occasional offpointing changes the light distribution at the focal plane of the detector. The inhomogeneous light distribution is due to aberrations that are not symmetrical around the normal of the primary mirror. This problem affects especially the guiding pixels that cannot yet be used for solar radius measurements.

The science performed with the LOI addresses three different subjects: p-mode seismology (Toutain *et al.*, 1997; Appourchaux *et al.*, 1998a, c; Gizon *et al.*, 1998; Appourchaux, 1998; Appourchaux and Toutain, 1998; Fröhlich *et al.*, 2001; Appourchaux *et al.*, 2000; Appourchaux *et al.*, 2002), g-mode detection (Fröhlich *et al.*, 1998; Appourchaux *et al.*, 2000, 2001; Appourchaux, 2003) and helioseismic analysis (Appourchaux *et al.*, 1998b, d). The latter is an interesting by-product that was not planned. It shows what can be achieved when instrumentation is only driven by Science.

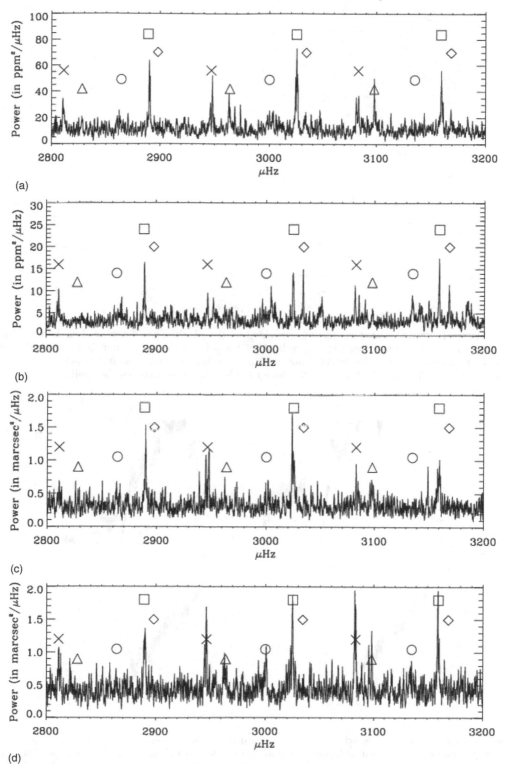

FIGURE 5.20. Power spectrum of the guiding pixels (a and b) and of the high voltages (c and d). The degree of the p-mode oscillation is indicated ($\diamond$ $l=0$, $\triangle$ $l=1$, $\square$ $l=2$, $\times$ $l=3$, $\circ$ $l=4$). The frequency resolution is 54 nHz. The spectra are smoothed to 10 points to enhance the p-mode signals.

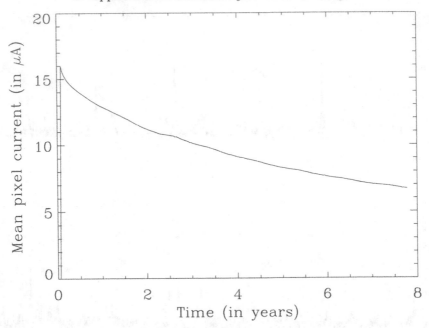

FIGURE 5.21. Variation of the current as provided by the pixels of the detector. The degradation shown here as an average over the 12 pixels affect them in the same manner. It is believed that the degradation is due to the detector and not to the entrance filter. The instrument can admit a reduction of throughput of a factor 8 before the electric noise becomes the limit.

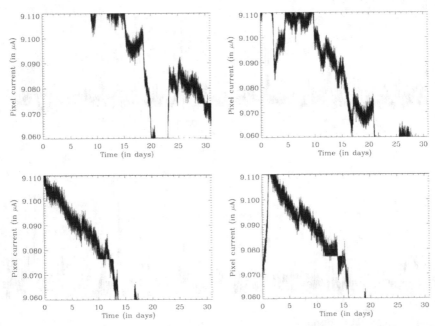

FIGURE 5.22. Signal measured by the four central pixels of the detector. On three of them, the *attractor* can easily be seen as locked value at around 9.75 μA. The attraction does not happen for each count but for some specific ones. This effect occurs also in the SPM. It is believed that it is due to the VFC locking onto these specific counts. The duration of the attractors is much shorter for the LOI than for the SPM because this latter does not degrade as fast as the LOI.

5.5. Discussion and conclusion

The LOI exceeded its design performances. In retrospect, there are three design features that could have been better thought about.

The number of pixels is for instance rather limiting the capability of the instrument above $l = 3$ due to the mode aliasing. Doubling the number of pixels would have push the limit to $l = 4$ only. This would have severely made the electronics more complex by having 24 individual read out lines instead of 12. At least, we could have also put the guiding pixels on a read out using VFC...

Last but not least, it would have been better to have a separated coarse pointing that would have allowed to use the telescope around its symmetry axis, and avoiding to have the secondary off pointed thereby destroying the symmetry of the telescope. At the time of design, a mechanism pointing the whole instrument towards the Sun required more space that was available for the whole LOI instrument. As far as the coarse pointing is concerned, this would require just a mere redefinition of the boundary conditions.

Since the end of 90s, the technology has evolved in such a way that it is now possible to have not only 24 pixels but about a million that can be read individually. These new detectors are called active pixel sensors (APS). They are obviously the future for high photon flux application such as for the Sun or for star trackers imaging bright stars.

Unfortunately, from the scientific point of view, there would be very little sense to fly such an Next Generation (NG) LOI... It would be very unlikely that we would be able to detect g modes with such an NG-LOI. In addition, everything that could be learnt using p modes would be better done with an instrument detecting the modes in solar radial velocity†.

This state of affair shows that instrument building must be indeed driven by Science. Even though I found quite sad that I will not be able to build another LOI, and that this instrument is one of a kind, I have to accept the outcome of my own analysis. In due time I will find other ways of detecting g modes but they may not be based on such an instrument as the LOI (Appourchaux, 2003).

Acknowledgements

SOHO is a mission of international collaboration between ESA and NASA.

I am grateful to Trevor Sanderson of the Solar and Heliospheric Science Division of ESA. He was the one who proposed to replace the "non-intelligence" of VIRGO by a well timed 'pulling of its plug'. We are also grateful to our colleagues of the Mechanics Section of ESTEC without whom my model of the cover could have not been validated: P. Coste, M. Eiden, M. York and M. Verin. Sincere thanks also to Matra-Marconi-Space (now Astrium, Toulouse) for the general management of the problem, to Saab for the implementation of the software patch and to P. Strada of the SOHO project for coordinating the software patch efforts. I am thankful to H.J. Roth for his faithful dedication to the VIRGO instrument and all tests performed on the spare.

The building of the LOI would have not been possible without the contribution of the following ESTEC personnel: J. Fleur, R. Scheper, A. Smit for the mechanical designs; J. Blommers, J. Postema, M.J. Kikkert, K.F. Frumau, for the machining and the assembling; K. Bouman, A. Fransen, J. Heida, L. Smit for the electronics fabrication; U. Telljohann, D. Martin, T. Beaufort for the electronic design; S. Lévêque for being Mr Propre, for detector testing and for various optical tests and integration

---

†The signal-to-noise ratio in the power spectrum is about a factor 10 higher in velocity than in intensity. The lowest detectable frequency is about $1000\,\mu$Hz in velocity, and $1800\,\mu$Hz in intensity.

activities; additional contributions were given by G. Gourmelon for the spectrometry, by B. Johlander for various irradiations, by J.M. Guyt for checking the contamination, by A. Zwaal for the cleanliness levels and by P. Brault for the bakings; we also thank R. Czichy for loaning us the Zygo interferometer, and D. Doyle for the technical assistance for this interferometer. The various part of the LOI instrument could not have been built without the help of the following individuals: T.E. Hansen at Ame, C. Shannon and T. Hicks at Queensgate, and P. Robert at Bertin.

Last but not least, the LOI would never have had a place in space without the efforts of the whole VIRGO team which are gratefully acknowledged. A special note to the dedication of Antonio Jiménez for operating the Virgo Data Center.

On a more personal note, I also thank my family for the unconditional support shown over the years. The family being my wife, Maryse, my two sons, Kevin and Thibault, and my two cats, Ah Kin and Myrtille. We succeeded.

REFERENCES

ANDERSEN, B., DOMINGO, V., JONES, A., JIMÉNEZ, A., PALLÉ, P., RÉGULO, C. & ROCA CORTÉS, T. 1988 In: *Seismology of the Sun and Sun-like Stars*, V. Domingo & E. Rolfe (Eds.), p. 385. ESA SP-286, ESA Publications Division, Noordwijk, The Netherlands.

APPOURCHAUX, T. 1998 In: *Structure and Dynamics of the Interior of the Sun and Sun-like Stars*, S. Korzennik & A. Wilson (Eds.), p. 37. ESA SP-418, ESA Publications Division, Noordwijk, The Netherlands.

APPOURCHAUX, T. 2001 In: *Helio- and Asteroseismology at the Dawn of the Millennium*, P.L. Pallé & A. Wilson (Eds.), *SOHO-10/GONG 2000*. p. 71. ESA SP-464, ESA Publications Division, Noordwijk, The Netherlands.

APPOURCHAUX, T: 2003 In: *Proceedings of SOHO 12/GONG+ 2002. Local and Global Helioseismology: The Present and Future*, H. Sawaya-Lacoste (Ed.), p. 131. ESA SP-517, ESA Publications Division, The Netherlands.

APPOURCHAUX, T., ANDERSEN, B., BERTHOMIEU, G., CHAPLIN, W., ELSWORTH, Y., FINSTERLE, W., FRÖLICH, C., GOUGH, D.O., HOEKSEMA, T., ISAAK, G., KOSOVICHEV, A., PROVOST, J., SCHERRER, P., SEKII, T. & TOUTAIN, T. 2001 In: *Helio- and Asteroseismology at the Dawn of the Millennium*, P.L. Pallé & A. Wilson (Eds.), *SOHO-10/GONG 2000*. p. 467. ESA SP-464, ESA Publications Division, Noordwijk, The Netherlands.

APPOURCHAUX, T., ANDERSEN, B., FRÖHLICH, C., JIMÉNEZ, A., TELLJOHANN, U. & WEHRLI, C. 1997 *Solar Physics*, **170**, 27.

APPOURCHAUX, T., ANDERSEN, B. & SEKII, T. 2002 In: *From Solar Min to Max: Half a Solar Cycle With SOHO*, C. Fröhlich & A. Wilson (Eds.), p. 47. ESA SP-508, ESA Publications Division, Noordwijk, The Netherlands.

APPOURCHAUX, T. & ANDERSEN, B.N. 1990 *Solar Physics*, **128**, 91.

APPOURCHAUX, T., CHAPLIN, W.J., ELSWORTH, Y., ISAAK, G.R., MCLEOD, C.P., MILLER, B.A. & NEW, R. 1998a In: *New Eyes to See Inside the Sun and the Stars*, F.-L. Deubner, J. Christensen-Dalsgaard & D. Kurtz (Eds.), *IAU 185*. p. 45. Kluwer Academic Publishers, Dordrecht, The Netherlands.

APPOURCHAUX, T. & CISLAGHI, M. 1992 *Optical Engineering*, **31**, 1715.

APPOURCHAUX, T., FRÖHLICH, C., ANDERSEN, B., BERTHOMIEU, G., CHAPLIN, W., ELSWORTH, Y., FINSTERLE, W., GOUGH, D., HOEKSEMA, J.T., ISAAK, G., KOSOVICHEV, A., PROVOST, J., SCHERRER, P., SEKII, T. & TOUTAIN, T. 2000 *ApJ*, **538**, 401.

APPOURCHAUX, T., GIZON, L. & RABELLO-SOARES, M.C. 1998b *A&A Sup. Series*, **132**, 107.

APPOURCHAUX, T., RABELLO SOARES, M.C. & GIZON, L. 1998c *IAU Symposia*, **185**, 167.

APPOURCHAUX, T., RABELLO-SOARES, M.-C. & GIZON, L. 1998d *A&A Sup. Series*, **132**, 121.

APPOURCHAUX, T., TELLJOHANN, U., MARTIN, D., FLEUR, J. & LÉVÊQUE, S. 1995a In: *Helioseismology, 4th SOHO workshop*, J. Hoeksema, V. Domingo, B. Fleck & B. Battrick (Eds.), p. 359, ESA SP-376, ESA Publications Division, Noordwijk, The Netherlands.

APPOURCHAUX, T. & TOUTAIN, T. 1998 In: *Sounding Solar and Stellar Interiors, IAU 181, Poster volume*, J. Provost & F.-X. Schmider (Eds.), p. 5. Kluwer Academic Publishers, Dordrecht.

APPOURCHAUX, T., TOUTAIN, T., JIMÉNEZ, A., RABELLO-SOARES, M., ANDERSEN, B. & JONES, A. 1995b In: *Helioseismology, 4th SOHO workshop*, J. Hoeksema, V. Domingo, B. Fleck & B. Battrick (Eds.), p. 265. ESA SP-376, ESA Publications Division, Noordwijk, The Netherlands.

APPOURCHAUX, T., TOUTAIN, T., TELLJOHANN, U., JIMÉNEZ, A., RABELLO-SOARES, M.C., ANDERSEN, B.N. & JONES, A.R. 1995c *A&A*, **294**, L13.

APPOURCHAUX, T.P. 1993. *SPIE*, **2018**, 80.

APPOURCHAUX, T.P., GOURMELON, G. & JOHLANDER, B. 1994 *Optical Engineering*, **33**, 1659.

APPOURCHAUX, T.P., MARTIN, D.D. & TELLJOHANN, U. 1992 *SPIE*, **1679**, 200.

BERTHOMIEU, G. & PROVOST, J. 1990 *A&A*, **227**, 563.

BONNET, R.M., LEMAIRE, P., VIAL, J.C., ARTZNER, G., GOUTTEBROZE, P., JOUCHOUX, A., VIDAL-MADJAR, A., LEIBACHER, J.W. & SKUMANICH, A. 1978 *ApJ*, **221**, 1032.

CHRÉTIEN, H. 1958, *Calcul des combinaisons optiques*, 4th ed. Paris: Libr. du Bac, 1958/59.

DAMÉ, L. 1988 In: *Seismology of the Sun and Sun-Like Stars*, V. Domingo & E. Rolfe (Eds.), p. 367. ESA SP-286, ESA Publications Division, Noordwijk, The Netherlands.

DELACHE, P. & SCHERRER, P.H. 1983 *Nature*, **306**, 651.

FRÖHLICH, C., ANDERSEN, B., CROMMELYNCK, B.G.D., DELACHE, P., DOMINGO, V., JIMÉNEZ, A., JONES, A., ROCA CORTÉS, T. & WEHRLI, C. 1988a In: *Seismology of the Sun and Sun-Like Stars*, V. Domingo & E. Rolfe (Eds.), p. 371. ESA SP-286, ESA Publications Division, Noordwijk, The Netherlands.

FRÖHLICH, C., APPOURCHAUX, T. & GOUGH, D. 2001 In: *Helio and Asteroseismology At the Dawn of the Millennium*, P.L. Pallé & A. Wilson (Eds.), SOHO-10/GONG 2000. p. 71. ESA SP-464, ESA Publications Division, Noordwijk, The Netherlands.

FRÖHLICH, C., BONNET, R.M., BRUNS, A.V., DELABOUDINIÈRE, J.P., DOMINGO, V., KOTOV, V.A., KOLLATH, Z., RASHKOVSKY, D.N., TOUTAIN, T. & VIAL, J.C. 1988b In: *Seismology of the Sun and Sun-Like Stars*, V. Domingo & E. Rolfe (Eds.), p. 359. ESA SP-286, ESA Publications Division, Noordwijk, The Netherlands.

FRÖHLICH, C., FINSTERLE, W., ANDERSEN, B., APPOURCHAUX, T., CHAPLIN, W., ELSWORTH, Y., GOUGH, D., HOEKSEMA, J.T., ISAAK, G., KOSOVICHEV, A., PROVOST, J., SCHERRER, P., SEKII, T. & TOUTAIN, T. 1998 In: *Structure and Dynamics of the Interior of the Sun and Sun-like Stars*, Korzennik, S. & Wilson, A. (Eds.), p. 67. ESA SP-418, ESA Publications Division, Noordwijk, The Netherlands.

FRÖHLICH, C., ROMERO, J., ROTH, H., WEHRLI, C., ANDERSEN, B.N., APPOURCHAUX, T., DOMINGO, V., TELLJOHANN, U., BERTHOMIEU, G., DELACHE, P., PROVOST, J., TOUTAIN, T., CROMMELYNCK, D.A., CHEVALIER, A., FICHOT, A., DÄPPEN, W., GOUGH, D., HOEKSEMA, T., JIMÉNEZ, A., GÓMEZ, M.F., HERREROS, J.M., CORTÉS, T.R., JONES, A.R., PAP, J.M. & WILLSON, R.C. 1995 *Solar Physics*, **162**, 101.

GIZON, L., APPOURCHAUX, T. & GOUGH, D.O. 1998 In: *New Eyes to See Inside the Sun and the Stars*, F.-L. Deubner, J. Christensen-Dalsgaard, & D. Kurtz (Eds.), IAU 185. p. 37. Kluwer Academic Publishers, Dordrecht, The Netherlands.

MACLEOD, H.A. 1986 *Thin-Film Optical Filters*, Institute of Physics Publishing, London, UK.

MARCO, J., CHOURREAU, A. & OSCAR, H. 1994 "Essai de longue durée sur revêtements de régulation thermique', CERT 439700'. Technical report, Centre d'étude de recherches de Toulouse, France.

RABELLO-SOARES, M.C., ROCA CORTÉS, T., JIMÉNEZ, A., ANDERSEN, B.N. & APPOURCHAUX, T. 1997a *A&A*, **318**, 970.

RABELLO-SOARES, M.C., ROCA CORTÈS, T., JIMÉNEZ, A., APPOURCHAUX, T. & EFF-DARWICH, A. 1997b *ApJ*, **480**, 840.

SCHERRER, P., HOEKSEMA, J., BOGARD, R. & *et al.* 1988 In: *Seismology of the Sun and Sun-Like Stars*, V. Domingo & E. Rolfe (Eds.), p. 367. ESA SP-286, ESA Publications Division, Noordwijk, The Netherlands.

TOUTAIN, T., APPOURCHAUX, T., BAUDIN, F., FROEHLICH, C., GABRIEL, A., SCHERRER, P., ANDERSEN, B.N., BOGART, R., BUSH, R., FINSTERLE, W., GARCIA, R.A., GREC, G., HENNEY, C.J., HOEKSEMA, J.T., JIMENEZ, A., KOSOVICHEV, A., CORTES, T.R. TURCK-CHIEZE, S., ULRICH, R. & WEHRLI, C. 1997 *Solar Physics*, **175**, 311.

VAN DER RAAY, H.B. 1988 In: *Seismology of the Sun and Sun-Like Stars*, V. Domingo & E. Rolfe (Eds.), p. 339. ESA SP-286, ESA Publications Division, Noordwijk, The Netherlands.

WOODARD, M. 1984 Short-period oscillations in the total solar irradiance. Ph.D. Thesis, University of California, San Diego.

6. Hipparcos and Gaia: the development of Space Astrometry in Europe

By M.A.C. PERRYMAN

Astrophysics Missions Division, ESA–ESTEC, Noordwijk, The Netherlands

The operational phase of European Space Agency (ESA)'s Hipparcos mission ended 10 years ago, in 1993. Hipparcos was the first satellite dedicated to the accurate measurement of stellar positions. Within 10 years, ESA's follow-on mission, Gaia, should be part way through its operational phase. I summarize the basic principles underlying the measurement of star positions and distances, present the operational principles and scientific achievements of Hipparcos, and demonstrate how the knowledge acquired from that programme has been used to develop the observational and operational principles of Gaia – a vastly more performant space experiment which will revolutionize our knowledge of the structure and evolution of our galaxy.

6.1. Introduction

Measuring star positions is an ancient science, but one now poised to make a remarkable contribution to 21st century astronomy, and our understanding of the evolution and present-day structure of our galaxy. Over the years 1600–1700, scientific motivation driving celestial cartography was inspired by navigational problems at sea, understanding Newtonianism, and comprehending the Earth's motion. In 1718, Halley determined the first stellar "proper" motions through a comparison of his star maps with those of the Greek Hipparchus, and in 1725 Bradley made the first measurement of stellar aberration, thereby confirming the motion of the Earth through space, the finite velocity of light and, because of his inability to discern a measurable parallax, the immensity of stellar distances. In 1753, Herschel had postulated the Sun was moving through space, and the first direct measurement of stellar distances were reported by Bessel, Henderson and Struve around 1838–1839. The late 1800s saw the first photographic maps of the celestial sphere, and vigorous debates about the size and structure of our galaxy. After a further 100 years in which ground-based astrometry battled against the effects of the Earth's atmosphere, and succeeded in demonstrating the importance of accurate astrometric measurements, astrometry escaped from the confines of the Earth's surface with ESA's launch of the Hipparcos satellite, dedicated to the precise measurement of star positions.

The success of Hipparcos has inspired other ideas for astrometry from space: in 2010, ESA will launch a follow-up mission, Gaia, which will revolutionize the field through a huge advance in accuracy, number of objects and limiting magnitude. Meanwhile National Aeronautics and Space Administration (NASA)'s Space Interferometry Mission (SIM) is a pointed space interferometer, due for launch around 2010, providing high-accuracy astrometry albeit using very different techniques to Hipparcos and Gaia, while providing enabling technology for terrestrial planet finder (TPF). Other space missions have recently been proposed but not approved: FAME (USA), DIVA (D), Amex (USA) and Jasmine (J).

In this Chapter I will explain what motivates the measurement of star positions, why these measurements have to be made from space, how the measurements are made and what the results will tell us.

6.2. Scientific motivation underpinning astrometry

The determination of precise stellar (or galaxy) positions alone has some scientific interest, essentially by providing a positional reference system for measuring the location of other objects (i.e. navigation), or a comparison of fluxes in different wavebands (radio, X-ray, etc). The change in star positions with time (over tens or hundreds of years, or over years if the measurement accuracy is high enough) provides (once the distance to the star is known) the velocity of the stars through space, and hence information on the kinematics of the constituent stars in our galaxy, and consequently on the dynamics of our galaxy and of star clusters, information on the distribution of dark matter, and evidence of our galaxy's formation. The change in a star's apparent position as the Earth moves around the Sun provides the "parallax", or distance to the star, the only direct way we have of measuring distances. It is these distances which allow the determination of stellar luminosities, information on the stellar distributions throughout space and, via suitable modelling, information on ages and stellar structure, and evolution. As our measurement accuracy improves, many more subtle effects appear: binary star orbits, the dynamical manifestation of extra-solar planets, relativistic light bending, the Sun's motion around the galaxy, astrometric microlensing, etc.

6.3. The measurement of stellar distances

The measurement of stellar space motion and distance (parallax) is rather straight-forward in principle: a number of relative angular positional measurements made over several years gives the motion of the star through space by virtue of its (essentially) linear motion with respect to the background stars, while different viewing perspectives as the Earth moves around the Sun provide the distance to the star via the phenomenon of parallax; that is the apparent annual elliptical motion of a nearby star with respect to background stars due to the Earth's motion around the Sun (Fig. 6.1). In practice, measurements must be spaced frequently and with different geometries with respect to Earth's orbit. Measurements must be highly accurate (and the instrument must be made very stable) since the effects sought are very small: typical stellar motions in the Solar neighbourhood amount to some 10–30 milli-arcsec/year, while even the nearest stars have a parallax (angular distance subtended by the Earth's orbital radius around the Sun)

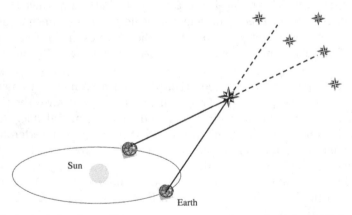

FIGURE 6.1. The principle of parallax measurements – the projected position of a foreground star with respect to background stars appears to change as the Earth moves in its orbit around the Sun. If this change in angle can be detected and measured accurately, then the distance to the star can be related to the orbital dimensions of the Earth–Sun system by simple trigonometry.

of less than 1 arcsec. Complications include the Sun's motion through space, the finite (and unknown) distances of the background stars being used as a reference system, the presence of binary and multiple stars, and the variable geometry of space-time due to the gravitational effects of bodies in the solar system (Fig. 6.2). But it is the very complexity of these space motions that makes their accurate measurement so rewarding scientifically.

A brief discussion of the immensity of stellar distances is appropriate. A distance of 1 parsec (pc) is defined to be the distance at which the mean Sun–Earth radius subtends an angle of 1 arcsec (1 second of arc, or 1/3600 degree), with $1\,pc = 3.26$ light years. The nearest stars have a parallax (π) of about 1 arcsec, and there are a few hundred stars within 10–20 pc. At $d = 100$ pc, $\pi = 0.01$ arcsec (10 milli-arcsec): a measurement accuracy of 1 milli-arcsec is needed to measure this to 10%. The galactic centre is at about 8–10 kpc, or $\pi \sim 100$ micro-arcsec: a measurement accuracy of 10 micro-arcsec is needed to measure this to 10%. Hipparcos reached accuracies of 1 milli-arcsec. Gaia will reach a few micro-arcsec, an accuracy not contemplated even two decades ago, while probing geometric cosmology ($z = 0.1$) will requires a few nano-arcsec.

It is worth trying to visualize and understand the difficulties confronted in measuring these tiny angles: the full moon subtends an angle of about 0.5 degree; 1 arcsec (ground) corresponds to measuring a distance of 0.5 mm at 100 m; 1 milli-arcsec (Hipparcos) corresponds to a distance of 2 m (the height of a person) on the Moon, or a golf ball viewed from across the Atlantic, or the angle human hair will grow in 10 s as viewed from a distance of 10 m. one micro-arcsec is the angular diameter of human hair viewed from 10,000 km, or a Bohr radius (5×10^{-11} m) viewed from 10 m. These analogies are themselves difficult to comprehend, but translated into instrument requirements and stabilities, they are extremely challenging. Some two hundred years ago, John Herschel tried to convey the immensity of stellar distances in the following way: *To drop a pea at the end of every mile of a voyage on a limitless ocean to the nearest fixed star, would require a fleet of 10,000 ships, each of 600 tons burthen.*

In developing instrumentation to make these highly precise measurements it becomes clear why it has been necessary to go to space to advance the field: much below a limit of about 1 arcsec, especially for large-angle measurements, a combination of atmospheric

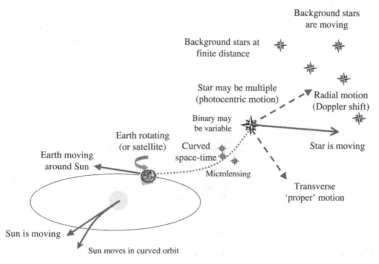

FIGURE 6.2. In practice, many effects complicate the accurate measurement of stellar positions, and more scientifically interesting effects become discernable as the accuracy improves.

phase fluctuations and refraction, and also instrumental thermal and gravitational stability, impose accuracy limits which have been impossible to surpass from the ground. In addition, space provides all-sky visibility, important for constructing a network of stellar positions (or reference frame) which extends uniformally over large angles on the sky.

A brief history of the Hipparcos project also indicates the stamina needed in space experiments (and in many other areas of science also): it was proposed as an outline concept to CNES (the French space agency) by a French scientist, Pierre Lacroute, in 1967; after studies and a decision that the project was too complex for the French national programme it was studied at a more European level, with major contributions from Erik Høg, Lennart Lindegren, Jean Kovalevsky, Catherine Turon and others – scientists fully involved in Hipparcos until its completion in 1997, and still deeply involved with the development of Gaia. With the involvement of ESA in late 1970s, Hipparcos was accepted as an approved mission in 1980. As a large collaboration between European scientists, ESA, and European industry, Hipparcos was launched from Kourou in French Guyana by an Ariane launcher in 1989, operated until 1993, and after a huge collaborative effort by the Hipparcos community, the final mission catalogues were published in 1997. It was a massive enterprise: some 200 scientists; around 30 industrial teams (probably greatly) in excess of a thousand people directly involved in the satellite development at some level; and an overall budget of 400 MEuros. Difficult advances in space sciences do not come cheaply or easily, yet the scientific rewards have been significant, and the project has opened the eyes of astronomers around the world to the new opportunities now accessible to improved measurements of the same type. European scientists and European

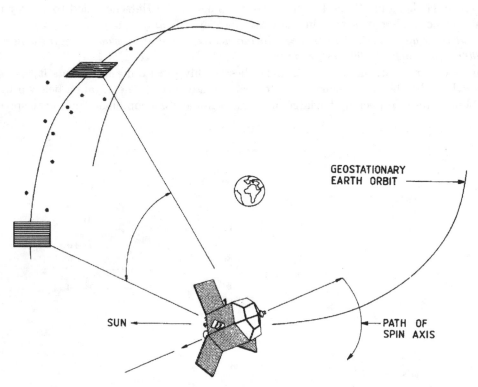

FIGURE 6.3. The measurement principle of the Hipparcos satellite. The two fields of view scanned the sky continuously, and relative positions of the stars along the scanning direction were determined from the modulated signals resulting from the satellite motion.

industry have acquired much new expertise and technical capabilities as a direct result of this programme, and scientific collaboration across Europe has gained substantially.

6.4. Measurement principles of the Hipparcos mission

A detailed explanation of how the Hipparcos satellite made its observations is beyond the scope of this chapter. The interested reader is referred to the published Hipparcos Catalogue, or to ESA's Hipparcos web site for further details, but here some basic principles will be summarized:

(a) The simultaneous observation of two widely separated viewing directions (Fig. 6.3) is central to the global measurement principle, providing an all-sky rigid reference frame. This was the main driver for the observations to be made from space, and proved to be one of the technically most difficult challenges of the Hipparcos programme (Fig. 6.4). To understand this, consider a situation in which all stars in one part of the sky have the same distance (e.g. if they are all members of a star cluster) and hence all have the same relative parallax motion. Then their distances could not be determined by measuring their relative motions with respect to each other. The solution (as proposed by Lacroute

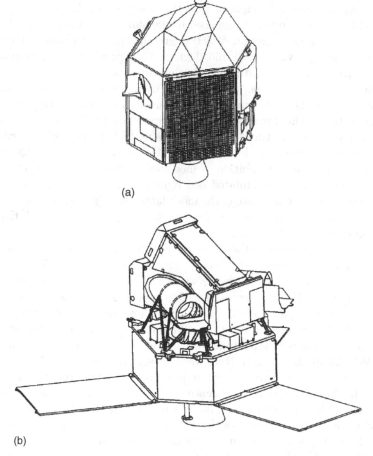

(a)

(b)

FIGURE 6.4. (a) Hipparcos satellite launch configuration with the solar arrays stowed, and (b) on-station configuration with the solar arrays deployed (and with the shade structure removed for illustration purposes).

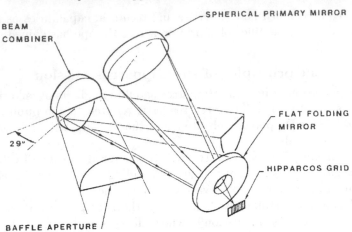

FIGURE 6.5. Configuration of the Hipparcos Schmidt telescope. Light entered from the two baffle directions, and was brought together at the beam combiner, which was figured as a Schmidt corrector. The combined light was reflected from the folding and spherical primary mirrors onto the focal surface where the modulating grids were located.

in the 1960s) was to have two viewing directions, separated by a large angle (say in the range 50−130 degree), and superimposed in the same focal plane. This "basic" angle cannot be close to a integer fraction of 360 degree for rigidity reasons: Hipparcos used 57 degree, Gaia uses (provisionally) 106 degree. This provides enough information to determine absolute parallaxes, and hence absolute distances.

(b) The two viewing directions are combined in the same focal plane in which the relative separations of the star images are measured (Fig. 6.5). These separations are made most accurately in the direction of the scanning motion; the separation of stars from two different fields are simply the separation measured in the combined focal plane plus the separation between the two viewing directions. As the Hipparcos detection system did not have detailed spatial resolution, a modulating grid was placed at the telescope focal plane, and the resulting modulated star light measured – angular separations were obtained from the difference in phase of the modulated star signals. Inclined slits provided information on the satellite attitude and the course star positions in both the along-scan and across-scan directions (Fig. 6.6).

(c) Measurements were repeated at many (some 100) different epochs, and at different orientations, over the observational lifetime (about 4 years). The different observations at different epochs were provided by the satellite's sky-scanning motion – basically a slow spin around a rotation axis, coupled with a precession of this spin axis around the Sun direction, maintaining a constant inclination to the Sun to provide thermal stability of the payload. Repeated measurements at different epochs and different scan directions provided the information necessary to allow the star positions, distances and motions to be disentangled. All of these parameters were determined on ground as part of a data analysis system, which took several years to develop, and which required state-of-the-art computational facilities on ground to execute. By-products of the measurements included information on double and multiple stars, and measurements of the space-time metric. Final accuracies were a combination of the instantaneous measurement accuracy, and the combination of these repeated measurements over several years.

(d) The Hipparcos Catalogue was built up from the observing programme which had to be defined (and the stars selected for observation) before launch. In addition to the resulting Hipparcos Catalogue of about 120,000 stars, the inclined slits of the modulating

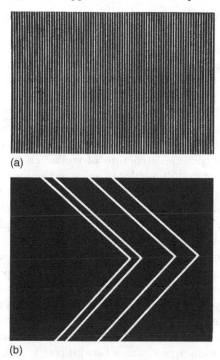

(a)

(b)

FIGURE 6.6. Optical micrographs of part of the main modulating grid for Hipparcos (a) and the star mapper grid (b). The grids represented one of the main technical challenges of the Hipparcos payload, and were of an extremely high quality – note that the repeat period of the upper grid is only about $8\,\mu$m (1.2 arcsec).

grid provided additional measurements, eventually compiled into the Tycho-2 Catalogue of some 2.5 million stars, of lower astrometric accuracy, but still much more accurate than those measured from the ground, and complete to about 12 mag. In addition, these measurements provided 2-colour photometry (roughly corresponding to B and V) at the measurement epochs – these were uniform, systematically calibrated over the sky, and provided information on stellar variability and coarse colours/temperatures. In principle, stellar photometry can be carried out from the ground, but Hipparcos provided much important information about the power and advantages of carrying out these measurements from space – in crowded regions, with global sky uniformity, and at many epochs coinciding with the astrometric measurements. This knowledge will be carried forward into the vastly more ambitious Gaia programme.

6.5. Scientific achievements of the Hipparcos mission

Before looking to the future plans, let me summarize what the Hipparcos mission achieved. Firstly, the Hipparcos Catalogue yielded a compilation of nearly 120,000 stars with accuracies of about 1 milli-arcsec in positions, parallaxes and annual proper motions. Most importantly, these measurements are on a common reference system, and the positions, proper motions and parallaxes are free of systematic effects at levels of a few tenths of a milli-arcsec. The catalogue provides an all-sky reference system with a density of about three stars per square degree, completeness between about 7.3 and 9 mag (depending on galactic latitude and spectral type), and with a limiting magnitude of about 12 mag. Catalogue annexes provide information on binaries and orbital systems, photometry and variability, and information on about 60 solar system objects. Light bending (relativity) was measured with an accuracy in γ of about 1 part in 10^3. The Tycho-2 Catalogue comprises about 2.5 million stars to 10–30 milli-arcsec, a magnitude limit (and completeness) of about 12 mag, and two-colour photometry (B and V) at about 100 epochs over 4 years (the Tycho-2 Catalogue replaced the Tycho Catalogue of 1 million stars released previously).

In terms of scientific applications, the Hipparcos mission results have yielded a stellar reference system to which all other major stellar catalogues have now been referred; accurate stellar distances (and therefore luminosities and structure of the Hertzsprung-Russell diagram) out to about 100 pc; the structure of stellar clusters (Hyades, Pleiades, ...) and associations; kinematics and dynamics of stars in the local disk population and the Solar neighbourhood; information on stars passing close to the Sun in the recent past or future (possibly responsible for perturbations of the Oort Cloud comets); evidence for halo structure pointing to clues about the formation of our galaxy; upper limits to the mass of nearby planetary systems; and many other fields.

Hipparcos represented the first steps in Space Astrometry, and experience with the design, operation, data analysis and scientific exploitation points to the way ahead. Meanwhile the rapid development of many areas of "high technology" means that a new and vastly more powerful space mission can now be contemplated, benefiting from the strengths and learning from the weaknesses. Hipparcos was immensely successful, but 120,000 stars is but a tiny fraction of our galaxy (which contains some 100 billion stars). The accuracy of 1 milli-arcsec was a breakthrough, but gives only 10% distance accuracies out to about 100 ps; distances out to 10 kpc would be a revolution, especially if combined with the observations of many more stars, and a much fainter limiting magnitude. The pre-defined observing catalogue was a necessary but serious limitation, and for a much larger and deeper catalogue star selection on ground is unfeasible, and cannot make provision for observation of variable stars, supernovae and rapidly moving (Solar System) objects. B,V photometry was very valuable but is clearly too limited: in the future it would be desirable to measure the positions and luminosities of very many more stars, but also determine their physical properties at the same time – hence much more comprehensive multi-colour photometry would give detailed astrophysical information (temperature, gravity, chemical composition and reddening). The third Cartesian component of a star's space motion (given by the radial velocity) is also needed for a deeper understanding of population kinematics and dynamics, binary stars, etc. Somewhat analogous to the issue of the photometry, radial velocities can in principle be obtained from the ground, by Doppler measurements. But making such observations for tens of millions of stars, in a common reference frame, and preferably at the same epoch as the astrometric measurements, is a different matter entirely.

As we shall see, a space mission measuring some 1 billion objects, obtaining completeness to 20 mag, and with a huge improvement in astrometric accuracy is indeed feasible, and ESA is working towards the launch of such a mission, Gaia, around 2010. Consequently, the scientific results of Hipparcos will be rapidly superseded. So was the 20 years of effort devoted to it of value? The answer, from someone involved in both missions, is a definite "yes": Hipparcos established and verified the principles of these measurements from space, and these principles are now no longer questioned – the situation was very different in the early 1980s. The scientific value of high-accuracy astrometry is also no longer questioned – and this represents a huge development from the 1980s when astrometry was still perceived by many in the wider scientific community as a marginal research field. This evolution means that many bright young minds, both scientific and technical, are embracing ESA's new mission with great fervour and excitement. The community also now embodies a great deal of "system knowledge" – of the design difficulties and considerations, of the data analysis complexities, and so on, which should guarantee that the elements needed to make a future mission successful will be in place.

6.6. Scientific objectives of the Gaia mission

The design of Gaia, now planned for launch in 2010, is based on the same basic principles as Hipparcos: two viewing directions separated by a large fixed angle, sky-scanning using the same basic prescription (rotation about a spin axis combined with precession around the Sun direction), relative positional measurements made in the combined focal plane, and a global data analysis on ground. But Gaia is motivated by the requirements to have a significantly improved accuracy, a significantly improved limiting magnitude and a significantly larger number of stars. In its current design, Gaia will observe some 1 billion stars down to a uniform magnitude limit of about 20 mag. On-board object detection will ensure that the catalogue is complete to these levels. Accuracies should come in at a remarkable 4–5 micro-arcsec at 12 mag and brighter, some 10–20 micro-arcsec at 15 mag, and some 0.2 milli-arcsec at 20 mag.

As a result, the Gaia Catalogue will provide comprehensive luminosity calibration: for example, distances to 1% for some 20 million stars out to 2.5 kpc; distances to 10% for 150 million stars to 25 kpc; rare stellar types and rapid evolutionary phases in large numbers; and parallax calibration of all distance indicators (e.g. Cepheids and RR Lyrae stars out to distances of the Large and Small Magellanic Clouds LMC, SMC, respectively). Resulting physical properties will include "clean" Hertzsprung–Russell sequences throughout the galaxy; Solar neighbourhood mass function and luminosity function, including the distributions of white dwarfs ($\sim$200,000) and brown dwarfs ($\sim$50,000); initial mass and luminosity functions in star forming regions; luminosity function for pre-main-sequence stars; detection and dating of all spectral types and galactic populations; and detection and characterization of variability for all spectral types.

One billion stars in three-dimensions will provide material for establishing the distance and velocity distributions of all stellar populations; the spatial and dynamic structure of the disk, and halo and their formation history; a rigorous framework for stellar structure and evolution theories; a large-scale survey of extra-solar planets ($\sim$10–20,000); a large-scale survey of solar system bodies ($\sim$100,000); and support to developments such as Very Large Telescope (VLT), James Web Space Telescope (JWST), etc. Beyond our galaxy, Gaia will provide definitive distance standards out to the LMC/SMC; rapid reaction alerts for supernovae and burst sources ($\sim$20,000); quasar detection, redshifts, microlensing structure ($\sim$500,000); and fundamental physical quantities to unprecedented accuracy, specifically measuring γ to 10^{-7} (compared to 10^{-3} at present).

Looking in a little more detail at the expected discoveries in some more "glamorous areas": for extra-solar planets, the Gaia astrometric survey will enable the monitoring of hundreds of thousands of FGK stars out to about 200 pc. The detection limits will correspond to about 1 Jupiter mass and for periods in the range of $P \sim 2$–9 years, with a complete census of all stellar types. Masses, rather than lower limits ($M \sin i$), will be determined, and the relative orbital inclinations of some multiple systems measurable – the detection of some 10–20,000 planets in total is equivalent to some 10 planets per day. For the nearer systems, the astrometric displacements are very significant: for example, for 47 UMa it amounts to about 360 micro-arcsec. Consequently, orbits for some 5000 systems are predicted, and masses down to some 10 Earth masses at distances of about 10 pc (if such planets exist). Studies of photometric transits suggest that some 5000 transiting exo-planets might be discovered from an analysis of the variability data.

Within our solar system, the deep and uniform (20 mag) detection of all moving objects is expected to result in some 10^5–10^6 new minor planets (some 65,000 are known presently). Physical diagnostics follow: taxonomy/mineralogical composition versus heliocentric distance; diameters for about 1000, masses for about 100; orbits some 30 times better than present, even after 100 years of ground-based observations; Trojan companions of Mars, Earth and Venus; and some 300 Kuiper Belt objects to 20 mag. For near-Earth objects (Amors, Apollos and Atens) some 1600 Earth-crossers larger than 1 km in size are predicted (compared to some 100 currently known, although many more, if not all, will have been measured from the ground by 2010), with a (size) detection limit of 260–590 m at 1 AU, depending on albedo.

6.7. Design considerations for the Gaia payload

With these considerations, Gaia has been designed to provide:

(a) Astrometry (to $V < 20$ mag) with the following requirements: completeness to 20 mag based on on-board detection; an accuracy of 10–20 micro-arcsec at 15 mag; obtained with a scanning satellite, with two viewing directions, and yielding a global sky coverage with optimal use of observing time; the measurements based on the Hipparcos principles, and the data analysis on ground using a global astrometric reduction.

(b) Radial velocity (to $V < 16$–17 mag, fainter would be desirable but is unrealistic). These measurements are motivated by the determination of the third component of space motion, perspective acceleration, dynamics, population studies, binaries and spectra (providing chemistry, rotation, etc.). The underlying principles are the measurement of Doppler shifts using a slitless spectrograph observing the Ca triplet between 848–874 nm.

(c) Photometry (to $V < 20$ mag); that is, for all objects for which astrometric measurements are made. The photometric measurements are driven by the use of multi-colour photometry to determine astrophysical diagnostics. The present design includes five broad filters in the main focal plain (to provide chromatic corrections and astrophysical diagnostics especially in crowded regions) with 11 medium-band filters in a dedicated focal plane which will provide ΔT_{eff}, $\log g$, (Fe/H), and interstellar extinction along the line of sight.

6.8. On-board object detection for Gaia

On-board object detection is a very specific feature of Gaia, and its successful implementation will be essential in yielding an unbiased sky sampling (in terms of magnitude, colour and angular resolution). Since no all-sky catalogue at the required Gaia resolution (0.1 arcsec) to $V \sim 20$ mag exists, the solution is to avoid the use of an input catalogue or pre-defined observing programme, and to detect objects as they enter either of the two viewing directions, with a good detection efficiency. A low false detection rate, even at very high star densities, is required, and the system must be able to discriminate false triggers such as cosmic ray events. The required on-board analysis and data-handling complexity is very considerable. But a successful implementation will provide for the detection of variable stars which are bright enough at the moment of observation (eclipsing binaries, Cepheids, etc.), the autonomous detection of previously unknown supernovae (some $\sim$20,000 during the Gaia lifetime are predicted), microlensing events ($\sim$1000 photometric and 100 astrometric events are predicted); and the detection of rapidly moving, and previously unknown, solar system objects, including near-Earth asteroids and Kuiper Belt objects.

6.9. Main features of the Gaia payload

In order to achieve these objectives, the Gaia payload and supporting spacecraft include many new features compared to Hipparcos. Amongst these are the larger mirrors and overall support structure made of silicon carbide, providing uniform structural stability, a low coefficient of thermal expansion, and a high Young's modulus (Fig. 6.7). The mirrors are rectangular, and elongated along scan. The combined focal plane suppresses the modulating grid and photon-counting photomultiplier detector of Hipparcos, replacing these with an array of CCDs in the focal plane, with a telescope focal length matched to CCD detector physical dimensions (Fig. 6.8). The CCDs provide much higher detection efficiency, a broader wavelength coverage, spatial resolution, higher dynamic range and a multiplexing advantage (i.e. all the stars in the focal plane can be observed simultaneously, with the full integration time corresponding to the focal plane passage, unlike the sequential time-partitioning of the observations which had to be employed for Hipparcos).

Altogether, the massive leap from Hipparcos to Gaia applies to accuracy (2–3 orders of magnitude, i.e. 1 milli-arcsec to 4 micro-arcsec); limiting sensitivity (4 orders of magnitude, i.e. $\sim$10–20 mag); and number of stars (4 orders of magnitude, i.e. 10^5–10^9). This instrumental improvement can be understood largely in terms of the larger primary mirror ($0.3 \times 0.3\,\mathrm{m^2}$ increasing to $1.4 \times 0.5\,\mathrm{m^2}$), and the improved detector efficiency, although many other details contribute to the overall improvement. A critical element is the corresponding control of all associated error sources, which include optical aberrations, chromaticity, accuracy of the solar system ephemerides, attitude control and many others. The radial velocity instrument must achieve the very faintest photon detection limits possible, and presents its own set of technical and scientific challenges (Fig. 6.9).

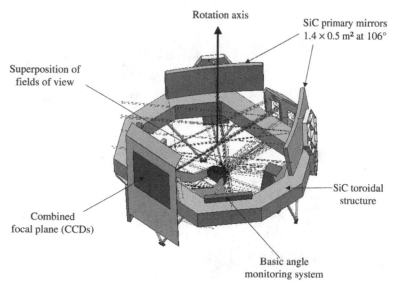

FIGURE 6.7. The heart of the Gaia payload is the two telescopes, corresponding to the two viewing directions, mounted on a silicon carbide optical bench structure. The combined fields are superimposed in the common focal plane. The satellite spins slowly around the indicated rotation axis, once every 6 h.

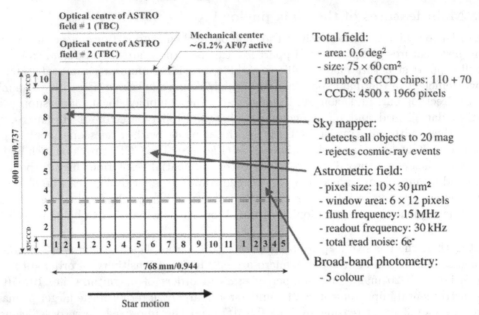

Optical centre of ASTRO field # 1 (TBC)

Optical centre of ASTRO field # 2 (TBC)

Mechanical center ~61.2% AF07 active

Total field:
- area: 0.6 deg²
- size: 75 × 60 cm²
- number of CCD chips: 110 + 70
- CCDs: 4500 x 1966 pixels

Sky mapper:
- detects all objects to 20 mag
- rejects cosmic-ray events

Astrometric field:
- pixel size: $10 \times 30\,\mu m^2$
- window area: 6 × 12 pixels
- flush frequency: 15 MHz
- readout frequency: 30 kHz
- total read noise: 6e⁻

Broad-band photometry:
- 5 colour

600 mm/0.737

768 mm/0.944

Star motion

FIGURE 6.8. The astrometric focal plane for Gaia. Star images pass from left to right, being first detected by the sky mapper, then information on these detections being passed to the main field, where small windows around the star images are isolated, read out and sent to ground. As the stars leave the field of view, their colours are measured in five broad photometric bands. All (CCDs) are operated in "time-delay integration" mode (i.e. read out in synchronization with the satellite scan motion) so that the stellar images are built up as the stars cross the focal plane. Their relative positions, at the measurement epoch, and projected in the along-scan direction, are determined from the image centroid positions determined on ground.

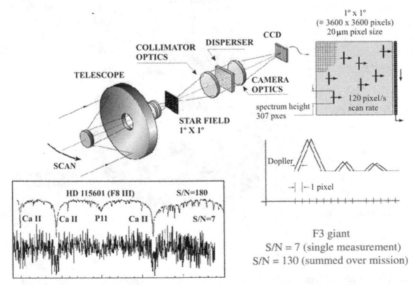

1° x 1° (# 3600 x 3600 pixels) 20 μm pixel size

CCD

COLLIMATOR OPTICS   DISPERSER

TELESCOPE

CAMERA OPTICS

STAR FIELD 1° X 1°

spectrum height 307 pxes

120 pixel/s scan rate

SCAN

Dopller

1 pixel

HD 115601 (F8 III)   S/N=180

Ca II   Ca II   P II   Ca II   S/N=7

F3 giant
S/N = 7 (single measurement)
S/N = 130 (summed over mission)

FIGURE 6.9. The radial velocity measurement concept for Gaia. A dedicated telescope produces slitless spectra, centred around the Ca line triplet at around 850–870 nm. The star spectra are read out, and sent to the ground.

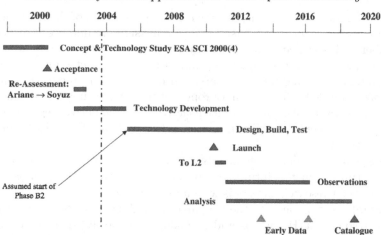

FIGURE 6.10. The overall schedule for the Gaia mission. At the present time (end 2003) the project is halfway through the technology development activities, and aiming for a start of the detailed design and implementation phase in early 2005. Following launch in 2010, the satellite will transfer to its L2 orbital location, and routine observations will start after instrument commissioning. During the operational phase, early catalogues will be released as appropriate, with the final catalogue expected to be completed not before 2018.

6.10. Development and schedule for Gaia

At the present time (late 2003), the Gaia project is in the middle of a number of important technical studies (2002–2004) which are aimed at proving or establishing the necessary technology required for the mission. The main activities comprise two parallel "system studies" (by Astrium and Alenia/Alcatel, respectively). Detailed technical activities include the CCD/focal plane development (Astrium and e2v, where the first CCDs have been produced); the silicon carbide primary mirror development (Boostec, where the primary mirror prototype is now under production); the high-stability optical bench prototype (Astrium and TPD Delft, which is nearing completion); payload data handling electronics (Astrium-D, where a breadboard development is underway); and the radial velocity instrument optimization (MSSL and Paris). Mission analysis aspects are under development by the European Space Operations Centre (ESOC). Also being studied are the FEEP thrusters, the high-gain antenna, the solar array deployment, the refocus mechanism, and ground verification and calibration.

Assuming that approval is given for start of the implementation phase in May 2005, launch will be targetted for mid-2010 (Fig. 6.10). The satellite will be launched by a Soyuz vehicle, and operated at the L2 Sun-Earth Lagrange point, 1.5 million km from Earth. An operational lifetime of 5 years is foreseen. A huge task, not addressed in the present contribution, is the preparation for the reception and automated analysis of the massive quantity of data expected from the mission. Serious effort on building this system, taking advantage of the Hipparcos experience but also major developments in data storage and computational power, has been underway for the past 3 years. The final Gaia Catalogue results are expected around the year 2018. Intermediate catalogues are expected to be released in advance of this date, as appropriate; for example, science alerts data (related to burst events like supernovae) will be released immediately.

6.11. Scientific organization of the Gaia project

ESA has overall responsibility for the execution and success of the Gaia mission. Like Hipparcos, Gaia is only feasible by bringing together the state-of-the-art expertise in

European space industry, and the huge knowledge and expertise of the ESA member state scientific community. The scientific aspects of the mission are under the responsibility of the 12-member Gaia Science Team. Within the current structure, there are 16 scientific working groups, focused on the payload optimization, specific astronomical objects (such as variable stars, double and multiple stars, and planets), and preparations for the data analysis. More than 200 scientists in Europe are now active in the working groups at some level. The Gaia scientific community is active and productive: regular science team and working group meetings are held, and an active archive of scientific working group reports is maintained.

Gaia presents many scientific challenges now and over the next few years, and many research opportunities in the future. Involvement by motivated and capable scientists starting out in their research careers is warmly welcomed.

For further information on both the Hipparcos and Gaia projects, the ESA www pages (http://www.rssd.esa.int/Hipparcos and .../Gaia, respectively) can be consulted as an entry into the extensive description and bibliography of each mission. The Hipparcos and Tycho Catalogues are also accessible from the Hipparcos www pages.

Acknowledgements

Both the Hipparcos and Gaia missions are very large scientific and technical undertakings which have extended over many years. This chapter is based on the collective work of these many colleagues.

7. Space Physics Instrumentation and Missions

By A. BALOGH

Space & Atmospheric Physics Group, The Blackett Laboratory, Imperial College, London, UK

Since the 1960s, instruments on Space Physics missions have changed and improved in many respects, but the basic physical parameters that we need to measure have remained the same. The requirement on any Space Physics mission is still to make measurements which are as extensive as possible of all the parameters of space plasmas: the distribution functions of all the constituents in the plasma populations; the DC and AC magnetic and electric fields; and the distribution functions of energetic particles species. All these parameters and distribution functions need to be measured with high spatial, directional and temporal resolution.

These lectures rely on extensive experience building magnetometers, energetic particle detectors, as well as on-board data processors and power supply and power management systems for space instruments. They provide an overview of the kind of instrumentation in which Europe has acquired considerable expertise over the years and which will be continued in future missions.

7.1. Space Physics and the space environment

7.1.1. *Introduction*

The first scientific space missions, at the birth of the space age in the late 1950s and in the 1960s, discovered that space in the solar system was filled with a complex medium, consisting of a fully ionized gas (plasma) that consists of electrons and ions (atoms from which some or all their electrons have been stripped). The plasma in the regions visited by these early spacecraft in the Earth's immediate space environment and in regions as far as the moon, Venus and Mars, was divided into distinct regions in space. As the regions, around the Earth and in the nearby interplanetary space were revisited by successive missions, it was also found that many of the properties of the plasma were variable in time. In these different regions, the plasma was found to be of different origin and had significantly different properties. The boundaries between regions were also variable. Since the pioneering days forty years ago, Space Physics, the study of plasmas in space, has become a recognized scientific discipline. Topics covered by Space Physics, defined in more detail below, include the different plasma regions in the heliosphere, the region that surrounds the Sun in which the Sun's extended atmosphere, the solar wind, controls the properties and dynamics of the medium.

In Space Physics, the acquisition of observations in the different regions in space often led the theoretical understanding of the phenomena. Due to the nature of space plasmas, observations are made locally, in the medium itself, and this has led to defining the methodology of the observations and the instruments used as *in situ* that can be translated as "on site" or "in the medium". In this chapter, this phrase will occasionally be used, although it is preferable to talk about Space Physics instrumentation and measurements. This definition aims at distinguishing measurements in space plasmas from remote sensing measurements and observations that use instruments (such as optical telescopes, for instance) that provide data about objects and their environments located away from the instruments themselves. As is often the case, this distinction between *in situ* and remote observations is somewhat artificial when the scientific objectives of the observations are considered. In many cases, *in situ* measurements provide important results that are also directly relevant to remote phenomena.

In this chapter, we study those instruments that make *in situ* measurements, either to characterize the space medium locally, or to make local measurements that can be interpreted in terms of processes and phenomena that occur elsewhere in the space environment. These instruments measure characteristic parameters of the space plasma, such as the magnetic field, the magnetic and electric components of waves in the plasma, the parameters of the plasma itself, such as its density, flow velocity and composition, and the flux intensities and composition of energetic particles.

A useful definition to describe the discipline has been formulated by National Aeronautics and Space Administration (NASA), primarily as a framework for funding missions and research, and it describes well the different topics covered by Space Physics. Using NASA's four headings, the regions and phenomena that are the subjects of *in situ* investigations are described below. Several recent books provide a comprehensive description of Space Physics phenomena (Baumjohann and Treumann, 1996; Cravens, 1997; Gombosi, 1998; Kivelson and Russell, 1995).

7.1.2. *The Sun and its atmosphere*

This topic includes all aspects of solar physics, and is therefore more suited to remote observations of the Sun. However, there are several types of *in situ* measurements (the composition of the solar wind plasma, magnetic fields, radio waves and energetic particles) that are directly relevant to the study of phenomena occurring on the Sun and its atmosphere. The most important of these for Space Physics is the solar wind.

For reasons that have not been completely understood, the atmosphere of the Sun, the corona, is very hot, always and everywhere in excess of a million kelvin. In many parts of the corona temperatures can reach more than 1.5 million K. At these temperatures, the outer corona is not stable and is always expanding into interplanetary space at speeds varying between $\sim$300 and $\sim$800 km/s. Although the first model of the solar wind was based on the high temperatures in the corona, it is now known that additional energy input in the form of waves is needed to accelerate the solar wind.

Combining remote observations of the solar corona that were obtained by the first X-ray solar telescope from NASA's Skylab 1 and *in situ* observations of the solar wind near the Earth, it was established that there are two kinds of solar wind, fast ($>$550 km/s) and slow ($<$450 km/s) that originate in different regions of the solar corona. Fast solar winds originate in coronal holes, so called because they are relatively cool ($\sim$1.2 million K) and were found to be darker in X-ray pictures of the corona. These regions also have magnetic field lines that are anchored in the solar surface, but extend into interplanetary space. The slow solar wind is emitted from coronal regions which are dominated by magnetic loops and are considerably hotter than coronal holes. The fast solar wind is considerably more uniform in all its properties than the slow solar wind.

At solar minimum activity, the polar regions are covered by large coronal holes with occasional extensions towards the solar equator. Fast solar wind streams expand from the polar coronal holes to fill the three-dimensional (3D) heliosphere, except for a band of about $\pm$20° around the solar equatorial plan that contains both fast and slow solar wind. Around the maximum in the 11-year solar activity cycle, most of the corona is tightly bound in complex magnetic loops that cover the whole solar surface. There are only small, transient coronal holes. The solar wind is generally slow and highly variable everywhere, at all solar latitudes. Frequent solar storms, resulting in coronal mass ejections (CMEs), inject considerable amounts of plasma and magnetic flux into the interplanetary medium, superposed on the solar wind. Successive large transient CME events at all heliolatitudes propagate away from the Sun and eventually form a significant obstacle to the penetration

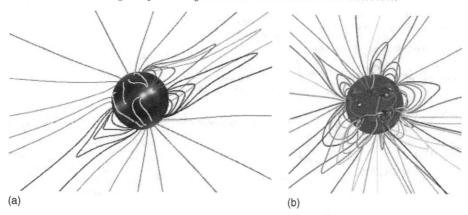

(a)   (b)

FIGURE 7.1. Magnetic field structures in the solar corona at solar activity minimum (a) and maximum (b), computed using a MHD model of the corona, but using ground-based measurements of the magnetic field on the solar surface (photosphere). (Courtesy of J. Linker and Z. Mikic.)

of cosmic rays into the inner heliosphere. The different appearance of the corona at solar minimum and solar maximum is illustrated in Fig. 7.1.

The two kinds of solar wind are illustrated in Fig. 7.2. These observations were made by instruments on the Ulysses spacecraft, the first to make measurements in the solar wind as a function of heliolatitude. The portion of the orbit that corresponds to the interval of measurements during spring 1995 is also shown in Fig. 7.2.

7.1.3. *The heliosphere*

The solar wind fills a vast volume around the Sun, the heliosphere. It carries the extended solar magnetic field; together, the plasma and the magnetic field form a highly structured, dynamic medium. The overall shape of the heliosphere is sketched in Fig. 7.3. Solar variability on all temporal and spatial scales introduces variability in the heliosphere on the same variety of scales, out to its boundary at a distance estimated to be of the order 100 AU. *In situ* measurements are essential to understanding the properties of the heliospheric medium.

Magnetic field structures in the solar corona at solar activity minimum (Fig. 7.1a) and maximum (Fig. 7.1b), computed using a magnetohydrodynamic (MHD) model of the corona, but using ground-based measurements of the magnetic field on the solar surface (photosphere).

Phenomena in the inner heliosphere, taken to be the volume from the Sun out to the orbit of Saturn, are much better known and understood than those at larger distances. The mainly dipolar component of the solar magnetic field is carried out into the heliosphere by the solar wind. As the magnetic field lines are stretched out, the rotation of the Sun (at a period ∼26 days) shapes them into a 3D spiral pattern. The Heliospheric Current Sheet (HCS) is a complex, 3D surface that separates magnetic filed lines directed towards and away from the Sun. Close to solar minimum, when the two hemispheres are filled with high-speed solar wind, magnetic field lines from these two hemispheres are directed, respectively, towards and away from the Sun, and the HCS has been likened to a ballerina's skirt, a wavy surface close to the solar equatorial plane. In the active phase of the solar cycle, the solar wind is mostly composed of slow streams, with occasional, interspersed high-speed streams, and the magnetic field of the Sun is more complex, the HCS becomes a very complex, 3D structure reaching to high heliolatitudes. The

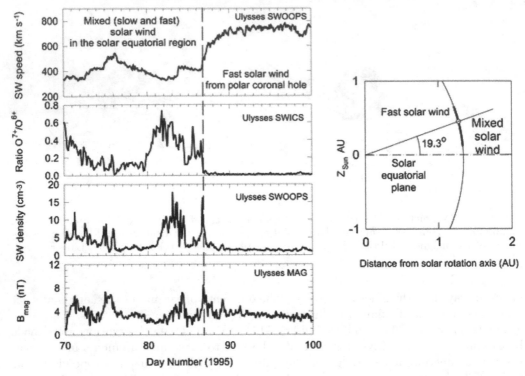

FIGURE 7.2. Illustrating the observations of fast and slow solar wind streams. In March 1995, the Ulysses spacecraft crossed the solar equatorial plane and observed the transition from a mixed (slow and fast) solar wind regime around the equator and purely fast solar wind streams poleward of about 20°. The right panel shows the trajectory of Ulysses that corresponds to the observations, the left panel shows, from top to bottom, the solar wind speed, the ratio of oxygen ions with charge $+7e$ and $+6e$ ($e =$ electronic charge), the density of the solar wind and the strength of the magnetic field. The electric charge state of the oxygen ion population is a proxy for the temperature in the corona where the solar wind originates.

schematic view of the inner heliosphere at solar minimum and solar maximum is illustrated in Fig. 7.4.

The dynamic interaction of solar wind streams of different speeds creates complex compressed and rarified regions in the heliosphere. Around solar minimum, the flow patterns from the Sun are relatively stable over many solar rotation periods. As the Sun rotates, fast wind streams compress the preceding slow wind streams to form the characteristic Corotating Interaction Regions (CIRs) that dominate the equatorial region of the heliosphere. The compressed solar wind regions develop shock waves at their leading and trailing edges that accelerate particles up to energies of several MeV. Around solar maximum activity, stable pattern of compression regions cannot be formed, but at smaller spatial scales, the dynamic interaction between solar wind streams of different speeds still takes place. At this time, the corona also becomes very disturbed and complex, with frequent emissions of CMEs that become the dominant factor in the dynamics of the heliosphere.

Solar wind and heliospheric magnetic field measurements are shown in Fig. 7.4 for (a) a CME in 2003 and (b) for a long-lasting series of CIRs in 1992–3. In the case of the CME observations, a main characteristic is the large-scale rotation observed in the magnetic field data that has been interpreted as a closed magnetic structure, a "magnetic cloud" (shown as the highlighted interval) that corresponds to a large coronal loop that was

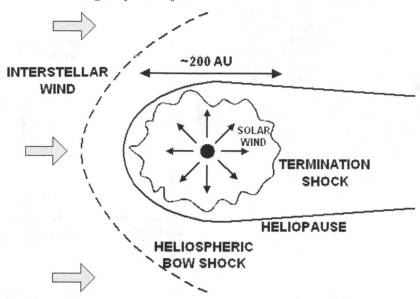

FIGURE 7.3. A schematic view of the heliosphere. The size of the heliosphere remains to be confirmed by the measurements expected from the Voyager mission, but the distance from the Sun to the first heliospheric boundary, the termination shock, is likely to be of order 100 AU (1 AU = 150,000,000 km).

ejected from the corona. The sharp increase in both magnetic field strength (a, b) and solar wind speed (c, d) ahead of the magnetic cloud indicates the passage of a shock wave that was generated by the CME as it compressed a slow-flowing solar wind stream ahead of it. The long series of CIRs shown in Fig. 7.4(d) were observed at a time of declining solar activity, as the observing platform, the Ulysses spacecraft, moved away from the ecliptic plane towards higher solar latitudes when, at the same time, the fast solar wind stream emerged from development of the large polar coronal hole. At mid-solar latitudes, the instruments on the spacecraft observed the CIRs formed by the interaction of the fast solar wind from the Sun's polar region with the slower solar wind streams from its equatorial region.

7.1.4. *Solar-terrestrial and solar-planetary relations*

Planets that have a sizable internal magnetic field (Mercury, the Earth, Jupiter, Saturn, Uranus and Neptune) present an obstacle to the flow of the solar wind. In the first approximation, the solar wind cannot penetrate into the regions occupied by the planetary magnetic fields. However, its dynamic pressure compresses the magnetic field on the sunward side of the planet and it then flows around the volume of space occupied by the magnetic field and plasma populations that are generated in the planet's proximity, for instance from its atmosphere. The enclosed space is the magnetosphere. The terrestrial magnetosphere is illustrated in Fig. 7.5; its main regions, containing different plasma and magnetic field regimes, are also shown in this figure and in Fig. 7.6. The magnetosphere is characterized by a compression of the dipole magnetic field of the Earth on the sunward side, on the anti-sunward side, the magnetic field lines are stretched out into a very long magnetotail.

The complexity of the different plasma regimes in the terrestrial magnetosphere is illustrated in Fig. 7.7. Plasma particles are found to have widely different energies and densities, covering many orders of magnitude in flux energy and flux density. *In situ*

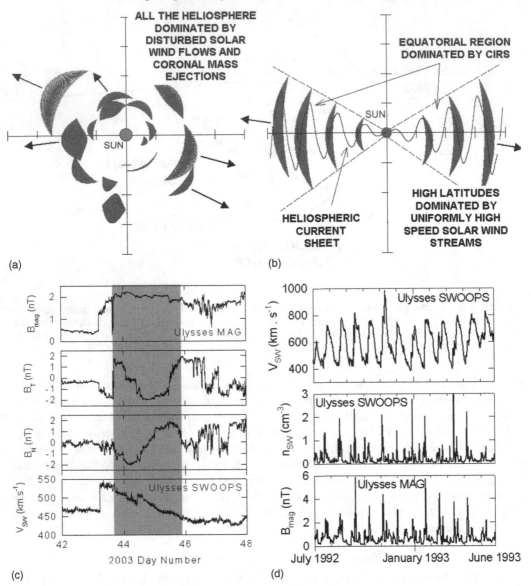

FIGURE 7.4. The structure of the inner heliosphere at solar maximum (a) and solar maximum (b) In the lower panels, heliospheric observations illustrate characteristic phenomena representative of the two phases of solar activity. (c) Typical measurements are shown that are associated with the passage of a CME (shaded interval). Ahead of the CME, the solar wind speed and the strength of the magnetic field jump at the passage of the shock wave driven by the CME. The CME itself is identified here as the large-scale rotation of the magnetic field, corresponding to a magnetic cloud. (d) The long series of CIRs observed in 1992–3 by the Ulysses spacecraft is shown. The alternating fast and slow speed solar wind streams cause periodic compressions of the solar wind and the magnetic field.

observing instruments need to take into account this wide range to cover the required measurements. In addition, the plasma populations are highly variable in time, the energy spectra shown in Fig. 7.7 only represents typical conditions.

Other planetary magnetospheres present both similarities and significant differences. Jupiter's magnetosphere is considerably larger than that of the Earth, due in part to

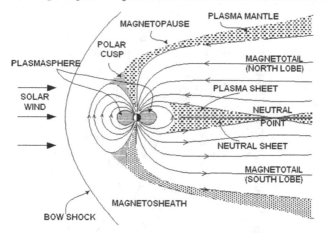

FIGURE 7.5. The different regions of the terrestrial magnetosphere. Although this sketch identifies the plasma regimes in the magnetosphere, in reality, constant dynamic changes cause the boundaries (bow shock, magnetopause, cusps) to be in constant motion, in response to changes in the solar wind and the magnetosphere's response to these changes.

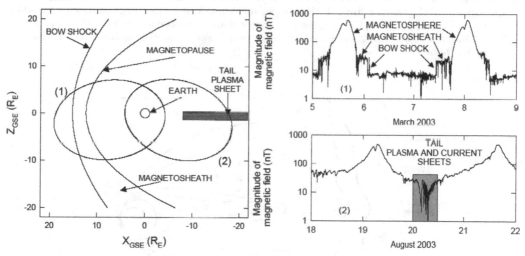

FIGURE 7.6. Measurements of the magnetic field along two different orbits of the four-spacecraft Cluster mission, one with its apogee in the solar wind, one in the magnetotail.

Jupiter's much larger magnetic field, but also to the intense source of plasma generated from the volcanic emissions of *Io*. The fast rotation of Jupiter (with a period ~10 h) imparts a centrifugal force to the relatively dense plasma in the equatorial plane: this results in a stretched-out plasma disk that drags out the dipole field of the planet, as illustrated in Fig. 7.8. The size of the magnetosphere coupled to the fast rotation creates a significantly different dynamic environment from that of the Earth.

7.1.5. Cosmic rays

Shock waves that surround supenovae accelerate particles to a broad range of very high energies. This high-energy particle population is one of the components of the interstellar medium in the galaxy (and beyond). Cosmic rays penetrate through the outer boundaries of the heliosphere and encounter the outward propagating solar wind and the

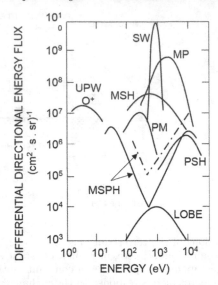

FIGURE 7.7. A schematic illustration of the energy spectra of the different plasma populations in the terrestrial magnetosphere and its vicinity. SW: solar wind; MSH: magnetosheath; MP: magnetopause; PM: plasma mantle; PSH: tail plasma sheet; MSPH: the range of spectra found in the inner magnetosphere; UPW O^+: upwelling, singly charged oxygen atoms from the ionosphere. (After a diagram representing the capabilities of the ion spectrometer on the Cluster mission, courtesy H. Rème.)

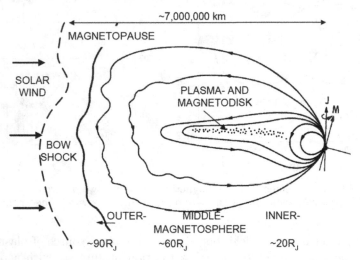

FIGURE 7.8. The sunward side of Jupiter's magnetosphere. Magnetic fields in the equatorial region are stretched out by the plasma disk formed by ions generated from volcanic emission from Io. The angle between the rotation axis of the planet and its magnetic field ($\sim 11°$) further distorts the magnetosphere. The irregular shape of the bow shock and the magnetopause are caused by the large size of the magnetosphere.

heliospheric magnetic field. The energy spectrum of cosmic rays measured at the orbit of the Earth (in the inner heliosphere) is shown schematically in Fig. 7.9; in reality, the energy spectrum continues to extremely high energies, at least to 10^{22} eV. As observed by detectors either on the Earth or in space, the intensity of the lower-energy cosmic rays is modulated in anti-phase with solar activity, as shown in Fig. 7.9. The shape of the energy

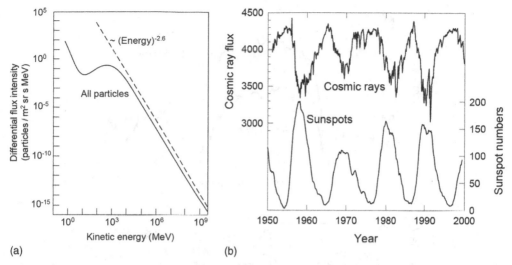

FIGURE 7.9. The energy spectrum of cosmic rays observed at the orbit of the Earth (a). The intensity of cosmic rays varies with the solar activity cycle (represented by the sunspot numbers) as a result of variable access through the outer heliosphere (b).

spectrum in the energy range up to 10^{10} eV, as shown in Fig. 7.9, is controlled by the heliosphere. It was this observation that originally led to the concept of the heliosphere as the volume of space in which the varying conditions created by different levels of solar activity controls the access of cosmic rays into the inner solar system. Although cosmic rays are primarily of astrophysical significance, their inclusion here is justified as their measurement, both Earth- and space-based, use *in situ* techniques; scientifically, their study is intimately linked with heliospheric structures and variability. A component of the cosmic ray spectrum, at moderate energies ($\sim$1–50 MeV), is constituted by neutral atoms penetrating into the heliosphere from interstellar space where they are first ionized by charge exchange and then accelerated, presumably at the heliospheric termination shock.

A special population of energetic particles, from 0.1 MeV up to several hundred MeV, has its origin on the Sun and in the heliosphere. Such particles are accelerated either near the Sun, as a consequence of solar flares and shock waves driven by CMEs, or by shock waves associated with CIRs. It is not quite clear, although it is likely, that all solar energetic particle events are accelerated by coronal shock waves formed by explosive events such as flares and CMEs. When observed in the heliosphere, or in Earth orbit, such events are characterized by an increase of energetic particles fluxes by several orders of magnitude over several days. The flux intensity versus time profile of these energetic particle events varies considerably, partly because of natural variability in the effectiveness of the shock waves during the acceleration process and partly because of propagation effects in the corona itself and in the heliospheric magnetic field. A relatively simple event is shown in Fig. 7.10(a), with a fast rise to high intensities and a slower decay to pre-event levels. A more complex event is shown in Fig. 7.10(b), in which the fast rise to peak intensities is followed by a period of sustained high fluxes. This was found to have been caused by large-scale magnetic structures associated with a CME following the energetic particle event that "stored" the high fluxes of energetic protons for more than 5 days in a magnetic "bubble" (indicated by the shading in the figure) that filled a significant fraction of the inner heliosphere.

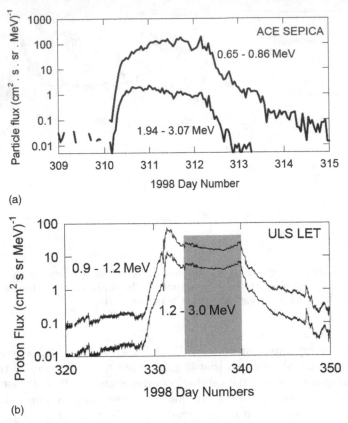

(a)

(b)

FIGURE 7.10. Solar energetic particle events observed in the heliosphere, near the orbit of the Earth (a) and at high solar latitude (b). The intensity versus time profile of these events is controlled by both coronal and heliospheric propagation effects.

A summary sketch of the cosmic rays and energetic particles populations, using energetic oxygen ions as an example, is shown in Fig. 7.11. Although protons (ionized hydrogen atoms) are the most numerous component of all energetic particle populations, ions of all elements (and most isotopes) are represented in space. The relative fluxes of the different elements and isotopes are important indicators of particle acceleration and propagation phenomena.

7.2. Measurements of magnetic and electric fields and waves

7.2.1. *DC magnetic field measurements*

The accurate *in situ* measurement of the magnetic field vector is of fundamental importance to all Space Physics missions. Space plasmas are always threaded by magnetic field lines that provide the basic structural organization in different plasma regimes, it is said that the magnetic field provides the "road map" in space plasmas. The properties of plasma, such as its temperature depend on the direction of the magnetic field and, in general, the temperature of the plasma measured perpendicular to the magnetic field and parallel to it are different, corresponding to different random velocity distributions of the plasma ions. More energetic particles are guided by the magnetic field; it is often possible to describe the directional energetic particle flux intensity in the form of pitch angle distributions that are rotationally symmetric around the local magnetic field. The

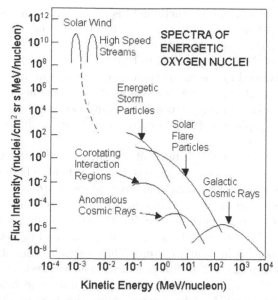

FIGURE 7.11. The energy spectra of energetic oxygen ions in the different populations of energetic particles in the heliosphere. (After a diagram by the WIND energetic particles instrument team.)

TABLE 7.1. Requirements on magnetic field measurements.

Environment	Range (nT)	Required accuracy (nT)
Earth's field and low-Earth orbit	0–45,000	~0.1–1
Magnetospheres	0–10,000	~0.05–0.2
Inner heliosphere	0–100	~0.05–0.1
Outer heliosphere	0–30	~0.01

required ranges and accuracies of magnetic field measurements appropriate to different plasma environments are shown in Table 7.1.

The magnetic field is a vector, usually denoted by **B**. Most magnetometers in space are dedicated to measuring the three orthogonal components of the vector, for example B_x, B_y, B_z in a specified Cartesian coordinate system. Two main types of magnetometers are used. The first, most frequently used type is the triaxial fluxgate magnetometer (FGM), the second is the vector helium magnetometer. The main characteristics of both these types are described below.

The performance of magnetometers is measured by their sensitivity (the smallest magnetic field they can measure) that is related to their noise level; their measurement range; and their frequency response. Magnetometers used on all space missions meet these requirements and have, in general, noise levels <10 pT in the frequency range 0.01–1 Hz. It is routinely possible to achieve accuracies better than 0.1 nT, within a frequency range from DC to 10 Hz. In and around the magnetosphere, magnetic fields to be measured range from a few nT to several thousand nT (the Earth's magnetic field at the equator, on the Earth's surface, is about 30,000 nT). However, in interplanetary space, at large distances from the sun, the magnetic field is usually 1 nT or less. It is nevertheless possible to build magnetometers which meet all the measurement requirements on current and foreseeable space missions. The measurement range is usually quite flexible; the dynamic range (ratio of the maximum field to be measured to the resolution of the instrument) is usually determined by the magnetometer electronics and the analogue-to-digital converter (ADC). On past space missions, 12-bit ADCs were used, currently 14–16 bit ADCs are routinely used, and more recently, on the latest space missions 20-bit ADCs are either used or being planned (20 bits correspond to a maximum range of 1 million times the digital resolution).

Magnetometers have, in general, a small offset, which means that the output values of the three magnetic sensors are not exactly zero in a zero field. These offsets are normally determined by calibration on the ground and have values of order 0.5–5 nT. However, the offsets are usually not constant in time, and an in-flight determination is necessary. On spinning spacecraft, the components of the offset vector in the plane perpendicular to the spin axis can easily be determined, as, in a slowly varying magnetic field, the average measured over a complete spin should be zero. The offset component along the spin axis is more difficult to determine, but techniques have been devised, valid in interplanetary space, which rely on the statistical properties of the magnetic field to determine the spin-aligned offset component. In the magnetosphere, this is more difficult to apply, and alternative methods are used; one such method is to rotate the spacecraft spin axis by 90°. On three-axis stabilized spacecraft, offsets are usually difficult to determine; in interplanetary space, again it is the statistical properties of the magnetic field which are used. In any case, it is important to remember that the offsets and their long-term stability are the most critical characteristics of space-borne magnetometers. The offset and other error sources in the transfer function of a typical magnetometer are shown schematically in Fig. 7.12.

The *triaxial* FGM consists of three sensors, each of which, together with the sensor-specific electronics, provides a voltage proportional to the value of the component of the magnetic field along its axis. The magnetic field vector is usually measured by an orthogonal arrangement of three sensors, as shown in Fig. 7.13.

A single-axis fluxgate magnetometer sensor with its electronic circuit is shown in Fig. 7.14. It consists of a magnetic core, which has a toroidal drive winding around it, and a coil former surrounding the sensor core and drive winding, around which another sense winding is placed. The two coils are effectively orthogonal, so that there is no magnetic coupling between them. Bipolar, symmetric current pulses in the drive winding are used to drive the core material deep into saturation around the hysteresis loop, at a frequency usually about $f_o = 15$ kHz. In the absence of an external field, the symmetry of the hysteresis loop ensures that there is no signal in the sense winding. However, in the presence of a non-zero component of an external magnetic field along the axis of the sense winding, the hysteresis loop is slightly displaced, leading to a non-symmetric magnetic signal which induces an alternating voltage in the sense winding, at a frequency $2f_o$. This signal, of order $<1\,\mu$V, is proportional to the component of the magnetic field along the axis of the sense winding. It is first amplified, then detected, using a synchronous detector. After some further amplification, the resultant voltage signal is fed back, through a transconductance (voltage-to-current) amplifier and the sense winding, as a feedback current counteracting the effect of the external field in the core. The voltage, representing the analogue value of the magnetic field component, is fed to an ADC that provides the digital output required for the data processing system.

Almost all magnetometers on space missions to date have used this design. The magnetometer on the four-spacecraft Cluster mission (Balogh *et al.*, 2001), for instance,

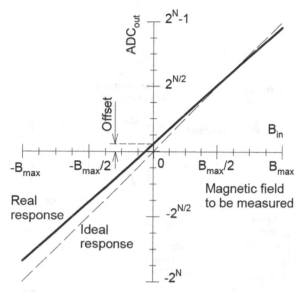

FIGURE 7.12. The response of a magnetometer to a given external magnetic field. The plot shows the output of the ADC versus the ambient field to be measured. The sources of the measurement errors are a zero offset and the non-linearity of the response of the sensor, sensor electronics and the ADC combined. These effects can be overcome by the calibration of the magnetometer and the application of the calibration values to the data. Note that the effects are exaggerated in this figure for clarity.

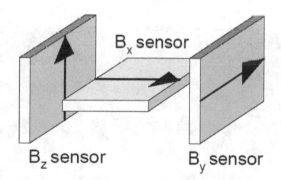

FIGURE 7.13. The schematic arrangement of three single axis fluxgate sensors to form an orthogonal triad. Each sensor measures one of the components of the magnetic field vector.

consists of two 3-axis fluxgate magnetometer sensors and the associated electronics box is shown in Fig. 7.15. For comparison, another, similar current magnetometer, on the Advanced Composition Explorer spacecraft is described by Smith *et al.* (1998).

It is possible to combine two single-axis sensors by placing two orthogonal sense windings around the same toroidal drive winding, as shown in Fig. 7.16. One of the sensors is thus constructed to measure two components of the magnetic field independently. The advantage of this design is the lower mass of the magnetometer sensor. This magnetometer has been successfully used on the Chinese Double Star two-spacecraft mission launched in 2004.

An alternative design of the fluxgate magnetometer uses the same sensors, but implements the electronics in a digital form. By replacing the analogue processing chain with

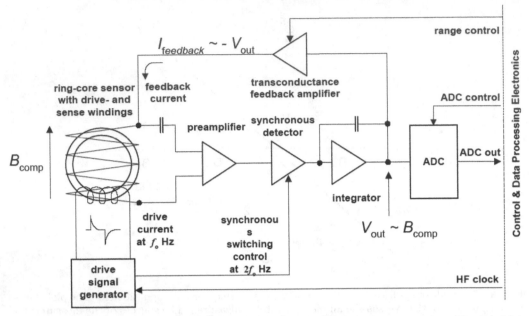

FIGURE 7.14. Functional block diagram of a single-axis fluxgate magnetometer. The toroidal sensor and its drive and sense windings are shown schematically on the left. The output of the magnetometer is a voltage proportional to the component of the magnetic field vector projected onto the axis of the sensor. The analogue voltage is converted to a digital value in the ADC for further processing and transmission to the ground through the spacecraft telemetry.

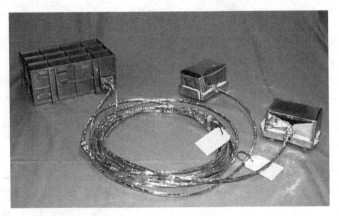

FIGURE 7.15. The two tri-axial fluxgate magnetometer sensors and the associated electronics box used on the four-spacecraft Cluster mission. The long cables connecting the sensors to the electronics box are used on the boom of the spacecraft. The electronics box contains the magnetometer sensor electronics, as well as the data processing and telemetry interface electronics, and the power supply unit used by the magnetometer.

a digital system, this signal processing is performed as a software function in the digital domain. As well as eliminating the temperature effects, this provides an inherent flexibility in the design, since the processing parameters can be modified by software. The basic design of the digital technique magnetometer is presented in Fig. 7.17. The fluxgate sensor output current is digitized immediately after the sense signal current is amplified and converted to a voltage; this stage is the same for both designs. Subsequently, the

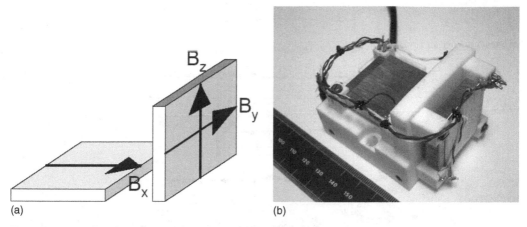

FIGURE 7.16. Combination of two single axis magnetometer sensors using a single core and drive and two sense windings (a); prototype of the three-axial sensor implementing the two-core solution used on the double star magnetospheric mission (b). The sensors are mounted on a ceramic frame for dimensional stability.

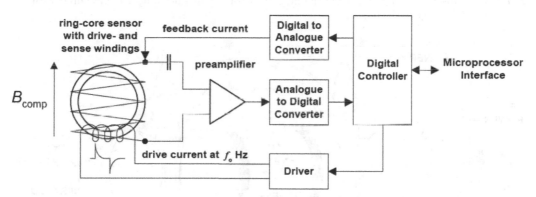

FIGURE 7.17. Functional diagram of a single-axis digital magnetometer. Replacing the analogue signal handling with digital processing in the digital controller improves the stability of the magnetometer.

filtering, synchronous detection and integration is performed by software, resulting in a calculated value for the feedback current which is used to null the ambient field. The feedback current is generated through a digital-to-analogue converter. This calculated value is therefore the field-proportional output of the magnetometer, and this value can be transmitted directly to the instrument Data Processing Unit (DPU).

Another type of magnetometer that has been used on some major missions (Pioneer 10 and 11, International Sun–Earth Explorer ISEE-3, Ulysses, Cassini Saturn Orbiter) is the Vector Helium Magnetometer (VHM). Its use has been restricted by (a) its relatively greater complexity, mass and power, when compared to the fluxgate magnetometers, and (b) its restricted availability: only the scientific team at the Jet Propulsion Laboratory have developed and used it. However, in principle, it out-performs the fluxgate types by its lower noise level, hence higher sensitivity, and by its greater stability (its offsets, in particular, have proved extremely stable on such missions as Pioneer and Ulysses). A sample plot of data covering 1 day of observations of the very weak interplanetary

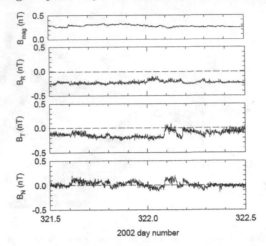

FIGURE 7.18. Measurement of the weak interplanetary magnetic field by the VHM magneto-meter on Ulysses for 1 day when the spacecraft was at $\sim$4.5 AU from the Sun. The measured components of the magnetic field (lower three panels) are very small, of order 0.1 nT or less. The fluctuations in the measured components represent natural (Alfvénic) fluctuations in the direction of the magnetic field, while the magnitude of the magnetic field is measured with an uncertainty of order 0.02 nT.

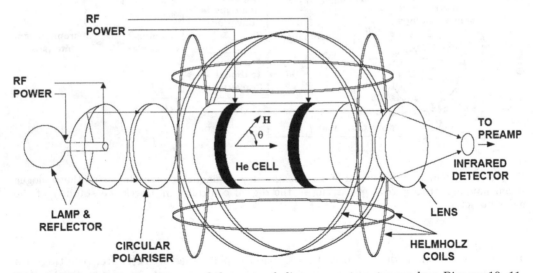

FIGURE 7.19. Schematic diagram of the vector helium magnetometer used on Pioneer 10, 11, ISEE-3, Ulysses and Cassini missions. For the Cassini Saturn Orbiter mission, the magnetometer has been configured to be used either as a scalar sensor (measuring the magnitude of the magnetic field) or as a vector magnetometer. (Figure courtesy of E.J. Smith.)

magnetic field when Ulysses was at 4.5 AU from the Sun is shown in Fig. 7.18 to illustrate the capabilities of the instrument.

The VHM, schematically illustrated in Fig. 7.19, is based on the response of an optically pumped metastable helium population (contained in a glass cell) in the presence of external magnetic fields (Smith *et al.*, 1975). Magnetic fields are detected by their effect on the efficiency with which metastable helium can be optically pumped. In optical pumping, a non-thermodynamic equilibrium distribution of atoms among the various energy substates is produced by incident polarized radiation. The circularly polarized light at 1.08 mm passing through the He cell induces optical pumping of the metastable He population in the triplet "ground" state. The change in optical pumping efficiency depends on the intensity of the magnetic field and its angle with respect to the optical axis of the sensor. When a rotating sweep field of ~300 Hz is applied using the Helmholtz coil that surrounds the optical cell, the pumping efficiency, and therefore the infrared (IR) throughput detected by the IR sensor vary as the vector sum of the sweep field and the steady ambient field in the sweep plane. Under these conditions, the signal in the IR detector contains a sinusoidal component at the sweep frequency whose magnitude is proportional to the ambient field strength and whose phase relative to the sweep waveform depends on the direction of the ambient field in the sweep plane.

Phase coherent detection of the output of the IR sensor in the magnetometer electronics produces voltages representing the ambient field components along the optical and transverse axes. These voltages are used to generate feedback currents in the sensor coils that null the ambient field on both axes. The feedback currents are highly linear measures of the two field components. Triaxial (vector) measurements are obtained by alternating the sweep field between two orthogonal planes that intersect along the optical axis. The performance of the VHM is measured by its very stable offsets, with variations of ~20 pT over several years, and its very low noise, ~3 pT rms/$\sqrt{\text{Hz}}$; the best current fluxgates have an offset stability of only about 0.2 nT, although their noise performance now approaches that of the VHM.

It is also possible to operate this magnetometer in the scalar mode, in which its response is made dependent, with very high accuracy and precision, on the magnitude of the external magnetic field. Combining the scalar capability of the helium magnetometer with the vector capability of a tri-axial FGM provides a very sensitive magnetometer system; such a system is currently flying on the Cassini Saturn Orbiter mission and is used to provide a very accurate mapping of the internal magnetic field of Saturn now that Cassini has been placed in orbit around the planet.

The weak magnetic fields to be measured *in situ* (other than in the inner magnetospheres of strongly magnetized planets: the Earth, Jupiter and Saturn) require that special care should be taken to minimize the magnetic background due to the spacecraft itself at the location of the magnetometer sensors. This is normally achieved in three successive steps. The first is to ensure that the spacecraft, its subsystems and its payload elements have individually small magnetic moments, if necessary by attaching small, compensating magnets to the intrinsically magnetic devices in the spacecraft (such as valves used for attitude manoeuvres). This step also includes minimizing the effects of current loops in the spacecraft power system (in particular in the wiring of the solar arrays). The second step involves placing the sensors at the end of a boom that can be deployed in flight, to remove the sensors from the immediate proximity of the spacecraft. The length of the boom employed is usually defined by mechanical considerations and its cost, but can vary from ~1 to >10 m for use in the outer heliosphere, such as on the Voyager and Cassini spacecraft). On a high-performance magnetospheric mission, such as Cluster, the boom needs to be about 5 m in length (as illustrated in Fig. 7.20).

Even these two steps are not usually sufficient to meet the requirements of the missions. A third step is needed. It consists in measuring and modelling the residual background magnetic field of the complete spacecraft in an appropriate facility, calculating the value of a compensating magnet to be placed on the spacecraft and then to re-measure and re-model the residual magnetic field at the location of the magnetometer sensor. Figure 7.21 shows the result of this third step as performed for one of the Cluster spacecraft; the DC background field at the location of the two magnetometer sensors was found to be ~0.1 nT, better the specification of 0.25 nT.

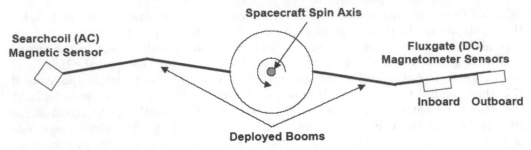

FIGURE 7.20. Boom mounting of DC and AC magnetometer sensors on the Cluster spacecraft to minimize the magnetic interference from the spacecraft and its systems. Each boom, when deployed, is about 5 m long from the outer circumference of the spacecraft.

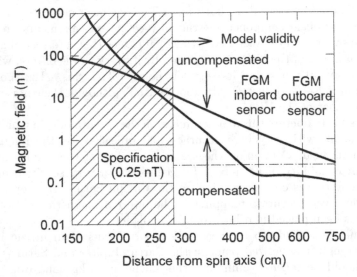

FIGURE 7.21. Magnetic compensation of the Cluster spacecraft. Although care had been taken to minimize the magnetic field of the spacecraft by magnetic compensation of specific units, the overall magnetic signature of the spacecraft at the location of the magnetometer sensors was still outside the specification. Measurements and modelling led to further magnetic compensation that achieved the low disturbance fields needed to meet the objectives of the mission. (Note the logarithmic scales on the graph.)

Data acquired in space by the magnetometer are transmitted through the telemetry. Data received on the ground need to be processed and calibrated before being ready for scientific analysis. A brief description is given below of the steps required to generate scientifically usable data from the raw telemetry data, based on the calibration procedures used for the magnetometers on the four-spacecraft Cluster mission. A first step is to "unpack" the telemetry data (that is transmitted in a special format) and to generate magnetic field vectors in physical units in the sensor coordinate system, called here FS. At the same time, it is necessary to generate the timing information and to determine (using timing data from the spacecraft) the phase of the spacecraft spin at which each measured vector was acquired. The calibration parameters then need to be applied to yield the magnetic field vector in physical units, in a coordinate system, denoted FSR, which is the magnetometer sensor's spin reference system, an orthogonal, right-handed coordinate system with its x-axis along the real spin axis of the spacecraft, and the

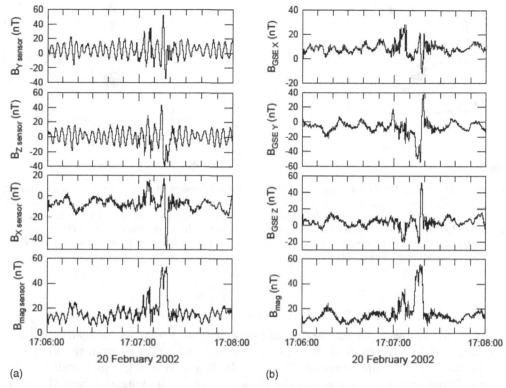

FIGURE 7.22. (a) Raw magnetic field data measured by the magnetometer on one of the Cluster spacecraft. The digital values transmitted through the telemetry have been simply converted to physical units (nT). The spin modulation of the measured magnetic field components in the spin plane (upper two panels) is clearly visible. The magnitude of the measured signal is also modulated by the spin, due to the offsets of the sensors. (b) The same data as in (a), but the signal has been despun and calibrated. This figure shows a typical Short Large Amplitude Magnetic structure (SLAM) associated with the quasi-parallel bow shock of the Earth's magnetosphere.

y-axis aligned with the sensor along the magnetometer boom. The defining equation of this transformation is:

$$B_{\mathrm{FSR}} = \underline{\underline{c}}^{\mathrm{cal}} B_{\mathrm{FS}} - o^{\mathrm{cal}} \tag{7.1}$$

The objective of the in-flight calibration is to determine the 3×3 matrix $\underline{\underline{c}}^{\mathrm{cal}}$ and the offset vector o^{cal} used in this equation for all measurement ranges. It is then necessary to carry out the transformation from the coordinate system that rotates with the spinning spacecraft, or to "despin" the vectors, and, using the spacecraft attitude data, generate a time series B_{GSE} in a selected geophysical system, normally in Geocentric Solar Ecliptic (GSE) coordinate system according to:

$$B_{\mathrm{GSE}} = \underline{\underline{c}}^{(\mathrm{att})^{-1}} \underline{\underline{c}}^{(\mathrm{spin})^{-1}} B_{\mathrm{FSR}} \tag{7.2}$$

The spacecraft spin is taken into account in the rotation matrix $\underline{\underline{c}}^{(\mathrm{spin})}$, corresponding to the spin-phase angle of sensors at the time of the measurement. The matrix $\underline{\underline{c}}^{(\mathrm{att})}$ represents the transformation into the spacecraft spin-aligned coordinate system from the GSE system. This matrix is determined from the spacecraft attitude measurements. An example of the "raw" data acquired in space and the corresponding processed and calibrated data is shown in Fig. 7.22.

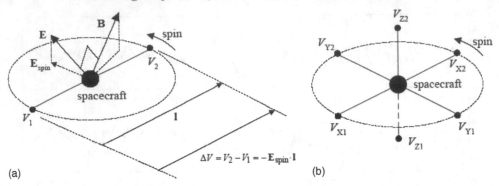

FIGURE 7.23. Principle of measurement of the electric field using long wire booms on a spinning spacecraft (a). The instantaneous potential difference between the isolated conducting spheres at the ends of the wire boom provides a measure of the component of the electric field vector along the direction of the boom. On some space missions, there are two or even three pairs of booms, to measure the three orthogonal components of the electric field vector simultaneously (b). (Redrawn after Pedersen *et al.*, 1997.)

7.2.2. *DC and AC electric field measurements*

Electric fields in space plasmas are weak; in many cases it can be assumed that there is no intrinsic DC electric field, $\mathbf{E} = 0$, only that field which arises from the motion of the plasma relative to the spacecraft, $\mathbf{E} = -\mathbf{v} \times \mathbf{B}$. Knowledge of the spacecraft velocity in some frame of reference, $\mathbf{v}_{SC}$ and the measurement of the magnetic field $\mathbf{B}$ can be used to calculate the electric field due to the motion of the spacecraft in that frame, $\mathbf{E} = -\mathbf{v} \times \mathbf{B}$. A measurement of the electric field at the spacecraft location, $\mathbf{E}_{\text{meas}}$, then gives the velocity $\mathbf{v}_p$ of the plasma with respect to that frame, from the relationship:

$$\mathbf{E}_{\text{meas}} + \mathbf{v}_{SC} \times \mathbf{B} = -\mathbf{v}_p \times \mathbf{B} \qquad (7.3)$$

The double-sphere method is the most frequently used technique that has been developed to measure electric fields in space. It uses long wire booms attached to a spacecraft as shown in Figs. 7.23 and 7.24. The booms are deployed from the spacecraft, once it reached its operational orbit, by centrifugal force, which also keeps them outstretched in flight. Spheres with a conducting surface are attached at the end of the booms, while the rest of the wire has an insulating cover. The coupling between the conducting spheres and the surrounding plasma is a very complex process which depends on the material and radius of the probe; its illumination by the sun (which generates photoelectrons and would cause the probe potential to become negative); on the Debye radius of the plasma (therefore its electron density and its temperature). In a very simplified manner, we can assume that the impedance between the two probes and the surrounding plasma is such that the electric potential of the probes is equal to that of the plasma. Therefore, if the potential is different at the locations of the two probes, at the end of the booms, the potential difference measured between the two spheres is equal to the electric field component along the wire boom multiplied by (approximately) the length of the boom (as shown in Fig. 7.23). This is because the electric field is derived from the electric potential Φ, according to $\mathbf{E} = -\nabla\Phi$, so that if the component of the electric field in the spin plane of the satellite (the plane swept out by the wire booms as the spacecraft spins) is $\mathbf{E}_{\text{spin}}$, then the potential difference between the probes is:

$$\Delta V = V_2 - V_1 = -\mathbf{E}_{\text{spin}} \cdot \mathbf{l} \qquad (7.4)$$

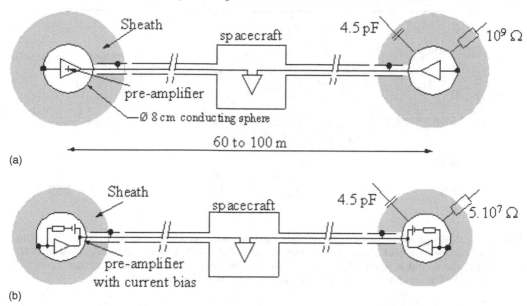

FIGURE 7.24. (a) A simplified schematic of a pair of wire booms and spherical sensors for measuring the potential difference between the sensors and, hence, the component of the electric field vector along the boom. The sensors have a conducting surface and are coupled to the surrounding plasma. The coupling is represented by the equivalent resistance and capacitance shown for the sphere on the right (it is the same for the sphere on the left). The plasma sheath surrounding the spheres represents the Debye-length effects in the plasma. (b) It is possible to improve the coupling of the spheres to the plasma by providing a small current to the spheres. (Redrawn after Pedersen *et al.*, 1997.)

where **l** is the vector distance separating the two probes. The two orthogonal components of $\mathbf{E}_{\text{spin}}$ can be measured simultaneously, if two pairs of wire booms are used in the spin plane, as shown in Fig. 7.23. This is the arrangement of the wire booms, for instance, on Cluster. If, as also shown in Fig. 7.23, a boom is used along the spin axis, then all three components of the electric field vector can be determined simultaneously, at least in principle. Note that for stability reasons, it is not possible to deploy wire booms along the spin axis – there is no centrifugal force to assist the deployment – so that on the few space missions on which such a spin-axis aligned boom was used, it was short and rigid or semi-rigid.

If the electric field is only slowly varying on the timescale of the spacecraft spin, then the signal is modulated by the spacecraft spin rate or angular frequency. If the electric field to be measured is constant in time (a field that can be assumed to be an electrostatic field) then:

$$\Delta V = -\mathbf{E}_{\text{spin}} \cdot \mathbf{l} = -|E_{\text{spin}}l| \cos(\omega_{\text{spin}}t + \varphi) \tag{7.5}$$

so that the electric field component measured by the double-probe is:

$$E \propto -\frac{\Delta V}{l} = E_{\text{spin}}\cos(\omega_{\text{spin}}t + \varphi) \tag{7.6}$$

This spin modulated signal is illustrated in Fig. 7.25. The envelope of the spin modulated signal is E_{spin} which can be recovered by the same despinning algorithm already used for recovering the magnetic field from the spinning signal.

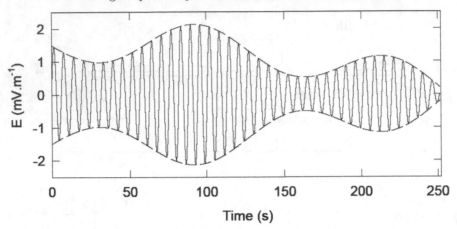

FIGURE 7.25. Typical signal from an electric field antenna on a spinning spacecraft. The component of the electric field along the antenna is measured, so that as the antenna spins with the spacecraft, the signal is modulated by the spin as described in the text. Slow variations of the spin-plane electric field are seen in the envelope of the modulating signal.

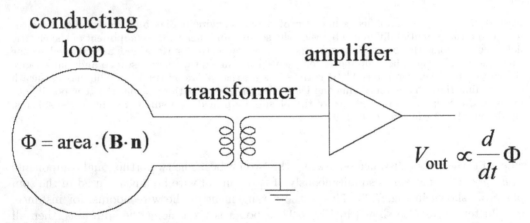

FIGURE 7.26. Schematic diagram of a magnetic loop antenna. For measurements in space, the diameter of the antenna is typically ~1 m, and it consists of a single rigid aluminium loop. The transformer is needed to match the impedance of the antenna to the input impedance of the sensing amplifier. The use of the transformer limits the low-frequency response of the loop antenna, so that it is sensitive to frequencies greater than a few hundred Hz. (Redrawn after Gurnett, 1997.)

7.2.3. *AC magnetic field measurements*

The magnetic field component of electromagnetic plasma waves is measured using magnetic antennas. There are two types in current use, the magnetic loop antenna (Fig. 7.26), and the search coil magnetometer (Fig. 7.27). In both cases, the time-dependent magnetic field vector $\mathbf{B}(t)$ is detected through the application of Faraday's law, by measuring the voltage induced in a sensing loop or coil, as a function of frequency, by the variation of the magnetic flux $\Phi = NA(\mathbf{B} \cdot \mathbf{n})$. Here N is the number of turns in the sensing coil, and A is the area of the sensing loop, while $\mathbf{n}$ is the unit normal to the area. The time rate of change of the component of the magnetic field vector parallel to $\mathbf{n}$ is thus obtained from the voltage induced in the coil. This means that if all three components of the derivative of $\mathbf{B}$ are to be measured, then, in principle, three simultaneous measurements along three orthogonal directions are needed.

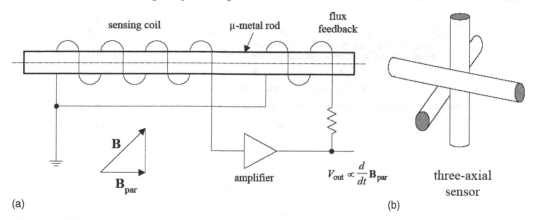

(a)

(b)

FIGURE 7.27. The schematic principle of the search coil magnetometer, sensitive to the frequency-dependent component of the magnetic field vector along the axis of the μ-metal rod around which a sensing coil of a large number of turns is wound. In order to operate close to its ideal, null operating point (to improve the linearity of the measurements) a feedback coil generates a magnetic flux in the rod opposing that generated by the signal. It is in effect this feedback signal, needed to null the signal, which is measured. On the right, a three-axial arrangement, consisting of three orthogonally placed single axis search coil sensors is shown, to measure simultaneously the three components of the AC magnetic field vector. (Redrawn after Gurnett, 1997.)

7.2.4. *Waves: combining AC magnetic and electric field measurements*

Characteristic plasma waves are generated in a wide range of environments that can be used as diagnostics of the state of the plasma and the processes that occur in it. Measurement of both the electric and magnetic components of waves is essential, over a wide range of frequencies, to cover the wave phenomena of interest. These measurements are provided by the electric field and AC magnetic field instruments described above. In addition, all active and dynamic plasma regimes, such as the solar corona and all regions of the Earth's magnetosphere, emit electromagnetic (radio) waves. These waves interact with the plasmas through which they propagate, a process in which both the waves and the plasmas are modified. *In situ* detection of these waves can thus remotely sense the properties of the plasma over large volumes of space.

An example of comprehensive wave instrumentation is shown in Fig. 7.28 as implemented in the payload of the WIND spacecraft (Bougeret *et al.*, 1995). There are three pairs of electric field antennas, two pairs of wire booms in the spin plane of the spacecraft, of 50 and 7.5 m in tip-to-tip length, and two rigid antennas, each of 5.28 m in length extending along the spin axis of the spacecraft. The three components of the AC magnetic fields are measured by a tri-axial search coil system as described above, located at the tip of a 12 m boom from the spacecraft to minimize artificial interference with the measurements. Following the pre-amplification of the signals from the sensors, the AC electric and magnetic measurements are combined in parallel signal processing units that cover different bandwidths and have different scientific diagnostic objectives. The five signal processing systems are a fast fourier transform (FFT) receiver (DC to 10 kHz), a broadband electron thermal noise receiver (4–256 kHz), two swept-frequency radio receivers (20 kHz to 1 MHz, and 1–14 MHz), and a time domain waveform sampler (up to 120,000 samples per second). The frequency ranges corresponding to the different receivers and their main measurement objectives are also shown in Fig. 7.28.

On the four-spacecraft Cluster mission, electric and magnetic fields are measured by a group of four instruments, using as sensors (i) a pair of $\sim$80 m wire booms in the

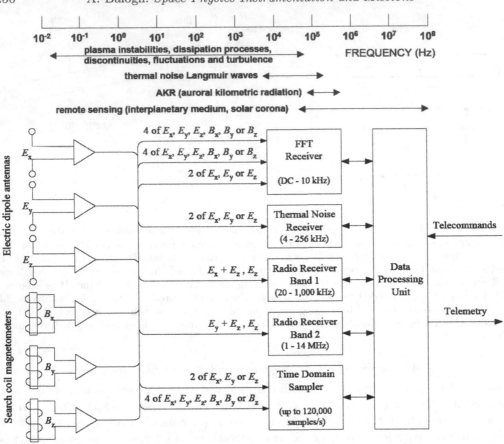

FIGURE 7.28. Block diagram of the WAVES instrument on the WIND spacecraft. At the top of the figure, the frequency ranges corresponding to the different scientific objectives are also indicated. (Redrawn after Bougeret *et al.*, 1995.)

spacecraft spin plane, with a modified version of the double-sphere sensors described above, and (ii) a tri-axial search coil magnetometer mounted at the tip of a 5 m boom. The four instruments take different combinations of AC magnetic and electric field signals from the sensors for further processing in the frequency domain to address different sets of characteristics of space plasmas in the terrestrial environment. In addition to these measurements that take the natural signals from the sensors, one of the instruments (WHISPER) also has an active (sounder) mode that can periodically transmit successive short ($\sim$1 ms) pulses of sinusoidal signal in the range 4–80 kHz, using the conducting outer braid of one of the pairs of electric antennas. In response to the transmitted signal, resonances are generated in the plasma surrounding the spacecraft. Following the transmission of the pulse at a given frequency, the response of the plasma is recorded over the same frequency and compared to the natural level of emission of the plasma at that frequency. Repeating the transmission at different frequencies explores all the characteristic frequencies in the plasma, as shown by the plot in Fig. 7.29. Of particular interest is the determination of the electron plasma frequency, $f_{\mathrm{pe}} \propto \sqrt{n_{\mathrm{e}}}$ where n_{e} is the electron density of the plasma. This instrument can therefore provide a precise determination of the plasma density, to complement and calibrate more conventional methods of density determination using particle detectors, described in the next section.

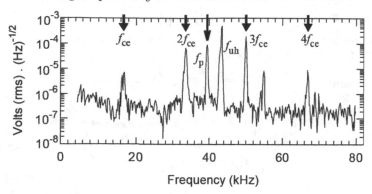

FIGURE 7.29. Measurement of characteristic frequencies in the terrestrial plasma using the WHISPER relaxation sounder on the Cluster spacecraft. The main resonances are at the gyrofrequency of the electrons (f_{ce}) and its harmonics, at the electron plasma frequency (f_p) and at the upper hybrid frequency of the plasma (f_{uh}); other peaks in the spectrum are harmonics of Berstein waves. The precise measurement of f_p allows an independent determination of the local plasma density. (Figure courtesy of Pierrette Décréau and the WHISPER team.)

7.2.5. *The electron drift instrument*

A special class of instruments (Paschmann *et al.*, 1977) exploits the fact that electrons, as all charged particles, move in a circular trajectory perpendicular the magnetic field vector $\mathbf{B}$ of magnitude B. The period of the circular motion, the gyroperiod, is:

$$T_g = \frac{2\pi m}{eB} \tag{7.7}$$

where m and e are the mass and the electric charge of the electron, respectively. In the frame of the left panel of Fig. 7.30, it is assumed that the electric field is zero, so that the drift velocity of electrons $v_d = \mathbf{E} \times \mathbf{B}/B^2$ is equal and opposite to the spacecraft velocity v_{sc}. If electrons are emitted from a spacecraft at time t_0 in two directions in the plane perpendicular to the local magnetic field, beam 1 returns to the electron detector mounted to the spacecraft after a flight time that is less than a full gyration, whereas beam 2 returns to the detector after a time longer than the gyroperiod. (Note that the relative dimensions of the trajectories and the distance travelled by the spacecraft are exaggerated to make the diagram clearer.) The two flight times are:

$$\Delta t_1 = t_1 - t_0 = T_g \left(1 - \frac{v_d}{v_e} \right) \tag{7.8}$$

$$\Delta t_2 = t_2 - t_0 = T_g \left(1 + \frac{v_d}{v_e} \right) \tag{7.9}$$

where v_e is the velocity of the electrons. The measurement of the flight times yields:

$$T_g = \frac{(\Delta t_1 + \Delta t_2)}{2} \tag{7.10}$$

and therefore the magnitude of the magnetic field can be calculated from Equation (7.5) as:

$$B = \frac{4\pi m}{e} (\Delta t_1 + \Delta t_2) \tag{7.11}$$

This measurement provides an accurate determination of the magnitude of the magnetic field and can be compared to the same quantity calculated from the vector measurement by the magnetometer. This comparison provides support in the determination of the offsets of the magnetometer. A sample comparison between measurements made by the

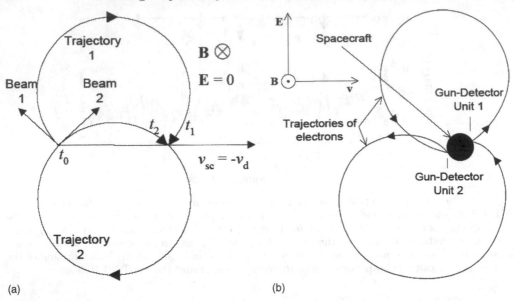

FIGURE 7.30. (a) The principle of the EDI on the Cluster mission. Electrons are emitted by the instrument and are detected as they return to the spacecraft after gyrating around the magnetic field direction. (b) Simulated trajectories of electrons in the presence of an external electric field. Note that the magnetic and electric fields used in the simulation are unrealistically high to assist with the clarity of the diagram. (Redrawn after Paschmann *et al.*, 1997.)

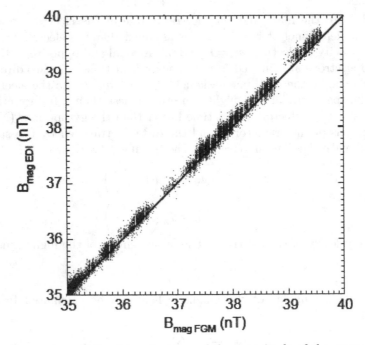

FIGURE 7.31. Comparison of the determinations of the magnitude of the magnetic field made by the magnetometer and the electron drift instrument on one of the Cluster spacecraft. The two measurements agree to within ∼0.2 nT in the weak fields in the terrestrial magnetosheath. (Figure courtesy of Edita Georgescu.)

electron gun experiment and the magnetometer on one of the Cluster spacecraft is shown in Fig. 7.31.

The implementation of the complete electron drift instrument (EDI) and its measurement capabilities (Paschmann et al., 1997) are considerably more complex than outlined above. Although the instrument's general principle of measuring the gyroperiod of test electrons fired by the incorporated electron guns remains valid, complexities arise from the targeting of the electron beams, their coding so that they can be recognized and identified by the detectors. Effects such as electron drifts arising from gradients in the electric and magnetic fields complicate the interpretation of the data. However, the scientific data that can be derived from the observable parameters lead to a unique characterization of the electric fields in the neighbourhood of the spacecraft.

Many of the instruments described in this section are described in more detail by Pfaff et al. (1998a).

7.3. Plasma measurements

7.3.1. *Introduction*

Space plasmas are low-density ion and electron gases which are usually electrically neutral (the positive charge density of the fully or partially ionized atoms balances the negative charge density of the electrons). In addition, the density is usually so low that particles effectively do not collide – this means that their mean free path, the average distance travelled by particles between collisions, is at least of the same order or even, in some cases, much larger than the physical scale of the system. Such a plasma is called collisionless. However, the particles interact in other ways, through the long range Coulomb force, so that the plasma can be treated as a fluid, even if a somewhat special one. However, given their extremely low densities, the usual macroscopic quantities that define gases such as density, pressure and temperature cannot be measured directly in space plasmas. Particles in the plasma need to be individually counted in almost all cases and, as described below, the macroscopic quantities are then calculated from accumulations of individually counted particles.

The subject of this section, plasma detectors, is the description of instruments that are used to detect individual particles, electrons or ions, which are the constituents of the plasma (Bame et al., 1986, Pfaff et al., 1998). The parameters that are usually measured are the fluxes of electrons and the different ion species as a function of direction of incidence and kinetic energy, as defined below. The simplest plasma detectors can measure the directional distribution and energy spectrum of electrons, and the most abundant ion species, H^+, or protons, and also usually of He^{++}, or alpha particles. Currently used plasma detectors can also distinguish between other ion species, and the more sophisticated ones can measure the charge state (the degree of ionization) of some important ion species, and can even distinguish among different isotopes of the most abundant species of ions. A brief review is given below of the basic operating principles of the different plasma detectors in current use.

7.3.2. *The plasma distribution function and its measurement*

In the description of space plasmas, the distribution functions of ions and electrons in phase space, as a function of location and time, play an important role. Phase space is the 6D parameter space made up of the three parameters that define position of the particle, usually denoted by the position vector $\mathbf{r}$, and the three components of its velocity or momentum vector, $\mathbf{v}$ or $m\mathbf{v}$. In principle, and usually in practice, electrons and each ion species have their own different distribution functions. We therefore define the

TABLE 7.2. Definition of some macroscopic quantities in the plasma using the distribution function.

Distribution function for species n	f_n
Mass density of plasma	$\rho = \sum_n m_n \int f_n d^3v$
Electric charge density of plasma	$\rho_e = \sum_n q_n \int f_n d^3v$
Bulk velocity	$V = \frac{1}{\rho} \sum_n m_n \int v\, f_n d^3v$
Current density	$j = \sum_n q_n \int v\, f_n d^3v$

distribution function for *each particle species* (electrons and the different kinds of ions). Such a distribution function, defined for each of the particle species separately, $f(\mathbf{r}, \mathbf{v}, t)$, is a function in phase space as already noted, and depends on the position vector $\mathbf{r}$ and the velocity vector $\mathbf{v}$ of the particles, as well as on time t. (In ordinary Cartesian coordinates, $\mathbf{r} = x\hat{\mathbf{x}} + y\hat{\mathbf{y}} + z\hat{\mathbf{z}}$ and $\mathbf{v} = v_x\hat{\mathbf{x}} + v_y\hat{\mathbf{y}} + v_z\hat{\mathbf{z}}$. The six variables, the components of the position vector $\mathbf{r}$ and of the velocity vector $\mathbf{v}$ are independent variables.) The distribution function $f(r, v, t)$ represents the *number density of the particles in phase space*, where it depends on the six independent variables, as well as on time. This means that the number of particles in an infinitesimal volume in this 6D space is:

$$\mathrm{d}n = f(r, v)\,\mathrm{d}r\,\mathrm{d}v \tag{7.12}$$

where, using Cartesian coordinates, we have $\mathrm{d}r = \mathrm{d}x\mathrm{d}y\mathrm{d}z$ and $\mathrm{d}v = \mathrm{d}v_x\mathrm{d}v_y\mathrm{d}v_z$. This shows that $f(r, v, t)$ has the units $\mathrm{m}^6/\mathrm{s}^3$.

Using standard techniques from statistical mechanics, the equations describing the plasma can be derived for the whole plasma based on the distribution functions of individual particle species. In particular, *moments of the distribution function* (as defined in Table 7.2) define the macroscopic quantities that describe the plasma, such as the mass density $\rho(\mathbf{r}, t)$, the electric charge density $\rho_e(\mathbf{r}, t) \approx 0$ (on scales larger than the Debye length), the bulk velocity vector $\mathbf{V}(\mathbf{r}, t)$, and the current density vector $\mathbf{j}(\mathbf{r}, t)$ in the plasma. In addition, the pressure of the plasma (more generally, the pressure tensor) can also be calculated from the distribution function. The differential equation for the distribution function is, in general, the Boltzmann equation.

The dependence of the distribution function on the position vector $\mathbf{r}$ can be dropped below, because for a plasma detector in space it simply corresponds to the position of the detector, or, on the position of the spacecraft on which the detector is mounted. Plasma detectors therefore measure, in the sense defined below, the distribution functions $f_n(v, t)$ for species n (electrons and ions) at the location $\mathbf{r}$ in space of the instrument/spacecraft.

A sample distribution function of solar wind protons measured by the plasma detector on the Helios mission is shown in Fig. 7.32. This is a 2D cut, in the v_x, v_y plane through the measured 3D distribution. Isointensity contours are shown, centred on the peak of the distribution function. In the solar wind, the velocity distribution of protons, due to their thermal velocity, is (approximately) symmetrical around the magnetic field direction. This type of distribution is usually approximated by two Gaussian (bell shaped) distributions, parallel and perpendicular to the magnetic field. The widths of the two Gaussians correspond to average thermal velocities of the protons parallel and perpendicular to the magnetic field. These two average thermal velocities are interpreted as the parallel and perpendicular temperatures in the solar wind plasma.

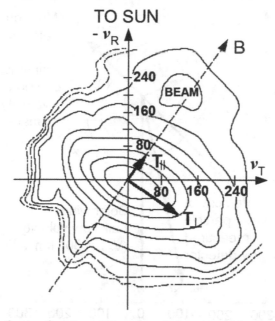

FIGURE 7.32. Ion distribution function of the solar wind measured by the plasma detector on the Helios spacecraft at 0.29 AU from the Sun. This distribution shows that the temperature of the plasma perpendicular to the magnetic field is higher than the parallel temperature, indicating that the plasma is heated locally. The distribution is shown after subtraction of the bulk flow velocity (781 km/s) of the solar wind.

An instrument measuring the solar wind in fact measures the total velocities, including the "bulk" or average velocity of the solar wind in the direction away from the Sun. A schematic view of a solar wind distribution is shown in Fig. 7.33. In the frame of reference of the instrument, the velocities that are measured are the combined bulk and thermal velocities of the protons. The distribution in Fig. 7.32 corresponds to the distribution in Fig. 7.33, with the bulk solar wind velocity subtracted from the R-component (in the direction of the Sun) of the measured velocities, to obtain the distribution function in the frame of reference moving with the solar wind.

Two examples of a 2D cut across the proton distribution function are shown in Fig. 7.34, as measured by the cluster ion spectrometer (CIS) instrument on European Space Agency's (ESA) Cluster mission. The distributions were measured just in front of the Earth's bow shock, in the solar wind. The figure shows the distributions in velocity space, in the x–z plane of the Geocentric Solar Equatorial (GSE) coordinate system. In the left panel, the undisturbed proton distribution of the solar wind is shown. The bulk speed of the solar wind is ∼600 km/s, and proton speeds are distributed around this value due to the (nearly Gaussian) thermal velocities of the particles. In the right panel, the distribution function is made up of two components: one is the incoming solar wind distribution, as in the left panel, the other is a distribution made up of hot ions propagating away from the bow shock upstream (against), the incoming solar wind flow. Such a distribution of hot ions is caused by a hot flow anomaly, a major disturbance of the bow shock by an interplanetary discontinuity. This figure illustrates the kind of measurements that are made in space to determine the plasma distribution function.

We define here a very simple plasma detector by its sensitive area and its orientation, or look direction, in space. Both these can be specified by the single vector $\mathbf{A}$, where $|\mathbf{A}| = A$ is the area, and the direction of the vector is taken to be perpendicular to

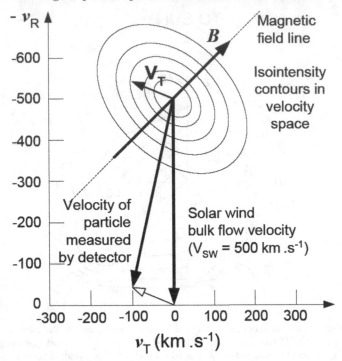

FIGURE 7.33. Schematic distribution function of the solar wind protons in the frame of reference of the measuring instrument. In this frame, the radial velocities have two components: one corresponding to the bulk velocity of the solar wind, and one corresponding to the thermal velocities of the solar wind protons.

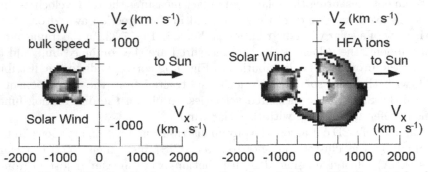

FIGURE 7.34. Plasma distribution functions in the solar wind from CIS. See text for detailed description of these distributions. (Figure courtesy of H. Rème, I. Dandouras and E. Lucek.)

the plane of the detector area. We assume that the detector will count plasma particles incident on it within an acceptance angle θ defined (using a simple collimator in this example) with respect to the normal to the area of the detector. This arrangement is illustrated schematically in Fig. 7.35. This assumes that the "detector" of area $\mathbf{A}$ is itself used to count the particles incident on it. In reality, the whole instrument, as described below for practical detectors, is a complex mechanical and electrical system, with an equally complex interpretation of what constitutes the sensitive area of the detector. However, the basic principles are well illustrated by this schematic example. The vector $\mathbf{u}_0$ in Fig. 7.35 is the unit vector along the "representative" or average arrival direction of the impacting particles, as will be defined below; in this example, it is also the unit vector along the velocity of particles with normal incidence on the detector.

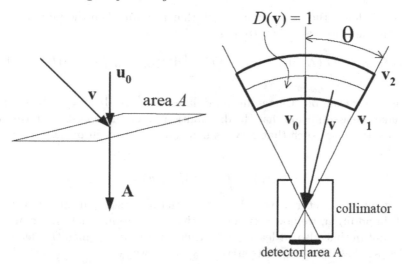

FIGURE 7.35. Definitions of the geometry of instruments for the measurement of the plasma distribution function. The form of detector shown is not representative of a large class of plasma instruments, with the exception of the Faraday Cup (see below), but is shown to illustrate the velocity selection process in all plasma detectors.

Given the definition of the distribution function, the primary quantity measured by a plasma detector can be defined as:

$$N - \int_T dt \int (\mathbf{A} \cdot \mathbf{v}) f(v,t) D(v) \, dv \qquad (7.13)$$

where N is the number of particles detected in a time interval T. In this formula, the vector $\mathbf{A}$ corresponding to the detector surface is taken to be along the inward directed normal to the area, and $D(\mathbf{v})$ describes the probability that a particle with velocity $\mathbf{v}$ incident on the detector will be counted. In the case of the simplified detector shown in Fig. 7.35, the function $D(\mathbf{v})$ is defined to be equal to 1 for velocities such that $v_1 < v < v_2$ within the directions bordered by the acceptance angle of the detector. For particles with velocity vectors that fall outside this volume, we have $D(\mathbf{v}) = 0$.

In general, the integral in velocity space (with respect to $d\mathbf{v}$) is taken over the hemisphere for which $\mathbf{A} \cdot \mathbf{v}$ is positive, in other words, we consider that the detector responds only to particles that are incident on one side of its sensitive area. An equivalent alternative is to set $D(\mathbf{v}) = 0$ for particles for which $\mathbf{A} \cdot \mathbf{v} < 0$ if the integration is taken over the whole velocity space. In the example in Fig. 7.35, the integration is performed for vectors within the area in velocity space (volume, in 3D) where $D(\mathbf{v}) = 1$.

As can be seen in the integral formula for the number of particles that are detected, the distribution function is not measured directly. The particle counting process involves its integration over velocity space as well as over time. The quality of plasma detectors is judged on their ability to make measurements of particle counts that can be interpreted, to a lesser or greater extent, in terms of an approximate determination of the distribution function.

We examine below the plasma detectors that make differential measurements; in other words, each measurement represents a sample of the distribution function. (For full details, see for example Vasyliunas, 1971 and Kessel *et al.*, 1989). If we assume that the distribution function remains approximately constant, $f(v_0, t)$, for non-zero values of

$D(\mathbf{v})$, and if we define the unit vector $\mathbf{u}$ such that $v = v_0\mathbf{u}$, then the number of particles counted by the detector can be written as:

$$N = \int_T \mathrm{d}t f(v_0,t) v_0 \int (\mathbf{A}\cdot\mathbf{u})D(v_0 u)v_0^3\,\mathrm{d}u = \int_T \mathrm{d}t f(v_0,t) v_0^4 \int (\mathbf{A}\cdot\mathbf{u})D(v_0 u)\,\mathrm{d}u \quad (7.14)$$

(because, using for example, Cartesian coordinates, $\mathrm{d}\mathbf{v} = \mathrm{d}v_x \mathrm{d}v_y \mathrm{d}v_z = v_0^3\,\mathrm{d}u$). In this formula, we make the assumption that the distribution function remains uniform in velocity space for velocity values such that $D(\mathbf{v})$ is non-zero, but can, in principle, vary in time. The expression:

$$G(v_0) = \int (\mathbf{A}\cdot\mathbf{u})D(v_0 u)\,\mathrm{d}u \qquad (7.15)$$

defines the geometric factor, a very important characteristic parameter of the plasma detector. If, in addition, we now assume that the distribution function remains approximately constant in time also, at least for the time interval over which the detector counts the particles, so that it can be written as $f(v_0)$, then we get:

$$N = TG(v_0)f(v_0)v_0^4 \qquad (7.16)$$

Thus, the number N of particles detected gives a sample value for the distribution function:

$$f(v_0) = \frac{N}{[TG(v_0)v_0^4]} \qquad (7.17)$$

This sample value of $f(\mathbf{v})$ is averaged over the volume in velocity space in which $D(\mathbf{v}) \neq 0$, so that a single count $N(v_0)$ provides a value for the distribution function for the representative velocity value $\mathbf{v}_0 = v_0\mathbf{u}_0$ at the time of the measurement (but averaged over the time interval T) and at the spatial location $\mathbf{r}$ of the detector. For a realistic determination of $f(\mathbf{v})$, many such measurements are necessary. This is usually achieved by varying $D(\mathbf{v}_0) = D(v_0\mathbf{u}_0)$, in other words, (a) by selecting v_0 from a range of values by changing the characteristics of the detector to be sensitive to different values of particle velocity and (b) by changing the orientation of the detector in space to detect incident particles from a series of different directions. For (a), the parameter v_0 is varied in the detector's sensitivity function $D(v_0\mathbf{u}_0)$, and for (b), it is $\mathbf{u}_0$ that is independently varied.

This is illustrated schematically in Fig. 7.36(b). The detector is made to be sensitive to increasing values of v_0, varying between $(v_0)_{\max}$ and $(v_0)_{\min}$, performing at each value a measurement the number $N(v_0)$ of incident particles per measurement interval. At the same time, the look direction (corresponding to $\mathbf{u}_0$) of the detector is changed, but much more slowly, so that the sweep in the magnitude of the velocity v_0 is performed while the detector's look direction changes relatively little during the sweep. The sweep in the look direction of the detector is usually achieved by using the spin of the spacecraft on which the detector is mounted. In Fig. 7.36(b), the achieved coverage of the velocity space is illustrated. In this example, we assumed for clarity that the sensitivity to the velocity parameter is stepped for each degree of the spacecraft spin (and therefore for each degree change in the orientation of the vector $\mathbf{u}_0$). The sensitivity is thus stepped up from its lower limit to its upper limit in 15 steps, while the look direction of the detector has changed by $15°$. Then the sensitivity sweep is restarted at the lower limit, for the next $15°$ of the spacecraft spin; this is repeated for a complete spin of the spacecraft, resulting in 24 sweeps in the 2D velocity space per $360°$ spin. The resulting schematic pattern of the points in velocity space where a determination of $f(\mathbf{v})$ is made in this example is shown in Fig. 7.36(b).

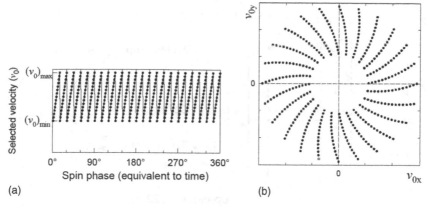

FIGURE 7.36. Principle of using the spacecraft spin, combined with stepped sensitivity to different velocities, for measuring the distribution function of space plasmas. Velocity selection is repeated periodically, as shown in (a). The detector on a spinning spacecraft looking in the radial direction samples the 2D velocity space as shown in (b).

Practical instruments, as described below, extensively use this technique for achieving a wide coverage of velocity space to determine, in detail, the distribution function of the plasma particles. In this schematic example, a number of simplifications were made for the sake of clarity. Plasma detectors of course have a 3D sensitivity (e.g. perpendicular to the spin plane shown in Fig. 7.36(a)); the sensitivity stepping is usually much faster when compared to the spin than in this example, and also has usually a much larger, often logarithmically spaced scale in v_0. But to a first approximation, almost all plasma instruments use the spacecraft spin for varying their look direction; this is one of the main reasons for using spinning spacecraft for space physics mission. Many new plasma sensors, based on the top-hat design described below, are simultaneously sensitive to a range of different directions. In that case, this feature is used for detecting incident particles in the third dimension, out of the spin plane of the spinning spacecraft.

On spacecraft that are three-axis stabilized, other ways need to be found for covering an adequate range of $\mathbf{u}_0$ vectors; this can be achieved either by the use of multiple sensors or by mechanically changing the plasma sensor orientation. In both cases, there is a penalty in the amount of payload resources needed, even if the directionally sensitive feature of the top-hat analyser type of particle detectors is used.

7.3.3. *Retarding potential analysers*

A classical plasma ion detector is the *Faraday Cup*, or *Retarding Potential Analyser* (RPA), used on a number of space missions, recently on NASA's WIND spacecraft, to measure the solar wind ion flux as a function of the energy-per-charge of the incident ions. The solar wind experiment on WIND is illustrated in Fig. 7.37 that shows the schematic operating principle of the Faraday Cup. It contains a series of wire-mesh, planar grids made from tungsten wire and, in this case, two collector plates. The velocity distribution function of ions is measured by applying a sequence of voltages to the "modulator" grid. If a voltage $V > 0$ is applied to the modulator grid, only particles with an energy per charge ratio $E/Q > V$ can cross the modulator and impact on the collector. If the number of impacting particles that satisfies this condition is $n(E/Q) \equiv n(V)$ then, if the voltage on the grid is stepped repeatedly between V_1 and V_2 (with $V_2 > V_1$) at a rate of several hundred times a second, the current generated by impacting particles on the collector varies between $en(V_1)$ and $en(V_2)$ at the same frequency. The purpose of the "suppressor"

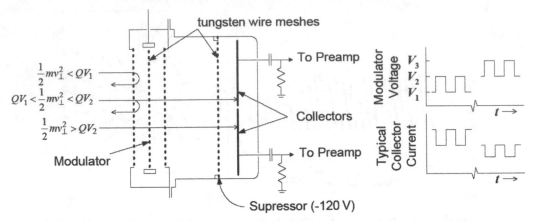

FIGURE 7.37. The Faraday Cup (also called the RPA) plasma detector. (Redrawn from Lazarus and Paularena, 1998.)

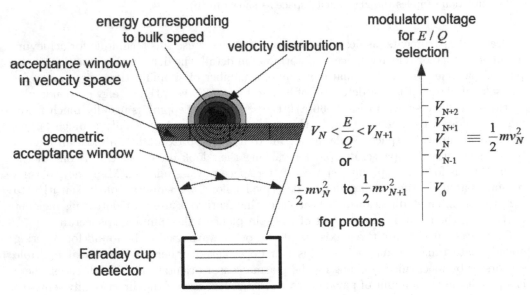

FIGURE 7.38. Application of the Faraday cup for the measurement of the solar wind ion distribution function. The stepping of the voltage on the modulator grid of the detector moves the window in velocity space across the velocity distribution of the solar wind. Note that the velocity distribution is a schematic representation of the type of real distribution shown in Fig. 7.32. (After a figure by A.J. Lazarus, MIT.)

grid is to eliminate the effect of secondary electrons generated by the impacting ions on the collector plates. The measurement of the modulated current on the collector therefore yields the number of particles in the energy per charge range $\Delta(E/Q) = V_2 - V_1$. Using a sequence of modulating voltage pairs such as $V_3 > V_2$, $V_4 > V_3$, further intervals of the full energy per charge range can be sampled by the instrument. This is illustrated in Fig. 7.38, where a schematic solar wind proton distribution function is sampled in energy per charge space (equivalent to velocity space for particles of a given charge) through stepping the voltage on the modulator. Figure 7.38 also illustrates the geometric acceptance angle of the Faraday Cup detector. As shown in this figure, the Faraday Cup's wide acceptance angle and large sensitive area is particularly suited to the measurement of the solar wind.

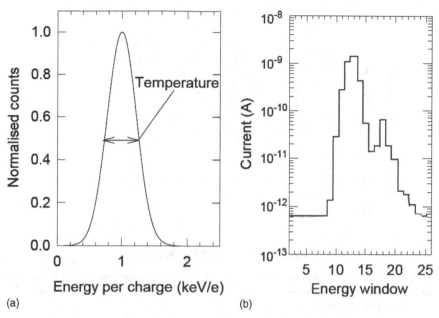

FIGURE 7.39. Simplified distribution of the solar wind protons measured by an instrument that measures the energy per charge of the incident particles (a). Actual distribution measured by a Faraday cup in the solar wind (b). The measured distribution shows two peaks, one for protons (charge = 1) and one for alpha particles (charge = 2).

7.3.4. *Principles and applications of electrostatic analysers*

The basic technique underlying almost all plasma detectors is the measurement of the (kinetic) energy per charge, E/Q of the incident plasma particles, E denotes the kinetic energy and Q, the electric charge of the particle. Similarly to the Faraday Cup described above, this technique also measures the energy per charge ratio, using the deflection of charged particles in an electric field that is usually orthogonal to the incident velocity of the particles. (In the Faraday Cup, the retarding potential creates an electric field that is anti-parallel to the incident velocity vector.)

The basic measurements made by the energy per charge instruments are illustrated, in the case of the solar wind, in Fig. 7.39. In Fig. 7.39(a), the counts per measurement interval are shown for protons that travel at ∼400 km/s, corresponding to the peak in the distribution. However, protons in the solar wind have a thermal speed of ∼50 km/s so that each incident proton has a speed that is the vector sum of the solar wind speed and its own thermal speed as already illustrated in Fig. 7.33. The resulting counts are usually fitted with a Gaussian curve, with the width of the distribution providing a measure of the temperature of the solar wind plasma. In the real solar wind, other particles are present, in particular alpha particles, with a relative abundance that is in general <10%. Figure 7.39 gives an example distribution measured by the Faraday Cup on the WIND spacecraft that shows a peak for both protons and alpha particles. In Fig. 7.40, four distributions are shown schematically, corresponding to the same relative proton-to-alpha counts, but with different temperatures of the plasma. It can be seen in the figure that if the temperature of the solar wind is high, the distributions corresponding to different species (with different energy per mass ratios), in this instance to protons and alpha particles, can significantly overlap, making it difficult to measure accurately the relative abundance and temperature of minor species.

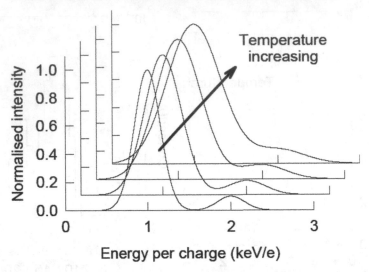

FIGURE 7.40. Typical but simplified energy per charge measurements in the solar wind, illustrating the effect of the temperature of the plasma on the overlapping distributions of adjacent species in energy per charge space.

We start with the simplest case, the deflection of charged particles in a parallel plate capacitor, as illustrated in Fig. 7.41. This geometry of the deflecting plates is not used in practical detectors in space. However, the principle of measuring the energy per charge ratio can be easily followed in this example and it applies to effectively all electrostatic deflection systems used for measuring space plasmas. Fig. 7.41(b) provides the basic geometrical arrangement showing that positively charged ions ($Q > 0$) will be deflected in the direction of the electric field, while electrons ($Q < 0$) will be deflected anti-parallel to the electric field. The deflection D is inversely proportional to the E/Q ratio, it is calculated as:

$$D = D(V, E/Q) = \frac{1}{4}\frac{L^2}{a}\frac{VQ}{E} = \frac{1}{4}\frac{L^2}{a}\frac{V}{E/Q} \tag{7.18}$$

where Q is the charge of the particle, V is the voltage applied to the deflecting plates and E is the energy of the incident particle. For a given charge, lower-energy particles will be deflected more than more energetic particles.

The lower panel in Fig. 7.41 illustrates a conceptual plasma ion detector (not used in practice) in which, for the voltage on the plates and the dimensions, ions with an energy to charge ratio of 0.8–2.0 kV entering through the collimator (to ensure that only particles incident normally are allowed in the detector) are recorded using a multichannel plate (MCP) detector. For protons, this corresponds to the energy range 0.8–2.0 keV, and for alpha particles (with an electric charge twice that of protons), the energy range is 1.6–4.0 keV.

Practical plasma particle detectors use curved plate electrostatic analysers. The geometry of the plates can be cylindrical, spherical or toroidal. In each case, the voltage on the plates, together with the geometry of the plates, selects particles in a small energy per charge range that follow the centreline of the plates to exit to a particle detector (usually a, materials control and verification program, MCVP). Stepping the voltage on the plates leads to a wide energy per charge range to be covered, usually from close 0 V to about 30 kV. Cylindrical plate analysers have the property, as shown

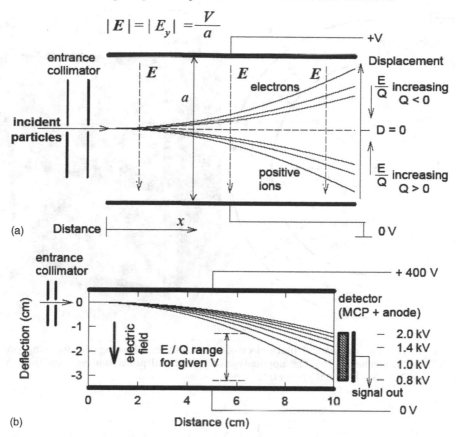

$$|E| = |E_y| = \frac{V}{a}$$

FIGURE 7.41. Basic principles of the energy per charge plasma analysers, using parallel plate electrostatic deflectors. Although indicating the principle of actual instruments, this geometry is not used in actual instrumentation.

in Fig. 7.42(a), that particles that are incident within a small range of normal incidence at the entrance aperture are focused after a trajectory of 127°. Even though current and future instruments preferentially use spherical geometries, the cylindrical analyser has been extensively used in a number of particle instruments in space and current missions, such as NASA's WIND and POLAR spacecraft that carry cylindrical plate electrostatic analyser instruments for a range of plasma ion and electron measurements. The HYDRA instrument (Scudder *et al.*, 1995) on the POLAR spacecraft, for instance, uses a large number of small, 127° cylindrical plate analysers to provide measurements in a range of directions of the plasma distribution function, as is shown in Fig. 7.42(b).

Spherical plate electrostatic analysers are widely used in plasma detectors. As illustrated in Fig. 7.43, for a given voltage difference V between the plates, ions with a charge Q and incident energy E follow a circular trajectory, if the condition:

$$\frac{E}{Q} \approx \frac{r_0 V}{2\Delta r} \qquad (7.19)$$

is satisfied, where $r_0 = (r_1 + r_2)/2$ is the radius of the centre line between the plates and $\Delta r = r_2 - r_1$ is the difference in radius between the outer and inner plates. Incident particles with an energy per charge near this value can still traverse the analyser, but the others collide with the plates and are absorbed, thus not reaching the detector. For a given analyser, at a particular voltage, ions therefore emerge to be detected within a

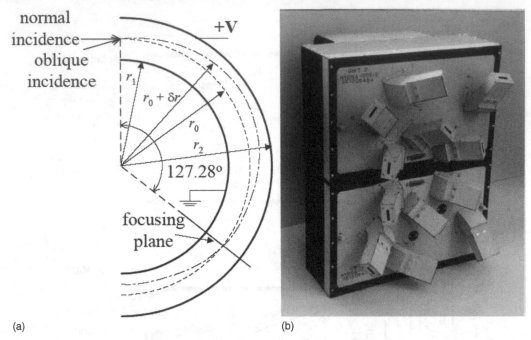

(a)   (b)

FIGURE 7.42. (a) The cylindrical plate electrostatic analyser, showing the focusing property of the cylindrical geometry. (b) The application of cylindrical plate analysers on the HYDRA instrument on NASA's POLAR spacecraft.

range:

$$\frac{r_0 V}{2\Delta r} \pm \Delta\left(\frac{E}{Q}\right) \tag{7.20}$$

where $\Delta(E/Q)$ is a function dependent on the energy per charge and the angle of incidence. The detected range of particles is obtained by calculations, by computer simulations (determining the trajectories of particles that cross the analyser, often called ray tracing, following the optical equivalent) and by calibration using ion sources. A schematic view of the sensitivity for a given voltage (and given geometry of the spherical analyser plates) is also shown in Fig. 7.43. As the voltage is stepped on the analyser plates, the instrument is made to select different E/Q values of the incident plasma particle flux.

For detecting electrons, the voltage V on the plate is reversed (with the inner plate at a higher potential). As the electrons are singly (negatively) charged particles, the E/Q technique (in all geometries of the electrostatic analyser plates) becomes simply an instrument to measure the energy spectrum of the electron fluxes incident on the detector, using successive voltage steps on the analyser plates.

The advantage of spherical plate analysers, used extensively since the 1960s, over cylindrical plate instruments is that particles incident in the plane defined by the aperture for normal incidence follow trajectories along a great circle, as shown for the quarter-sphere (quadrispherical) analyser geometry in Fig. 7.44. In this way, particles incident at different polar angles (but with an energy per charge ratio selected by the voltage on the analyser) exit the plates at locations that correspond to the polar angle of incidence. At the exit of the analyser, a semicircular particle detector is used. If this detector is "position-sensitive", that is it records the location of the impact of the particle at the exit of the quadrispherical plates, the instrument is able to resolve not just the energy per charge of the incident particles as selected by the analyser, but also their polar

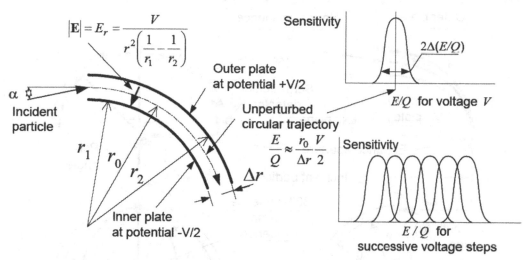

FIGURE 7.43. Principle of the spherical plate electrostatic analyser. The analyser allows the transmission of ions with a defined value of energy per charge between the plates, dependent on the voltage V applied to the plates. On the right, the schematic sensitivity curve of the analyser is shown to select the detection over extended ranges of the energy per charge of the particles.

angle of incidence, thus contributing to the requirement that instrument should determine directional properties of the plasma distribution function. This property, enabling quadrispherical plasma analysers to cover a wide range of incident angles, close to 180° simultaneously, has made them the prime choice for space instrumentation, in particular in the "top-hat" analyser described below.

The further development of quadrispherical analysers, the top-hat geometry, combines the advantages of these analysers with a 360° coverage in the polar direction. The top-hat analyser was first proposed by Carlson et al. (1983) and is illustrated in Fig. 7.45. This type of analyser is constructed as follows. Using a pair of hemispherical analyser plates, a circular section of the outer plate is cut out of it and raised, around the axis of symmetry. This geometry then effectively forms a quadrispherical analyser that has been extended over 360°, with the entrance aperture formed between the raised, top-hat section and the outer plate of the analyser. In this way, particles can enter the electrostatic analyser over a range of 360° in polar angle (in the plane perpendicular to the rotation axis of the analyser) and within the azimuthal acceptance angle that covers a range of 3–10° in the direction along the analyser axis. As shown in Fig. 7.45, particles that enter around the directions defined by A, B and C effectively exit the analyser, after a 90° path between the plates, to points a, b, and c, respectively.

At the exit of the analyser plates, a circular detector, as shown also in Fig. 7.45, can record the incident particles that selected for their mass per charge ratio by the electrostatic analyser. This detector can be built in circular sections in the form of independent detectors (MCPs), for example in 24 sections as shown in the figure. The top-hat analyser in this example can therefore resolve the incident particle flux in 24 directions in the plane perpendicular to the axis of the detector. When such an analyser is mounted with its rotation axis perpendicular to the spin axis of a spinning spacecraft, as shown in Fig. 7.46, the instrument can resolve particles incident over the complete 4π solid angle in a complete spin of the spacecraft. If at the same time, the voltage is stepped on the analyser plates at a rate which is much faster than the spin rate, thus stepping the energy per charge range of the detected particles, then the complete distribution function (as a function of E/Q of the incident particles) can be measured in each spin of the spacecraft.

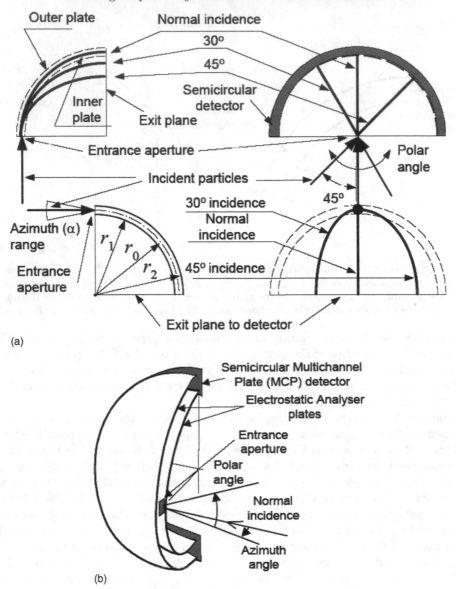

FIGURE 7.44. The quadrispherical plate electrostatic analyser. (a) Shows the projected drawings of the quadrisphere, with sample particle trajectories at different polar angles of incidence; (b) Shows the perspective view of the plate geometry.

7.3.5. *Plasma composition measurements (high-resolution measurements of plasma ions)*

Plasma ion detectors described so far can determine the plasma distribution function in terms of angular dependence, but the different particle species are not distinguished except in terms of their energy per electric charge ratio E/Q. In the solar wind, where most of the kinetic energy of all the particles (at least to a first order approximation) arises from the streaming velocity at $\sim$30–800 km/s, the energy scales with the atomic number of ions, so that alpha particles, for instance, have four times the kinetic energy

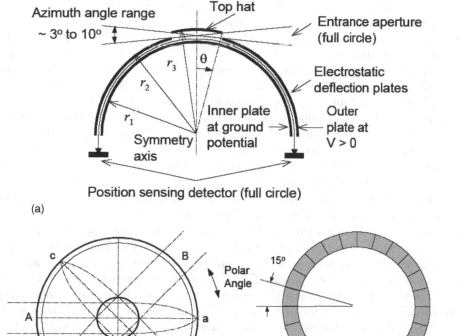

(a)

(b)   (c)

FIGURE 7.45. Schematic diagram of the quadrispherical top-hat electrostatic analyser. As shown in the top panel, ions enter through the gap between the top hat (circular cap) and the hemispherical analyser are deflected by the electric field between the plates through 90° onto a circular position sensing detector (illustrated in (c)). Particles entering over a full 360° in polar angle are detected by this analyser (three sets of typical particle trajectories are shown in the lower left diagram).

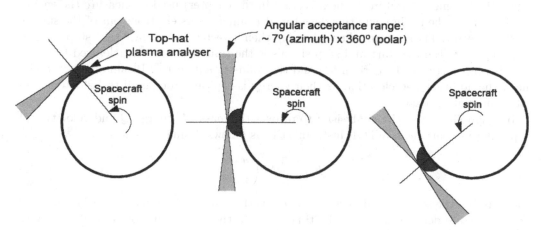

FIGURE 7.46. Mounting a top-hat plasma analyser with its symmetry axis perpendicular to the spacecraft spin on a spinning spacecraft provides an angular coverage of 4π steradian in each spin period of the spacecraft.

of solar wind protons and twice the electric charge, so that their E/Q ration is twice that of protons. The same is true for fully ionized carbon and oxygen, so that these three elements, when fully ionized, cannot be distinguished in the solar wind on the basis of their energy per mass ratio. Similar ambiguities concerning the elemental composition of the measured plasma ions can arise in other space environments. Although protons are usually the most abundant species in most plasma regimes, the characterization of the plasma cannot be fully complete without distinguishing between some of the more relatively abundant elements. In addition, a range of plasma diagnostics becomes possible if both elemental composition and the relative charge states of the heavier ions are determined. This is the case, for instance, in the solar wind, where the charge states of the heavy ions provide an important amount of information on the source regions of the solar wind in the solar corona.

The determination of the constituents of the plasma involves resolving independently the particles' mass and electrical charge. The measurement of their energy then becomes equivalent to the measurement of their velocity. A new class of plasma ion instruments was introduced in the early 1980s that combined the measurement of the energy per charge measurement with an independent measurement of the particle mass. As the technique involves the measurement of the velocity of the particle through the measurement of the time the particle takes to travel a given distance, these instruments are known under the generic name "time-of-flight" or TOF instruments (for a review, see Wuest, 1998). The first of such instruments, the solar wind composition spectrometer (SWICS) is illustrated in Fig. 7.47 (Gloeckler *et al.*, 1992). The principles of the technique can be described using this instrument as the generic example of the TOF instruments. The energy per charge ratio E/Q of incident ions is selected using a curved plate electrostatic analyser. The selected ions are then accelerated by an electric field generated by a calibrated high voltage, typically set at $\sim$20–30 kV. Particles thus accelerated are incident on a very thin carbon foil ($\sim$2 µg/cm^2). The particles cross the carbon foil, but scatter electrons as they pass through the foil. The scattered electrons are detected using a standard particle detector, such as a MCP. The time at which the signal is generated by the scattered electrons (the "start" time) corresponds to the particle to be detected crossing the carbon foil. The plasma particle then travels a set distance, usually $\sim$10 cm, and impacts on a detector which, in the case of the SWICS instrument, is a silicon solid-state detector (described in more detail in the next section in the chapter) used to measure the energy of the ion. As the ion impacts this detector, it again scatters electrons out of the surface of the detector. The electrons themselves are then detected to generate the "stop" signal. The time between the start and stop signals is the time that the particle used to travel the distance between the carbon foil and the solid-state detector. Plasma ions generally need the additional acceleration as their energy is typically too small to be detected in the solid-state detector.

In a simplified way, the analysis of to recover the mass M, charge Q and velocity v of a plasma ion particle in a TOF instrument is as follows. Using the energy:

$$E_{\mathrm{ss}} = \frac{1}{2}M\left(\frac{d}{\tau}\right)^2 \tag{7.21}$$

measured in the solid-state detector, where d is the distance between the carbon foil and the solid-state detector and τ is the time between the measured stop and start signals, the mass of the ion can be calculated as:

$$M = 2E_{\mathrm{ss}}\left(\frac{d}{\tau}\right)^{-2} \tag{7.22}$$

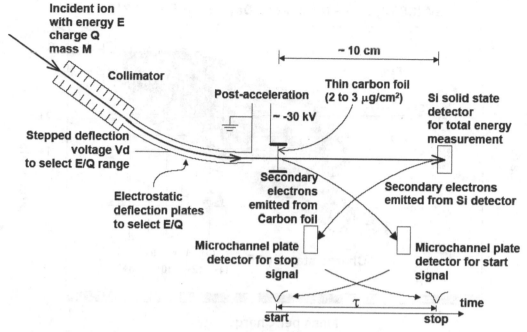

FIGURE 7.47. The TOF plasma ion analyser SWICS flown on the Ulysses mission. This instrument was the first to carry out detailed observations of the elemental and charge state composition of ions in the solar wind. (Redrawn after Gloeckler *et al.*, 1992, courtesy of G. Gloeckler.)

The energy measured in the solid-state detector is also the sum of the original energy E of the particle and the additional energy gained by the accelerating voltage V, so that:

$$E_{\mathrm{ss}} = QV + E = Q\left(V + \frac{E}{Q}\right) \tag{7.23}$$

from which the incident particle's charge can be determined:

$$Q = E_{\mathrm{ss}}\left(V + \frac{E}{Q}\right)^{-1} \tag{7.24}$$

where E/Q is the quantity determined by the voltage on the electrostatic plates. The mass-per-charge ratio is

$$\frac{M}{Q} = 2\left(\frac{d}{\tau}\right)^{-2}\left(V + \frac{E}{Q}\right) \tag{7.25}$$

The incident energy of the particles is calculated from $E = Q(E/Q)$, and therefore their incident velocity is $v = \sqrt{2E/M}$.

The corrections that need to be made to this simplified analysis are related to a small energy loss of the incident particle through the carbon foil and a more important fraction of the energy that is lost through the surface layer of the solid-state detector. These two effects are determined in the pre-flight calibration of the instrument and are taken into account in the analysis of the data returned from space.

An example of results from the Ulysses SWICS instrument is shown in Fig. 7.48. This instrument is capable of determining the elemental composition of solar wind ions as well as the charge state of partially ionized ions up to the atomic number of iron. Partially ionized ions can provide important information on the source regions of the solar wind;

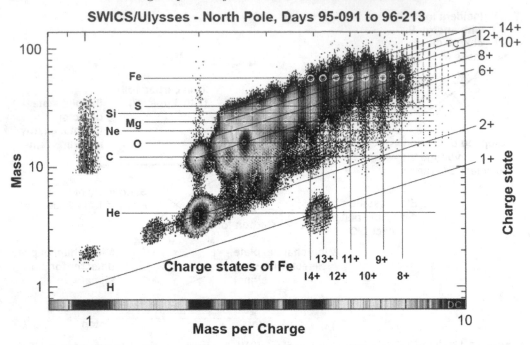

FIGURE 7.48. The elemental distribution and the charge states of ions in the fast solar wind measured by the SWICS instrument on Ulysses. (Figure courtesy of G. Gloeckler and R. von Steiger.)

relative ratios of ionisation for ions such as oxygen can indicate the temperature in the solar corona where the corona and the solar wind become collisionless.

A further development of the TOF plasma analyser, the CODIF instrument (Rème et al., 2001) on the four Cluster spacecraft, is shown in Fig. 7.49. The instrument is shown schematically in a half cross-section. In this case, the electrostatic analyser is a toroidal top-hat design; the toroidal geometry for the plates is a variant on the quadrispherical design, with improved geometrical properties over the spherical design focusing the ions at the exit of the analyser plates on to the TOF section below. Following the selection of ions with a given E/Q value through the selection (and successive stepping) of the voltage on the toroidal analyser, particles are post-accelerated by a voltage of 25 kV (in fact, variable between 17 and 25 kV), and then enter the TOF section through a thin carbon foil. Secondary electrons generated in the carbon foil are deflected to the inner part of the circular microchannel plate that acts as an electron multiplier. Electrons that emerge from the MCP are detected using the inner semi-transparent grid to provide the start signal for the TOF analysis. The ions that traverse the TOF section impact on the outer part of the MCP; the electrons emerging from the MCP are again detected, first by the signal generated in the outer semi-transparent grid to provide the stop signal for the TOF analysis, followed by the sectored plate that provides information on the position of the impacting ions. In the composition and distribution function analyser (CODIF) instrument, unlike in the SWICS instrument described above, the energy of the ions is not measured after the post-acceleration section. However, the mass per charge of the ions can be obtained from the same Equation (7.25), using the quantities defined for the SWICS instrument described above. The determination of the position of the incident ion on the MCP is a key measurement by the CODIF instrument, in order to determine the angular distribution of the incident ions. This, together with the energy per charge and the mass

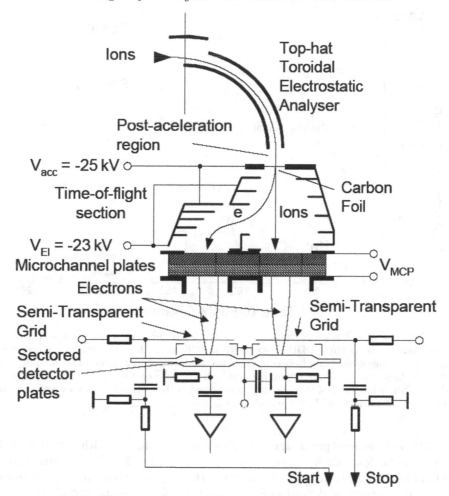

FIGURE 7.49. A simplified half-cross section of the CODIF instrument of the CIS experiment on the four-spacecraft Cluster magnetospheric mission. The diagram shows the geometry of the top-hat electrostatic analyser plates, followed by the TOF section for determining the atomic mass of magnetospheric ions and their charge states. (Redrawn from Rème *et al.*, 2001, courtesy of H. Rème and I. Dandouras.)

per charge determination leads to the determination of the distribution function of the plasma population of interest in the magnetosphere. Fig. 7.50 illustrates the capabilities of the CODIF plasma analyser, showing the relative abundance of the different main ion species that emerge from the Earth's ionosphere into the magnetosphere.

Although TOF plasma instruments using curved plate electrostatic analysers are capable of resolving the atomic number of ions and their charge states, their mass resolving capability $M/\Delta M$ is limited to about 10–20, mainly due to energy losses and scattering of the ions in the carbon foil. In the case of the CODIF instrument described above, the mass per charge $\Delta(M/Q)/M/Q$ resolution is ~ 0.15 for protons and ~ 0.25 for low energy, singly charged oxygen ions. This resolution meets the requirements of the investigation for the study of the distribution functions of the important ions in the Earth's plasma environment.

Another type of instrument, using a linear electric field (LEF) has been developed to increase the mass resolving capabilities of plasma instruments. This type is based on the

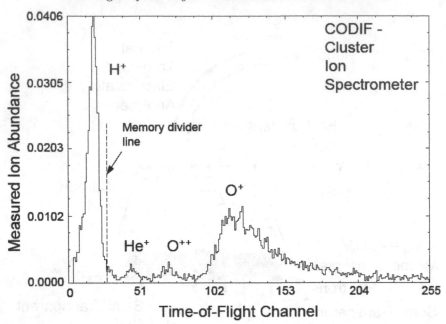

FIGURE 7.50. The distribution of "upwelling" ions, from the terrestrial ionosphere, measured in the magnetosphere by the CODIF instrument on Cluster. The TOF channel number is proportional to the measured time between the start and the stop signals. The memory divider line indicates that the most abundant ions, H^+, have 20% of the instrument event memory devoted to them, while 80% of the memory is reserved to less abundant ion events detected by the instrument. (Courtesy of H. Rème and I. Dandouras.)

determination of the mass per charge (M/Q^*) of plasma ions, not with the original charge Q of the incident ion, but with the charge Q^* the ion retains after passing through a carbon foil of a TOF instrument. Instruments of this type use the characteristic motion of particles in an electric field defined as $E_z = -kz$, where k is a constant. An ionized particle of mass M and charge Q^* entering a region with such a linear field moves according to the equation:

$$M\frac{d^2z}{dt^2} = -Q^*kz \tag{7.26}$$

Rearranging yields an equation of simple harmonic motion:

$$\frac{d^2z}{dt^2} + \frac{kQ^*}{M}z = 0 \tag{7.27}$$

with solutions of the form:

$$z = A\sin\left(\sqrt{\frac{kQ^*}{M}}t + B\right) \tag{7.28}$$

where A and B are constants of integration determined form the initial conditions. A particle entering the field region at $z = 0$ will return to the same plane after a flight time of:

$$\tau = \pi\sqrt{\frac{M}{kQ^*}} \tag{7.29}$$

independently of the energy of the ion and its direction of travel when it first enters the linear field region. Sample trajectories of particles in the LEF are illustrated

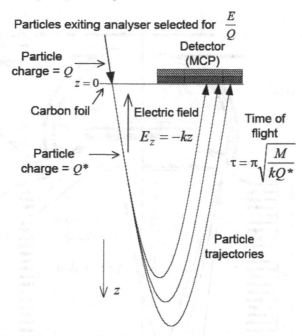

FIGURE 7.51. The principle of the LEF technique for high-resolution mass per charge determination.

in Fig. 7.51. Instruments of this type have been developed for a number of missions since 1990.

Ions passing through a carbon foil at the entrance of a TOF system lose their original charge state through charge exchange in the foil. Most incident ions emerge as neutral particles (∼70%), with a smaller number (10–25%) as singly charged positive ions and an even smaller number as singly charged negative ions. The charge state of the emerging particle depends on the particle species and its energy. Instruments using the LEF technique determine the mass per charge M/Q^* of the ions (not their energy), but in practice, $Q^* = +1$ for the particles that respond to the LEF, so the mass of the ion can be determined. Neutrals follow straight trajectories as they are insensitive to the electric field and negatively charged ions are accelerated by the electric field away from the entrance plane, to which the positively charged ions return.

In TOF instruments of the Ulysses SWICS type (as described above) the total energy of the particle is measured after it has been accelerated, and therefore the original charge state Q as well as the mass M of the incident ion can be determined. LEF instruments using electrostatic deflection techniques for E/Q selection and measurement, on the other hand, measure only the mass M of the incident ions in addition to their energy per charge, but are capable of much greater mass resolution, $M/\Delta M$, in the range 50–100.

The Cassini Plasma Spectrometer (CAPS) group of instruments includes the ion mass spectrometer (IMS) that uses the LEF technique for high-resolution mass resolution of space plasma ions (McComas et al., 1990). The schematic diagram of the IMS instrument is shown in Fig. 7.52. It consists of a toroidal top-hat electrostatic analyser for selecting incident ions according to their energy per charge ratio and also using the top-hat geometry for determining the angles of incidence of incoming ions within an ∼8° × 160° acceptance angle of the analyser. (Although such symmetrical top-hat analysers have, in principle, a 360° acceptance angle perpendicular to the axis of the analyser, in the case

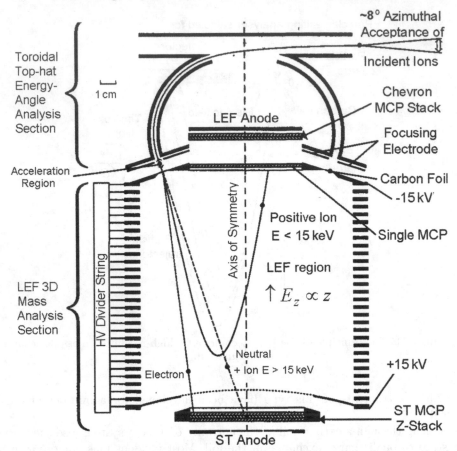

FIGURE 7.52. Schematic diagram of the IMS in the CAPS investigation on the Cassini Saturn
Orbiter mission (after McComas *et al.*, 1990).

of the IMS this is limited by the mounting of the instrument within the CAPS complex
on the Cassini spacecraft.) At the exit of the toroidal deflection plates, the selected particles are accelerated by a voltage set at 14.5 kV and cross a thin carbon foil into the
LEF section of the instrument. Ions undergo charge exchange processes in the carbon
foil and emerge mostly as neutral particles, with a small fraction of the ions retaining a
single positive electric charge. The LEF (directed upward in Fig. 7.52) is generated by
aluminium rings attached to a voltage divider around the perimeter of the cylindrical
LEF section that generates an electric potential $\propto z^2$ so that the vertical electric field is
$\propto -z$. The total voltage across the LEF section is ~30 kV. The singly charged ions with
energy less than ~15 keV are reflected in the vertical LEF and after a flight time $\propto \sqrt{M}$
are detected by the upper detection system. The start signal for the TOF measurements
are generated by secondary electrons from the carbon foil impacting on the lower MCP.
Neutral particles and singly charged ions with energy >15 keV, on the other hand, impact
the lower detection system.

An alternative design based on the LEF principle is the CELIAS/MTOF (Hamilton
et al., 1990; Moebius *et al.*, 1990) instrument on the ESA/NASA solar and heliospheric
observatory (SOHO) spacecraft. The energy per charge selection of incident ions is performed in an electrostatic deflection system of complex geometry that allows both a wide
window in energy per charge (about a factor 5) and a wide acceptance angle. Both these

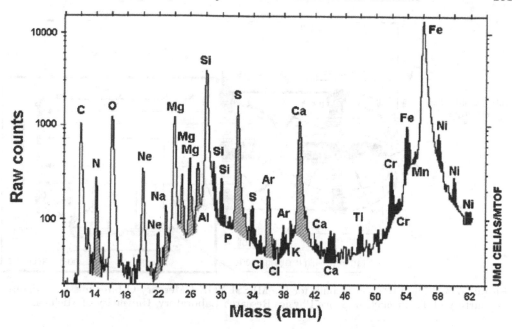

FIGURE 7.53. Mass determination of solar wind ions by the CELIAS/MTOF instrument on the SOHO space mission. The instrument is able to resolve the elemental and isotopic composition of the solar wind ions up to Ni.

are significantly larger than the more selective top-hat geometry for the electrostatic deflection system. The need for such a wide-band acceptance for the energy per charge selection on this mission arose because SOHO is a three-axis stabilized spacecraft, so that the often-used technique of the spacecraft spin for covering a wide range of acceptance angles cannot be used. The LEF for the TOF section is generated between two plates, one (at ground potential) is in the shape of a V, the other (at a positive potential), is a hyperbolic surface. Particles enter the linear field region through a carbon foil through the vertex of the V-shaped plate and return to the same plate, impacting on a MCP detector, after a flight time as given above, proportional to $\sqrt{M/Q^*}$, where Q^* is the charge of the ion after crossing the carbon foil. A sample result from the MTOF instrument, shown in Fig. 7.53, illustrates the mass resolving capabilities of this technique as it is able not just to resolve the elements but also their isotopes.

A new type of plasma ion instrument has been developed recently that combines a wide angular acceptance angle with the capability to measure the energy per charge and the mass per charge ratios of incident particles. The fast imaging particle spectrometer (FIPS) was developed for NASA's MESSENGER mission, a Mercury orbiter due to arrive to the planet in 2011. The operating principle of the instrument (Zurbuchen *et al.*, 1998) is shown schematically in Fig. 7.54. The wide acceptance angle is required because MESSENGER is a three-axis stabilized spacecraft, and the top-hat electrostatic analysers have an intrinsically narrow polar field of view that provide wide angular coverage only when combined with the spin on spinning spacecraft (as shown in Fig. 7.45). The solution (Koehn *et al.*, 2002), as shown in Fig. 7.54, is to use a more complex electrostatic deflection system that preserves the information on the direction of incidence of particles over a wide range of incidence. A cylindrically symmetric set of deflection plates, combined with collimators, steers the incident particles through to a post-acceleration section, followed by a TOF section.

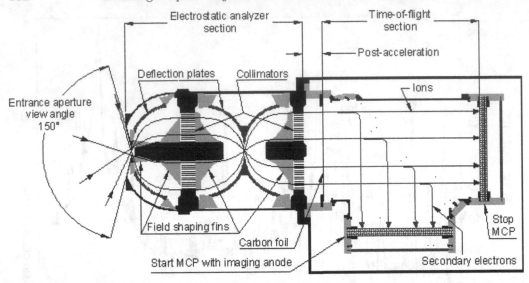

FIGURE 7.54. The FIPS sensor built for the MESSENGER Mercury Orbiter mission. (Figure courtesy of T. Zurbuchen, Space Physics Research Laboratory, University of Michigan.)

7.3.6. *Electron multiplier detectors: the channeltron and the microchannel plate (MCP)*

A very important part of any plasma particle instrument is the actual detector that indicates, by the production of a recognizable electrical pulse, the impact of the particle to be detected. By far the most common particle detector (for the low-energy particles in the plasma population) is the multichannel plate detector, or MCP as it is commonly called. The functioning of the MCP is based on its predecessor, the electron channel multiplier (also called the channeltron). The principle of both devices is illustrated in Fig. 7.55. A channeltron or one of the elements of the MCP consists of a hollow tube of glass that has electrical contacts at both ends, with its inner surface made to be a semiconductor through reduction in a hydrogen atmosphere. A high voltage (of order 1–4 kV) is applied between the two ends of the tube, thus establishing a strong electric field in the tube. An incident particle or photon of sufficient energy creates, on impact on the inner, semiconductor surface of the tube, secondary electrons that move (and are accelerated) under the influence of the electric field in the tube. The secondary electrons, in turn, produce further electrons by impacting on the surface of the tube, thus creating an electron cascade, with, eventually, a very large number of electrons exiting the tube, as a result of the first impacting particle. The gain of these devices (i.e. the number of electrons that exit the channel, for each incident particle) is very high, of order 10^5. It depends on the physical characteristics of the semiconducting layer which coates the inner surface of the channels, as well as on the voltage applied (i.e. the strength of the electric field) across the channel. The electron cascade, on leaving the tube reaches a conducting plate, the anode, which thus becomes negatively charged. This charge pulse, corresponding to a voltage signal, is then amplified, and represents the detection of a particle. The resistor connecting the anode to ground allows the anode to be discharged to detect further particles.

    Channeltrons are small, but individual devices and are often used in groups. In the Ulysses solar wind plasma instrument, the electron analyser has seven chaneltrons, while

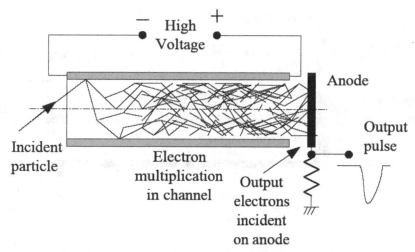

FIGURE 7.55. Principle of operation of the channel electron multiplier (channeltron) and of one element of a MCP.

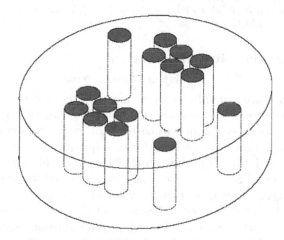

FIGURE 7.56. Schematic diagram of a MCP. The size of the individual channels is exagerated in this diagram and only a few of them are shown. Both the front and the rear surfaces are metallized, to provide an electrical connection to the high voltage applied to the MCP. The typical thickness of the MCP disc is ~1–2 mm, the diameter of the channels is usually in the range 10–100 μ.

the ion analyser has 16, to help resolve the direction of impact of the particles to be detected. MCP detectors consist of a very large number (>10,000) of very small diameter tubes, fused together in the form of a plate, as illustrated schematically in Fig. 7.56 (where only a few of the channels are shown explicitly, for the sake of clarity). In such a case, the whole of the top and bottom plates are coated with a conducting surface, and the high voltage is applied between the top and bottom plates. MCPs (as Fig. 7.57) offer considerably larger detecting areas than single channeltrons; in addition, MCPs can also be produced in any shape to provide the best matched detector surface area to any particular plasma detector.

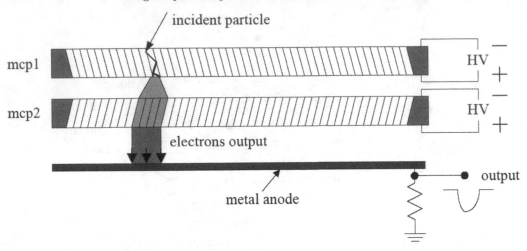

FIGURE 7.57. The so-called chevron arrangement of two mcps for increased electron gain. Each channel of an MCP has a gain $\geq 10^5$ Mcp channels are usually inclined with respect to the normal to the plate. By placing two mcps one behind the other, not only is the combined gain increased substantially, but this arrangement minimizes the possible effects of ion contamination, which occurs when an ion is generated within the channel, it is accelerated back into the channel by the electric field, and when it collides with the channel wall, it starts another electron cascade, resulting in an unwanted additional pulse.

7.4. Energetic particles instrumentation

7.4.1. *Principles of measurement*

Energetic particles in space generally originate in processes that are different from those that lead the formation of space plasmas. The latter are formed by heating a neutral gas by a variety of processes through some form of ionizing interaction between the gas and a wave or radiation field, such as, for example, the Ultraviolet (UV), extreme UV (EUV) or X-ray radiation from the Sun that is responsible for creating the plasma in the Earth's ionosphere. Such collections of ionized particles are referred to as thermal plasmas, characterized by their thermal velocity distributions and, in the case of the solar wind, by their bulk velocity. Plasma particle energies are usually less than $\sim$30–40 keV. Energetic particles are accelerated, in general from a seed population in the high-energy tail of the plasma distribution, by electric fields associated with shock waves in the plasma, and by other, less well understood processes and mechanisms associated with magnetic reconnection. Energetic particles therefore form the so-called non-thermal component of the space plasma environment.

There is also a difference in the techniques used for detecting non-thermal, energetic particles, compared to the plasma measurement techniques described in the previous section. The use of electrostatic deflection, the basic element of all plasma detectors, can only use voltages up to about 30 kV; this limits its use to particles with an energy per charge ratio of the same value.

Energetic particles are detected through the by-products of their interaction of with matter. An energetic particle travelling through any material, gaseous, liquid or solid, loses energy mainly by colliding with electrons orbiting the atoms of the material and producing a track of ionized atoms and free electrons along its path. What happens to the ion–electron pairs depends strongly on the material and its physical properties. Different particle detectors make use in different ways of the ion–electron pairs to generate a signal characteristic of the energy lost by the particle. A simplified formula for the rate

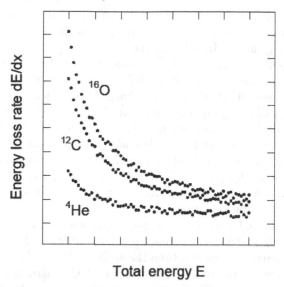

FIGURE 7.58. The rate of energy loss as a function of total energy for He, C and O energetic ions. This is the basis for determining the elemental composition of a flux of high-energy particles, whether of cosmic rays or particles accelerated near the Sun.

at which energetic particles lose energy in matter through collisions with electrons can be written as:

$$-\frac{dE}{dx} = K\frac{z^2}{v^2}\left(\ln\frac{K'v^2}{1-v^2/c^2} - \frac{v^2}{c^2}\right) \tag{7.30}$$

where the rate of energy loss, as a function of path length $-dE/dx$ is given usually in eV/μm in solids; z and v are, respectively, the electric charge and velocity of the incident particle; K and K' are constants characteristic of the material; and c is the velocity of light in free space. The range of the particle in the material can be obtained by integrating the above formula. At moderate energies, the logarithmic term varies only slowly, so that the rate of energy loss is approximately proportional to the square of the electric charge of the particle, and inversely proportional to its energy. The simplified formula above breaks down at very low and very high energies.

Further simplifying the formula, we have:

$$-\frac{dE}{dx} \approx K_1\frac{z^2}{v^2} = K_1\frac{mz^2}{2E} \tag{7.31}$$

where m is the mass of the energetic particle, we can see that the rate of energy loss for an ion of atomic number A and electronic charge z (in units of the electronic charge of the electron or proton e), can be calculated as a function of the rate of energy loss of protons, as:

$$\left(\frac{dE}{dx}\right)_{A,z} \cong Az^2\left(\frac{dE}{dx}\right)_{proton} \tag{7.32}$$

Note that this is an approximate formula only, and that a number of corrections are required in practice, dependent on A and z. Nevertheless, this is the basis for distinguishing not only among the different ion species (as shown in Fig. 7.58), but also their degree of ionization and atomic mass number, and even the different isotopes of a given ion species. (In general, unlike plasma ions, energetic particles are fully ionized and their electric charge is equal to their atomic number.)

Figure 7.58 also indicates the principles of energetic particle detectors. As will be described below, all energetic particle instruments are based on the measurement of both the rate of energy loss $-\mathrm{d}E/\mathrm{d}x$ and the total energy E of the incident particles. These two measurements (together with the measurement of the direction of incidence), in principle, provide all the information needed on incident energetic particles.

For energetic protons and other positively charged ions, each collision represents only a relatively small loss of energy and momentum. Their trajectory remains therefore an approximately straight line. Incident energetic electrons also tend to lose their energy by the same process, although as the projectile and the target (the electrons in the silicon lattice) have identical masses, the path of the electrons is more random. Furthermore, electrons also lose energy by passing close to the nuclei of the atoms in the material, emitting a photon representing the electron's energy loss. This radiation is known as *brehmsstrahlung*, or breaking radiation.

Particles with very large kinetic energies, moving at speeds close to that of light in free space, can lose energy in matter by emitting Cerenkov radiation. This phenomenon occurs if the velocity of the particle is greater than the speed of light, c/n in the material, where n is the refractive index of the material. Detection of Cerenkov light sets a lower limit on the velocity of the particle; this effect can be used in some particle detectors, usually those that are aimed at observing high-energy cosmic rays, to eliminate lower energy particles from the detection process. Using a set of Cerenkov detectors with different refractive indices allows further discrimination, on the basis of their velocities, among high-energy particles.

An energetic particle detector counts the number of particles incident on it, as a function of the detector itself (its response function), and as a function of the differential intensity of the particle flux which is present at the location of the detector. In other words, the number of particles detected is:

$$N_i = \int J\left(\mathbf{r}, E, t, \theta, \varphi, i\right)\,\mathrm{d}t\,\mathrm{d}S\,\mathrm{d}\Omega\,\mathrm{d}E \qquad (7.33)$$

where N_i is the number of particles of species i counted in an interval Δt and in an energy interval ΔE (over which the integral is taken); J is the differential flux of particle species i, at a location given by the position vector $\mathbf{r}$, at energy E, at time t, in a direction defined by the angles θ and φ in some given coordinate system. The integration is also performed on the angles θ and φ over the solid angle $\Delta\Omega$ and a surface area ΔS, characteristic of the detector, defined by its *geometric factor*, in units of cm^2 sr. J is normally measured in units of either $\mathrm{cm}^{-2}\ \mathrm{s}^{-1}\ \mathrm{sr}^{-1}\ \mathrm{keV}^{-1}$ or $\mathrm{cm}^{-2}\ \mathrm{s}^{-1}\ \mathrm{sr}^{-1}\ \mathrm{MeV}^{-1}$. The dependence of J on the location r is usually taken out of the integral, as r depends on the location of the measurement in space, therefore on the orbit of the spacecraft carrying the detector.

The most frequently used detector in space instruments for measuring the fluxes of energetic particles is the silicon detector. It consists of a silicon wafer, of thickness from a few μm to a few mm, and of surface area from a few mm^2 to several cm^2; the front surface is either a diffused p-n junction, or a metal-semiconductor surface barrier junction; the back surface has an ohmic connection (see Fig. 7.59). In normal operation, a large reverse voltage is applied to the detector, so that the depletion layer characteristic of reverse biased semiconductor junctions penetrates the whole thickness of the detector. An incident energetic particle loses energy according to the above formula, and creates electron-hole pairs, one pair for each 3.6 eV energy lost. In the electric field created by the reverse voltage across the detector, the electric charge is collected by the electrodes and produces, after suitable amplification, a voltage pulse that is proportional to the energy lost by the particle in the detector. (This operation is similar to the classical ionization

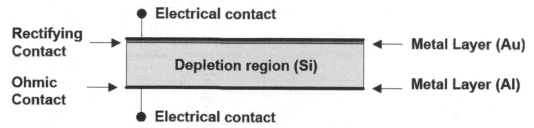

FIGURE 7.59. The construction of a planar solid state (silicon) energetic particle detector. Reverse biasing the detector that is equivalent to a diode creates a charge-free region inside the detector in which impacting high-energy particles can create electron-hole pairs to be collected by the electrical contacts on the detector.

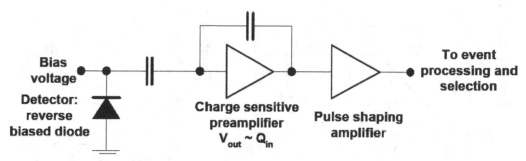

FIGURE 7.60. Front-end electronic circuit that collects the charge deposited by a high-energy particle in the solid state detector and turns it into a voltage pulse that is proportional to the collected charge and therefore to the energy lost by the particle in the detector.

chamber that used the charge generated in a gas container by a passing energetic particle to detect its energy loss in the chamber.)

The silicon used in solid state particle detectors is either extremely pure, close to its intrinsic conductivity (i.e. only very marginally n-type), or, as this is difficult to achieve for thicker detectors, is diffused with lithium, to achieve close to intrinsic conductivity. Surface barrier detectors can have a thickness up to a few hundred microns; thicker detectors are usually of the lithium-drifted types.

The typical front-end electronic circuit used with a solid state detector is shown schematically in Fig. 7.60. The detector itself is equivalent to a reverse-biased diode. When a particle impacts on the detector, an electrical charge signal is generated by the electron-hole pairs in the detector. This signal is first transformed into an equivalent voltage signal by a charge-sensitive preamplifier. The voltage signal is then shaped into a pulse with an amplitude that is proportional to the charge created by the passage of the particle in the detector or, equivalently, by the energy deposited (lost) by the particle in the detector. Figure 7.61 illustrates the energy loss versus total energy for a surface barrier detector. Protons with an incident energy up to 1 MeV are absorbed in the detector and the pulse response of the output circuit shows the proportionality of the signal to the (total) energy lost by the particle as it is stopped in the detector. Protons with an incident energy greater than 1 MeV penetrate through the detector and generate smaller amplitude pulses as their energy exceeds the maximum energy that is generated by a proton that is just stopped in the detector. However, as is indicated in Fig. 7.61 to illustrate the effect, a penetrating particle of ~2.8 MeV generates the same signal as a stopping particle of energy 0.15 MeV; similarly, the stopping particle of energy 0.2 MeV

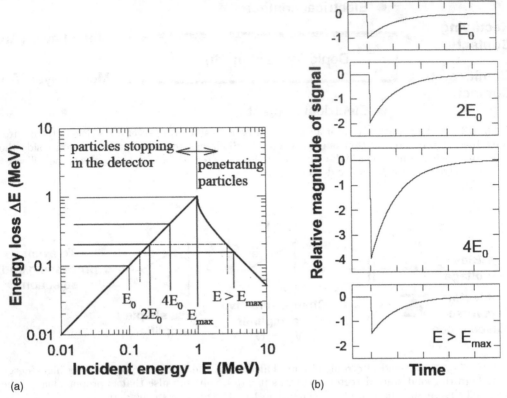

FIGURE 7.61. The energy loss versus incident energy curve for a given thickness ($\sim 30\,\mu$) silicon detector is shown in (a). In (b) the resulting voltage signals are shown for four incident energies. The signal is proportional to the total energy of the incident particle if it is absorbed by the detector. Penetrating particles give a signal that is proportional to the energy lost in the detector.

produces the same signal as the penetrating particle shown with energy ~ 3.2 MeV. With a single solid state detector there is therefore an ambiguity concerning the energy of the particle that generates a given signal.

7.4.2. *Energetic particle telescopes*

In order to overcome the inherent ambiguity of measuring the energy of particles with a single detector, silicon detectors are normally combined into stacks, as shown in Fig. 7.62. The direction of incidence of the particles is restricted to a cone around the axis of the stack by active (as in the figure) or passive shielding. As particles travel through the stack, they lose energy in successive detector elements, and provide a set of signals (or absence of signals) in the detectors, from which the incident energy of the particle and even its species) can be deduced. Such a stack is customarily called a particle "telescope".

A typical energetic particle telescope consists of a stack of three detectors, as shown in Fig. 7.62, based on a simplified schematic of the low energy telescope (LET) on the Ulysses mission. The acceptance cone is defined by a circular aperture in metal, and/or an active or passive shielding cylinder surrounding the stack. An active shielding is also a particle detector, usually made of a scintillator material. This active shielding detector is viewed by a photomultiplier tube or photodiodes, so that particles detected crossing the shield will not be counted as valid particles, even if, after penetrating the shield,

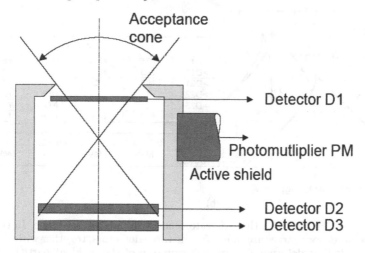

FIGURE 7.62. A typical stack of three solid state detectors surrounded by an anti-coincidence shield. Such an arrangement of detectors is usually called a solid state energetic particle telescope. This schematic is based on the LET on the Ulysses spacecraft.

they deposit enough energy in the detectors of the stack to generate a signal. The active shield operating in this way is called an anti-coincidence detector. In many simple particle telescopes, the shield is simply a metal cylinder, so that all but the most energetic particles (of which there are only a few, compared to the population of particles which the telescope is intended to detect) are stopped from penetrating through to the silicon detectors in the stack. In addition, the third detector in the stack is also used in anti-coincidence: this means that if a particle produces a signal in this third detector, it is either because it is incident from the back of the telescope or because its energy exceeds the energy range to be analysed in the first two detectors of the telescope.

The operation of the telescope in Fig. 7.62 is further illustrated in Fig. 7.63. This figure shows the energy loss versus energy curves for the stack of three solid state detectors for protons. Only protons with sufficient energy to penetrate detector 1 can generate a signal in detector 2 and only protons of sufficient energy to penetrate both detectors 1 and 2 can generate a signal in detector 3. A simple way to illustrate the determination of the energy of the incident particles is through the use of discriminators. These are electronic circuits following the primary amplification chain that provide a digital signal at their output if the amplitude of the pulse generated in the solid state detector exceeds a preset value. The levels set for the six discriminators in this simple example are shown on the energy loss versus energy curves. There are two discriminators, D11 and D12, associated with detector D1, three discriminators, D21, D22 and D23 associated with detector D2, and a single discriminator, D3 is associated with detector D3. The table in Fig. 7.63 shows that four adjacent proton energy channels can be defined, from 0.1–4.0 MeV, by simultaneously detecting the digital combination of the presence (coincidence) and absence (anti-coincidence) of indications of signals at the output of the discriminators. It is noted that a signal at the output of detector D3 indicates that the particle penetrated through both detectors 1 and 2 and therefore its energy is greater than 4 MeV; this a signal is used to reject the particle event. Similarly, if there is a signal in the anti-coincidence scintillator shield shown in Fig. 7.62 above, the event is rejected as it indicates that the particle arrived in the detectors from outside the acceptance cone of the particle telescope.

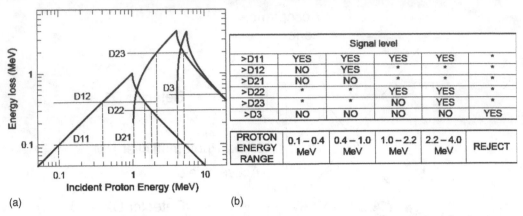

(a)   (b)

FIGURE 7.63. The operation of a three-detector energetic particle telescope. In (a), the energy loss versus incident energy curves are shown for the three detectors, together with the (schematic) discriminator levels that determine a "yes-no" response of the incident particles as they cross the stack. (b) Table showing the principle of energy discrimination according to the response in the discriminators.

Although greatly simplified compared to the operation of actual particle telescopes, this description of the principles of operation of solid state telescopes shows how detailed measurements of signals derived from the energy deposited in the stack of solid state detectors can be used to determine precisely the energy of incident particles. In similar instruments used in space, there are in general a larger number of discriminators; in addition, through a high-resolution determination of the amplitude of signals in pulse height analysers (a specialised version of analogue-to-digital converters), very refined energy loss and total energy measurements are possible. This makes it possible to determine different particle species, as shown in Fig. 7.64, taken from ground calibrations of the LET on the Ulysses mission (Marsden *et al.*, 1984).

7.4.3. *Position sensing particle telescopes*

The energy loss of particles penetrating through a detector depends on the angle of incidence. If, as shown in Fig. 7.65, the thickness of the detector is d, and the path length of a particle incident at an angle a is d', then the energy loss in the detector for the obliquely incident particle is greater than the energy loss of a normally incident particle:

$$\int_0^{d'} \left(\frac{\mathrm{d}E}{\mathrm{d}x}\right) \mathrm{d}x > \int_0^{d} \left(\frac{\mathrm{d}E}{\mathrm{d}x}\right) \mathrm{d}x \tag{7.34}$$

At the same time, incident particles of higher energy are absorbed in the detector than for normal incidence. This effect is also illustrated in Fig. 7.65. The energy of the incident proton that is just absorbed in the detector at an angle of incidence of 30° is 1.155 MeV, rather than the 1.0 MeV at normal incidence. As the energy of the obliquely incident particle increases, the difference also increases: an energy loss of 0.1 MeV corresponds to a particle of energy 5.0 MeV at normal incidence, but to almost 7 MeV at 30° incidence. It is clear that, for a precise energy loss determination, the angle of incidence also needs to be determined.

The generally adopted solution to determine the angle at which particles are incident within the acceptance cone of the telescope is illustrated schematically in Fig. 7.66. Specially designed solid state detectors are used for this purpose that are able to indicate

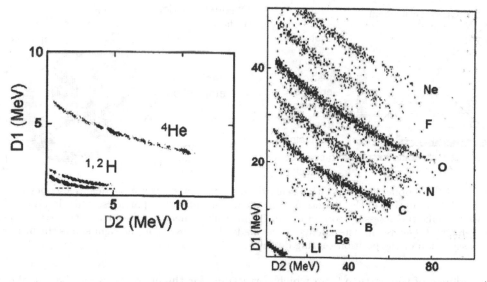

FIGURE 7.64. Two plots of the energy loss in the two front detectors of the LET on Ulysses (the particle telescope illustrated schematically in Fig. 7.62). Plotting the energy loss in Detector D1 versus the energy loss in Detector D2 allows to distinguish between different particle species. (Figure courtesy of Dr. R. Marsden, ESA.)

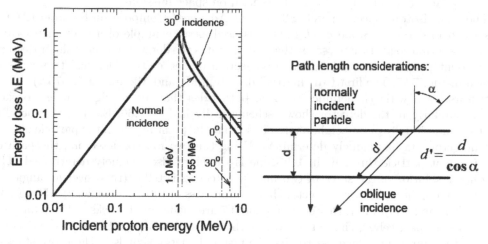

FIGURE 7.65. The effect of oblique incidence of particles on the path length through the detector and on the energy loss in the detector.

the location of the impact of the particles; these are called position sensitive detectors. The principle of operation of such detectors is also illustrated in Fig. 7.66. Both the front and back layers are metallized in strips, at 90° orientation with respect to each other. The position of the impact of a particle is detected on the front layer as a distance from the left side of the detector (the "x" position) and, on the back layer, as a distance from the bottom (as seen in Fig. 7.66) of the detector (the "y" position). The x, y data thus measured locates the particle in the plane of the detector. Using a stack of two such position sensing detectors, as in Fig. 7.66, allows the path of the particle (assumed to be a straight line) to be determined through the telescope. This, in turn, provides the angle

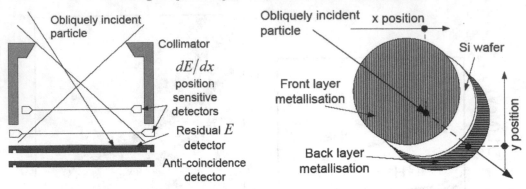

FIGURE 7.66. Application of position sensitive detectors in energetic particle telescopes for the precise determination of the direction of incidence of individual particles. This determination allows to make the necessary corrections to the measured energy loss so that very precise energy loss versus total energy measurements can be made. The drawing on the right shows the principle of construction of the position sensing detector.

of incidence of the particle from which corrections for the energy loss can be calculated. The residual energy of the particle can be measured by one or more solid state detectors placed below the position-sensitive detectors. An anti-coincidence detector below the residual energy detectors determines the upper limit of the energy range measured by the telescope. The telescope drawing in Fig. 7.66 represents schematically the construction of a very large class of actual instruments used on space missions.

The Solar Isotope Spectrometer (SIS) on the Advanced Composition Explorer (ACE), described in detail by Stone *et al.* (1998), is used as an example of a high energy- and mass-resolution solid-state telescope that incorporates the determination of the trajectory of incident particles. The schematic cross section of one of two identical telescopes is shown in Fig. 7.67. The first two "matrix" detectors M1 and M2 are thin ($\sim$75 μ) silicon detectors, each with 64 metallic strips on both sides that are at right angles to each other, similarly to the detector shown schematically in Fig. 7.66. Each of the metallic strips is individually pulse-height analysed, so that the point at which the particle crosses the detector can be uniquely determined. Using the two matrix detectors, the path of the incident particle through the telescope is therefore also uniquely determined. The angular resolution of the stack of two matrix detectors is 0.25° (rms) over all angles of incidence. Further solid-state detectors have a thickness of 100 μ (T1 and T2), 250 μ (T3), 500 μ (T4) and 750 μ (T5). Detectors T6 and T7 are made up of stacks of three and five detectors, respectively, giving an equivalent thickness of 2650 μ and 3750 μ; the outputs of the stacks are added together to give the residual energy signals in these two stacks.

The capability of the SIS telescope to distinguish among the different species of particles is well illustrated in Fig. 7.68 by the measurements made during the intense, solar flare generated energetic particle events in October 2003. In general, using ground calibrations, the capability of the SIS telescope has been charted as shown in Fig. 7.69.

7.4.4. *Directional flux measurements*

Energetic particle telescopes measure the flux of the different species of particles as a function of energy within the acceptance cone of the telescopes. A full description of particle fluxes as a function of the direction of incidence requires the use of techniques that allow the telescopes to sweep through a range of angles of incidence. This means that either multiple telescopes are used to cover the required angular range, or that the telescope is made to look successively in different directions.

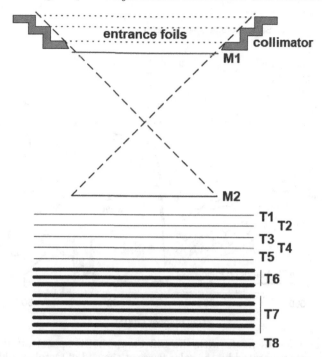

FIGURE 7.67. Schematic construction of the SIS on the Advanced Composition Explorer spacecraft. The two position sensing "matrix" detectors M1 and M2 determine the direction of incidence of energetic particles in the telescope, and detectors T1–T8 determine the rate of energy loss versus total energy for a large class of energetic solar particles, so that not only different particle species can be distinguished, but also the isotopic composition of the particles that compose a energetic solar particle event. (Figure courtesy of Dr. E. Stone, CALTECH.)

On spinning spacecraft this is achieved by using the spin to rotate the look direction of telescopes over an angular range of 360° per spin. This technique has been used on many space physics missions. The LET system on the ISEE-3 spacecraft (Balogh *et al.*, 1978), shown in Fig. 7.70, is used here as an example of how to achieve a wide range of angular coverage with good resolution. Three identical particle telescopes were used, with their acceptance cones centred at 60°, 30° and 135° with respect to the mounting plate of the system.

The schematic cross section of one of the telescopes of this system is shown in Fig. 7.71, together with the energy loss versus incident energy of the two detectors in each telescope. Each telescope incorporated a light rejection collimator; an electron rejection magnet; a detector stack; and an amplifier chain for each detector.

The detector stack in each telescope contained detector A, of surface area 1 cm^2, and 33 μ thick; detector B, of surface area 2 cm^2 and 150 μ thick. The instruments normally operated by detecting particles satisfying the condition of a signal in detector A and no signal in detector B, in other words, counting the flux of particles which were absorbed in detector A and therefore did not generate a signal in detector B. For protons this condition corresponded, as shown in Fig. 7.71, to incident energies between 35 keV and 1.6 MeV. The overall energy range was divided into eight logarithmically spaced energy channels. The discriminator levels for detector A (DA1–DA8 in the figure) were in fact set at $\sim$2 keV lower than the values of the energy levels defining the energy channels. The reason for this is that an incident energetic proton loses about 2 keV in the front ("dead") layer of the detector before it enters the active volume of the detector in which

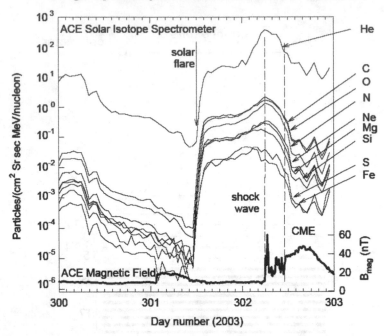

FIGURE 7.68. Measurements made with the SIS on the ACE spacecraft during the intense energetic particle events generated by the solar flares at the end of October 2003. The capability of the instrument to distinguish among the different species is well illustrated in this plot.

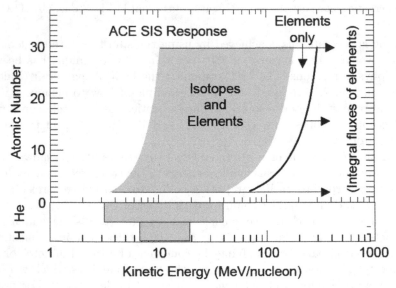

FIGURE 7.69. The capability of the SIS on the ACE spacecraft to distinguish among particle species (elemental composition) and also their isotopic composition. As illustrated, Isotopes and elements can be distinguished in the telescope (Fig. 7.67) in the energy range from about 10–100 MeV/nucleon. (Figure courtesy of Dr. E. Stone, CALTECH.)

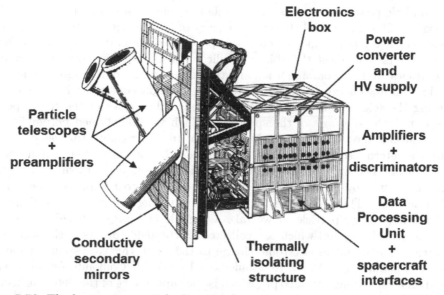

FIGURE 7.70. The low-energy particle directional spectrometer telescope on the ISEE-3 space-craft. The instrument had three identical telescopes with their acceptance cones angled with respect to the spacecraft spin axis. The instrument was mounted on the spacecraft on a plane perpendicular to the spin axis (spin plane). Each telescope had an acceptance cone of 30° full width.

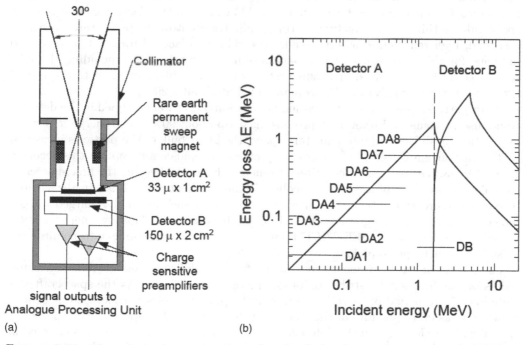

FIGURE 7.71. The schematic construction of each of the three telescopes on the ISEE-3 low-energy particle telescope (a) and (b) the energy loss versus incident energy curves for the two detectors A and B in the stack, together with the discriminator levels to allow the energy spectrum of particles to be measured.

the electron-hole pairs generated as the particle loses energy are collected to provide the output signal. This difference between the energy of the incident proton and the signal it generates as it is stopped in the detector is particularly important at low energies.

Further effects that affect the ability of silicon detectors to measure accurately low energy particles are light contamination and the intrinsic noise of the signal preamplifier. For the ISEE-3 low energy telescope, minimizing the light contamination was achieved by pointing the telescope away from the sun (hence the angles of the telescope units with respect to the spin axis of the spacecraft which was maintained perpendicular to the spacecraft-sun direction). In addition, the light rejecting collimator was designed to minimize the internal scattering of photons, and also had a special black, non-reflective internal coating. The first (or pre-) amplifier which detects and measures the amount of charge generated by the particle in the detector needs to be specially designed for low intrinsic noise. This can be easily seen by evaluating the amount of charge generated by a low-energy particle. If we take a particle of energy $\sim 36\,\mathrm{keV}$, it generates $\sim 10,000$ electron hole pairs which are collected in the detector. As each electron-hole pair contributes one unit of electronic charge to the total, this corresponds to an electric charge of about $1.6 \times 10^{-19} \times 10^4$ Coulomb $= 1.6 \times 10^{-15}$ Coulomb. As this charge is deposited on a capacitance of ~ 300 pF, which is the capacitance of the detector itself, the voltage signal can be calculated as charge/capacitance, or $1.6 \times 10^{-15}/3 \times 10^{-10}$, about $5\,\mu\mathrm{V}$. This very small signal has to be amplified accurately. Amplifier noise increases with (absolute) temperature, so a lower noise level in the amplifier can be achieved by operating it at a temperature as low as practicable. In the case of the ISEE-3 low-energy particle experiment this was achieved by thermally decoupling (isolating) the telescope units from the rest of the electronics. The telescopes were mounted on a special plate (machined from magnesium, for low mass), held away from the electronics unit and the spacecraft by thin, low conductivity struts, while the front of the plate was covered by a special, high reflectance material (called secondary surface mirrors, extensively used in space for this purpose) so that in equilibrium, the whole plate, including the telescopes, could be maintained at a temperature of about $-30°$, compared to the rest of the spacecraft (internally) which was maintained at about $+20°$.

Other particles (such as α-particles and heavier ions) that were stopped in the detector could also be counted. However, as protons make up about 90% of the energetic particle population at all times in interplanetary space, the great bulk of the particles measured by this instrument consisted of low-energy protons. A significant exception this occurred when ISEE-3, renamed the International Cometary Explorer, became the first spacecraft to make a close flyby of a comet, Giacobini–Zinner, in September 1995. In the vicinity of comets, there is an abundance of water-group ions, such as singly charged oxygen molecules and atoms, energized or "picked-up" in the solar wind. This low-energy particle telescope instrument on ISEE-3/ICE then became the first instrument to confirm the existence of such "pick-up" ions.

The success of this instrument came mainly from its ability to make detailed measurements of the directional distribution of low-energy proton fluxes. As the spacecraft was spinning with a period of 3 s (i.e. at a rate of 20 rpm), the telescopes swept out conical viewing areas corresponding to their pointing directions with respect to the spacecraft spin axis. Particle events in the different energy channels and telescopes were counted in accumulation intervals corresponding to eight angular sectors of 45° each in each complete spin of the spacecraft. The sectors were fixed with respect to the spacecraft–sun direction, as shown in Fig. 7.72. The accumulation period contained an integer number of spins; particles were counted in each sector for one-eighth of the accumulation period. This way, during each accumulation period, 24-directional samples (eight each from the

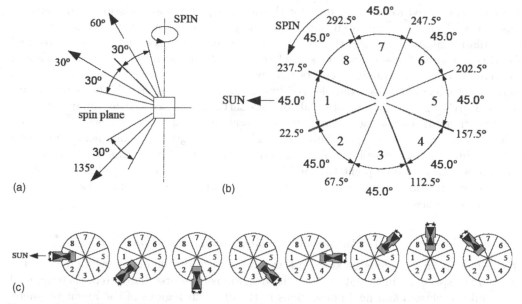

FIGURE 7.72. The schematic arrangement of the three telescopes of the ISEE-3 low energy particle instrument (a), and (b) the definition of the eight angular sectors in the spacecraft spin plane in which the detected particles were separately recorded, to allow the directional distribution of energetic particles to be determined. The (c) further clarifies the definition of the angular sectors as the spacecraft was spinning and the telescopes viewed different ranges of direction in space.

three telescopes) of the particle flux was obtained for each of the eight energy channels in each telescope, so that a total of 192 samples of the particle distribution function was measured in an accumulation period. The accumulation period contained either five or six spin periods, so that the distribution function sample was obtained in 15 or 18 s. This high time resolution still represents, more than 25 years after the launch of this mission, a record.

This very high time resolution was obtained at the cost of not distinguishing between protons and other ion species in this energy range. However, as already mentioned, protons dominate the overall distribution function, so that the approximation can be allowed, if the objective is to look for very high time resolution to study the dynamic effects of the low energy populations of interplanetary particles, for instance near interplanetary shock waves.

7.5. Space physics missions and their payloads

The objective of the first space missions, from 1957, was the exploration of the space environment using *in situ* instruments. Since then, many dozens of spacecraft explored the Earth's magnetosphere, the heliosphere and the fields and particles environment of planets and comets in the solar system. In this section, a few, selected missions and their payloads are described that have specially contributed to our current understanding of space.

Together with its payload of *in situ* instruments, the most important attribute of a Space Physics mission is its orbit. A mission's scientific objectives are expressed in terms of the observations that need to be carried out in specified regions of the magnetosphere, heliosphere or near other planets. Given a target region, celestial mechanics, the capabilities of launchers and the amount of fuel (for on board propulsion) that can be carried on the spacecraft need to be considered for the definition of a space mission. This activity, mission analysis, is an essential part of the preparation for any new mission. In general, mission analysis requires the use of highly sophisticated computer codes developed for this purpose. These calculations need to take into account not only basic celestial mechanics, but also a range of effects that perturb the trajectories of spacecraft. Such perturbing effects are, for example, residual atmospheric drag and higher order terms of the Earth's gravitational potential for low-Earth orbits, as well as lunar and solar perturbations. In this section, the most important orbits for space science missions are briefly summarized. Some special missions and orbits are given as examples: the properties of elliptical magnetospheric orbits and the orbit of the Cluster mission; missions at the Sun–Earth Lagrange point L1, the Ulysses mission over the solar poles; the Voyager and Pioneer missions to the outer planets and the outer heliosphere; the Helios mission to the inner heliosphere.

7.5.1. *Earth-orbiting, magnetospheric missions*

There have been numerous missions in Earth orbit to study the magnetosphere since the start of the space age in the late 1950s. These missions have first discovered, then mapped in detail the plasma and field populations in the different regions of the magnetosphere. Understanding the dynamics of the magnetosphere in response to variable solar wind conditions has been the objective of most Earth-orbiting Space Physics missions since the 1970s. What was quickly recognized is that observations made by any single spacecraft along its orbit were inherently ambiguous: were the changes recorded due to the motion of the spacecraft through different plasma regions or were they due to the temporal variations in the magnetospheric plasma at the instantaneous location of the spacecraft? The large scale variability of the magnetosphere could be well characterised by statistical studies of the observations made by either the same spacecraft as it repeatedly crossed the same regions or by the many spacecraft that visited the different regions of the magnetosphere. As a result, magnetospheric missions conceived and developed since the mid-1970s have used two or more spacecraft to make simultaneous observations, either in different regions of the magnetosphere, or relatively closely spaced to follow temporal and spatial changes on small scales.

All magnetospheric missions, whether single- or multi-spacecraft missions, use the natural rotation of the orbits with respect to the magnetosphere, as shown in the case of a highly eccentric orbit in Fig. 7.73. As the magnetosphere has a constant orientation with respect to the Earth–Sun direction through the year, elliptical orbits around the Earth that remain fixed in inertial space appear to "rotate" as indicated in the figure for the 12 months of the year. This property of spacecraft orbits is exploited to explore different regions of the magnetosphere and the nearby interplanetary medium through the year.

The first dual spacecraft mission was the ISEE programme (Ogilvie *et al.*, 1978). (The third spacecraft in this programme, ISEE-3, is described below.) The ISEE-1 spacecraft was built by NASA, while the ISEE-2 spacecraft was supplied by ESA. ISEE-1 and -2 were launched together in October 1977 and followed the same highly elliptic Earth orbit (23 Re by 270 km). Both spacecraft carried a full complement of *in situ* instruments: magnetometer, plasma particle analysers, energetic particle detectors and plasma wave instrument. The inter-spacecraft distance could be controlled by manoeuvring ISEE-2, so that measurements made at the location of the two spacecraft could be compared to separate temporal and spatial changes along the orbit. In particular, near the bow shock, the distance between the two spacecraft ranged between 10 km and 5000 km, so that the structure of the bow shock and its dynamic features could be studied. The orbits of the

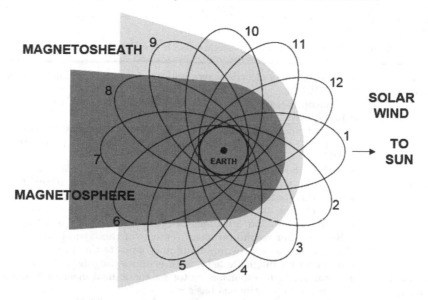

FIGURE 7.73. The natural "rotation" of spacecraft orbits around the Earth's magnetosphere through the year. As the magnetosphere is always oriented along the Sun–Earth line, orbits that have a constant orientation in inertial space appear to rotate around the Earth through the year. One orbit is shown in this sketch for each month.

spacecraft were subject to solar and lunar perturbations, and eventually they re-entered the Earth's atmosphere in September 1987.

The Global Geospace Science programme (GGS) was first proposed in the early 1980s by NASA. Its objective was to improve the understanding of the flow of energy, mass and momentum in the solar-terrestrial environment (Acuña *et al.*, 1995). For this purpose, four spacecraft were to be launched to study key regions of the magnetosphere. Eventually, three spacecraft were launched in the framework of this programme:

• The GEOTAIL spacecraft, a collaborative project between the Institute of Space and Astronautical Science (ISAS) of Japan and NASA, was the first to be launched in July 1992. Its objective is to study the dynamics of the Earth's magnetotail over a wide range of distances, from the near-Earth region, at about eight Earth radii (Re) from the Earth to the distant tail, out to about 200 Re.

• The WIND spacecraft, shown in Fig. 7.74, was launched in November 1995 into a complex, variable orbit, using a series of lunar swingbys to keep the spacecraft in front of the magnetosphere, in the upstream solar wind, to monitor the solar input to the magnetosphere (Fig. 7.75). The technique of double-lunar swingby has been devised to counteract the natural, seasonal rotation of the orbit exploited by other magnetospheric missions.

• The POLAR spacecraft was launched in February 1996, to obtain data from both high- and low-altitude perspectives of the polar regions of the magnetosphere, in particular to study the aurora and phenomena associated with this active part of the magnetosphere.

The GGS programme was also due to include a spacecraft in equatorial orbit around the Earth. This component of the programme was cancelled, although a small German spacecraft, Equator-S, was eventually launched in December 1997 to monitor the equatorial region of the magnetosphere. However, this spacecraft was short-lived and failed in May 1998, due to severe radiation damage inflicted by solar storms in April 1998.

TABLE 7.3. The scientific investigations on the WIND spacecraft.

Radio and plasma waves	• Low-frequency electric waves and low-frequency magnetic fields, from DC to 10 kHz. • Electron thermal noise, from 4 kHz to 256 kHz. • Radio waves, from 20 kHz to 14 MHz. • Time domain waveform sampling, to capture short duration events which meet quality criteria set into the WAVES data processing unit (DPU).
Solar wind experiment	• High time-resolution 3 D velocity distributions of the ion component of the solar wind, for ions with energies ranging from 200 eV to 8.0 keV. • High time-resolution 3 D velocity distributions of subsonic plasma flows including electrons in the solar wind and diffuse reflected ions in the foreshock region, with energies ranging from 7 eV to 22 keV. • High-angular resolution measurements of the "strahl" (beam) of electrons in the solar wind, along and opposite the direction of the interplanetary magnetic field, with energies ranging from 5 eV to 5 keV.
Magnetic field investigation	• Accurate, high resolution vector magnetic field measurements in near real time on a continuous basis. • A wide dynamic measuring range, from $+/-0.004$ nT up to $+/-65,536$ nT, in eight discrete range steps. • Measurement rates up to 44 vector samples per second for analysis of fluctuations.
Energetic particle acceleration, composition, and transport	• Energy spectra of electrons and atomic nucleon of different charge and isotopic composition, from hydrogen to iron, over an energy range extending from 0.1 to 500 MeV/nucleon. • Isotopic composition of medium energy particles (2–50 MeV/nucleon) in solar flares, in the anomalous component and in galactic cosmic rays, extending up to $Z = 90$. • Determination of angular distributions of these fluxes.
Solar wind ion composition study, the "Mass" sensor, and suprathermal ion composition study	• Energy, mass and charge composition of major solar wind ions from H to Fe, over the energy range from 0.5 to 30 keV/e (SWICS). • High mass-resolution elemental and isotopic composition of solar wind ions from He to Ni, having energies from 0.5 to 12 keV/e (MASS). • Composition, charge state and 3D distribution functions of suprathermal ions (H to Fe) over the energy range from 8 to 230 keV/e (STICS).
3 D plasma analyser	• The 3 D distribution of plasma and energetic electrons and ions over the particle energy range from solar wind to cosmic ray energies, a few eV to several McV. • Energy resolution of 0.20 ($\Delta E/E$) and angular resolution from 3 eV to 30 keV; and energy resolution of 0.3 ($\Delta E/E$) and angular resolution of $22.5° \times 36°$, for particles from 20 keV to 11 MeV. • Perturbations to the electron distribution function, in wave-particle interactions.

The European Space Agency's four spacecraft Cluster mission was also conceived in the 1980s (Escoubet *et al.*, 1977). The spacecraft and payload were completed for a launch on the first test flight of Ariane-5 in June 1996. However, the explosion of the Ariane launcher shortly after lift-off destroyed the four original Cluster spacecraft. Fortunately, the spacecraft and complete payload were rebuilt (see illustration in Fig. 7.78) and launched in two pairs by the Soyuz–Fregat launcher on 16 July and 9 August 1990. The orbits of the four Cluster spacecraft are polar (with an inclination of the orbit plane close to 90°), with an apogee of about 19.4 Re and a perigee of about 4 Re. The key feature of

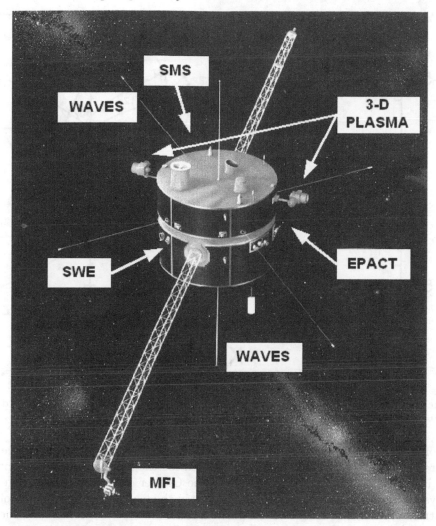

FIGURE 7.74. The WIND spacecraft of NASA's GGS programme, launched in 1995. The location of the *in situ* instruments is indicated. (MFI: magnetometer; EPACT: a set of energetic particle telescopes to measure solar energetic particles; 3 D Plasma: plasma ion and electron detectors; SWE: solar wind ion and electron detectors; SMS: a group of detectors for measuring the solar wind composition; WAVES: plasma and radio wave investigation.)

the Cluster mission is the non-planar configuration of the group of four spacecraft with inter-spacecraft distances that can be controlled by commanding the onboard propulsion systems of the spacecraft. This is illustrated for a particular orbit of Cluster in Fig. 7.76.

Cluster can therefore not only separate temporal and spatial variations on the scale of the separation of the four spacecraft, but, by combining measurements at the four locations, new parameters of the plasma can be determined. In particular, using the times of boundary crossing (bow shock or magnetopause) at the four spacecraft, the vector normal to the boundary can be determined, as well as the speed with which the boundary is moving. Combining the magnetometer measurements, wave vectors can be determined; using the magnetic field measurements in Ampere's law, the current density vector in the plasma can also be estimated. The way the four-point measurements are used, and the

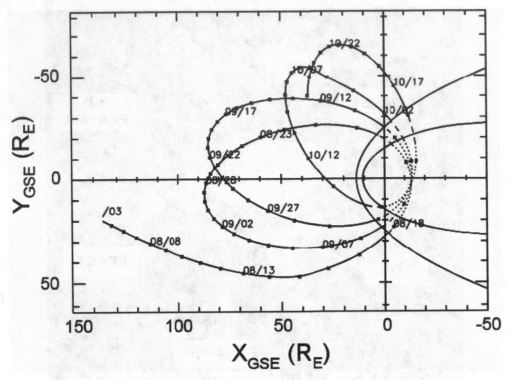

FIGURE 7.75. The orbit of the WIND spacecraft from August to October 1999. This mission uses the double-lunar swingby technique to keep the orbit of the spacecraft sunward of the magnetosphere.

range of phenomena that can be investigated, depend on the separation distance of the spacecraft. The distances that have been selected for the mission and implemented are shown in Fig. 7.77. The particular feature of the Cluster orbit is that the apogee is on the dayside (towards the Sun) in February and is in the magnetotail in August, following the natural "seasonal rotation" of the orbit. Orbit manoeuvres for changing the inter-spacecraft separation are usually performed in May–June and in November to set them ready for the regions of main interest to the mission: the cusps of the magnetosphere and the magnetotal.

Cluster carries a comprehensive package of *in situ* instrumentation, listed in Table 7.4. The complexity of the spacecraft can be judged from the photograph showing it when it was partially assembled. The operation of the payload on the four spacecraft is co-ordinated according to a Master Science Plan, set normally 6 months in advance, to ensure that instrument modes are optimized to the magnetospheric regions to be swept by the orbit.

7.5.2. *Heliospheric missions*

While many Earth-orbiting spacecraft in eccentric orbit can observe the solar wind at least part of the year, truly heliospheric missions are usually launched into escape orbits to give them an orbit fully in the heliosphere. Missions in the three classes of heliospheric orbits are presented below. The first class consists of spacecraft at the Lagrangian point L1 of the Sun–Earth system (strictly speaking, the Sun–Earth–Moon system). This L1 point orbit (at about 1,500,000 km ahead of the Earth towards the Sun) is based on a special solution of the three-body problem by Lagrange. In general, solving the equations

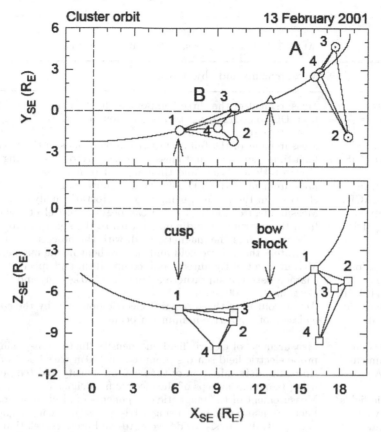

FIGURE 7.76. Projections of a portion of the Cluster orbit in the Solar–Ecliptic co-ordinate system, as the four spacecraft were inbound from their apogee on the dayside. On the scale of this plot, orbits cannot be distinguished, the inter-spacecraft distances were ∼600 km, or 0.1 Re. The configuration of the Cluster tetrahedron is shown in two points on the orbit, enlarged by a factor 30.

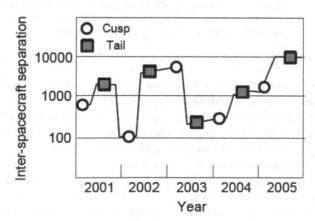

FIGURE 7.77. The typical inter-spacecraft distances in the Cluster tetrahedron during the mission.

TABLE 7.4. The Cluster scientific investigations.

Instrument	Measurements and objectives
Magnetometer (FGM)	Measurement of the three components of the magnetic field vector, from DC to 30 Hz, using two boom-mounted tri-axial fluxgate magnetometers.
Cluster ion spectrometer (CIS)	Measurement of the full 3-D distribution of ions in the plasma with high time-, spectral- and mass-per-charge resolution, using electrostatic analysers and time-of-flight techniques.
Plasma electron analyser (PEACE)	Measurement of the 3-D velocity distribution of electrons in the plasma, using two electrostatic analysers.
Electron drift instrument (EDI)	Measurement of the drift of a weak beam of test electrons emitted from the instrument and which return to the spacraft after one gyration around the magnetic field, with the objective to determine separately the electric field and the gradient in the magnetic field.
Suprathermal ions and electrons instrument (RAPID)	Measurement of the directional, composition and spectral characteristics of suprathermal ions (2–1500 keV/n) and electrons (20–400 keV).
Spacecraft electric potential control (ASPOC)	Active control of the spacecraft electric potential by the controlled emission of a positive indium ion beam.
Electric fields and waves instrument (WEC/EFW)	Measurement of electric field and density fluctuations, using spherical probe electric field sensors mounted on 100 m wire booms from tip-to-tip; determination of the electron density and temperature using Langmuir sweeps of the electric field signals.
AC magnetic fields and EM waves measurements (WEC/STAFF)	Measurement of the magnetic components of electromagnetic fluctuations up to 4 kHz, using a tri-axial search coil magnetometer, and spectral analysis, to derive auto- and cross-correlation coefficients of the spectral matrix, using three magnetic and two electric components of the waves.
Active plasma sounder instrument (WEC/WHISPER)	Measurement of the total plasma density from 0.2 to 80 cm^{-3}, using a resonance sounding technique; survey of one electric component of plasma waves from 4 to 80 kHz.
Plasma waves instrument (WEC/WBD)	High resolution frequency/time spectra of plasma waves: electric and magnetic fields in selected frequency bands from 25 Hz to 577 kHz.
Digital processor of electric and magnetic waves observations (WEC/DWP)	Electronic processing, analysis and control of the WEC instruments measuring magnetic and electric waves from DC to over 100 kHz; onboard computation of wave-particle correlations.

of motion for three gravitationally interacting bodies is nearly impossible. (There is a mathematically rigorous general solution, but it is not useful in practice, as the solution converges very slowly.) However, given some special assumptions, several classes of special solutions have been obtained. One such special solution is the one derived by Lagrange in the late 18th century; this solution has now found some important applications to space missions. The five Lagrange points are shown in Fig. 7.79.

The L1 orbit is quite special and ideal for monitoring the sun and the solar wind close to the Earth. However, it is not possible, or even desirable to place a spacecraft at the precise L1 point whose location in fact wobbles somewhat, as it is a null point for the gravitational potential of the sun, on the one hand, and the Earth–Moon system, on the other hand, and the centre of mass of the Earth–Moon system oscillates as the

FIGURE 7.78. One of the Cluster spacecraft under construction in the large assembly building of Astrium GmbH in Friedrichshafen, Germany. The large silver spheres are the fuel tanks that allow orbital manoeuvres to control the separation of the four spacecraft. The upper platform shows some of the equipment and payload already installed. In the background, on the left, another Cluster spacecraft is prepared for assembly.

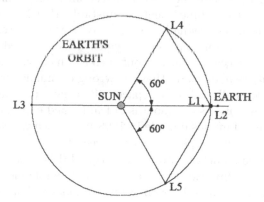

FIGURE 7.79. The relative positions of the five Lagrange points of the Sun–Earth system. Note that the distances of L1 and L2 from the Earth are exaggerated, for clarity. The actual distances are only 1% of the Sun–Earth distance. The other three points are correctly indicated.

Moon orbits the Earth. The three spacecraft sent there were in fact placed into a "halo" orbit around the L1 point itself, so that the spacecraft actually circles the L1 point at distances of several hundred thousand kilometres. This is in any case necessary, because for communications (both for sending commands to the spacecraft and for receiving telemetry from it) the ground station antenna should not be pointed directly at the sun, as the sun is a very strong and variable radio source, so that the communication link would be swamped by solar radio noise.

Three important space missions, ISEE-3 in 1978, then SOHO in 1995 and ACE in 1997 have used a special orbit which allowed them to remain at an approximately constant

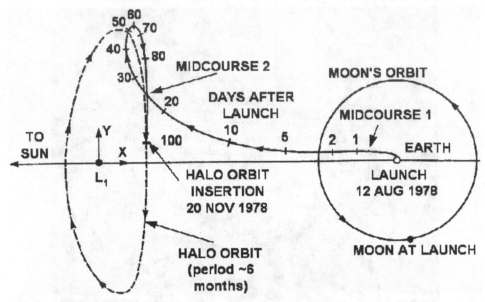

FIGURE 7.80. The orbit of the ISEE-3 spacecraft from launch into its halo orbit around the L1 point. The L1 point is situated at about 1.5 million kilometres from the Earth towards the Sun.

distance between the Sun and the Earth. The transfer trajectory and mission orbit in a halo orbit around the L1 point for the ISEE-3 spacecraft is shown in Fig. 7.80. A problem with the halo orbit is that it is fundamentally unstable, that is, the spacecraft needs to carry out orbit corrections occasionally, using its onboard thrusters, to remain in that orbit. This is a relatively simple manoeuvre which does not need much fuel. It is also simple to command the spacecraft to carry out such a manoeuvre; however, it was during such an orbit maintenance manoeuvre that the wrong commands were sent up to SOHO in June 1998, which eventually led to the spacecraft being temporarily lost. Fortunately, the mission was recovered after some months of frantic and expensive search and careful in-orbit recommissioning, but this episode in 1998 has shown the very great importance of the quality control of ground operations of spacecraft.

The first mission to be launched into the L1 orbit, ISEE-3, achieved in the end three unrelated objectives (including two that were originally unplanned). The ISEE-3 spacecraft was launched in 1978, first into an orbit around the Lagrange point L1 of the Sun–Earth system, therefore about 1.5 million kilometres in front of the Earth, for uninterrupted measurements in the solar wind. SOHO is now in the same orbit, launched in 1995, for uninterrupted observation of the Sun; an even more recent spacecraft in that orbit is NASA's, the ACE, launched in 1997 to study the details of the ionic composition of the solar wind.

In 1982, after 4 years of highly successful observation of the solar wind and energetic particle events associated with the solar maximum, it was decided to send ISEE-3 into the distant magnetospheric tail, close to the L2 Lagrange point. Having explored the tail, ISEE-3 was then made to fly very close by the Moon (with a closest approach of 90 km) to perform a sling-shot manoeuvre to send it to rendezvous, in September 1985, with the comet Giacobini–Zimmer. Having survived the cometary rendezvous, the spacecraft was then left to drift further away from the Earth in a heliocentric orbit at 1 AU, to continue making measurements of the solar wind into the 1990s, until it moved to such a distance

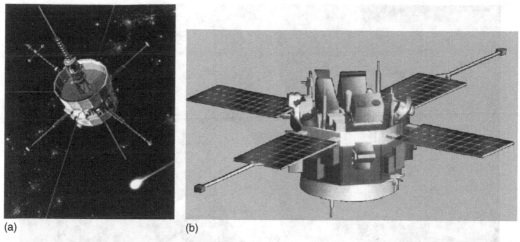

(a)   (b)

FIGURE 7.81. The ISEE-3 spacecraft (a) was the first to be placed, in 1978, into an orbit around the Lagrangian point L1 between the Sun and the Earth. Its monitoring role in that unique orbit is now performed by the ACE spacecraft (b), launched in 1997.

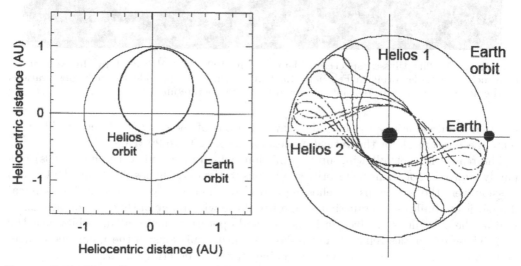

FIGURE 7.82. The ecliptic projection of the orbits of the two Helios spacecraft, launched in the early 1970s. Observations made by these spacecraft that remained operational to the mid-1980s remain unique from the inner heliosphere.

(more than 150 million kilometres) that even NASA's largest dishes could no longer pick up its signals. Pictures of the ISEE-3 and ACE spacecraft are shown in Fig. 7.81.

The second class of heliospheric missions is in orbit around the Sun (although, technically, the L1 missions are also in orbit around the Sun). There have been only two such missions to date: the Helios and Ulysses missions, although several are now in planning. The two Helios spacecraft, Helios-1 and -2 (built in Germany, launched by NASA and operated jointly), were launched on 10 December 1974 and 15 January 1976, respectively, into heliocentric orbits with aphelion near 1 AU and perihelion near 0.3 AU, as illustrated in Fig. 7.82. To this day, they remain the only spacecraft to have explored the inner heliosphere. Their observations remain unique in that important region, where solar wind streams of different speeds first interact. Despite dating back to the 1970s,

FIGURE 7.83. The Ulysses spacecraft, launched in 1990, was deflected by Jupiter into a high-inclination solar orbit in February 1992. It continues to provide the only observations on the heliosphere at high solar latitudes and is due to operate until 2008.

the Helios data continue to be used in investigations of the solar wind and are likely to remain a benchmark for theoretical models for another 10 to 20 years.

The reason for not following up this mission (which operated to the mid-1980s) was partly a question of scientific priorities: missions to the outer heliosphere and the Ulysses mission were the next steps in heliospheric research, and partly because of the technical difficulties to build spacecraft that support well the wide range of thermal conditions, as well as the difficulties involved in launching and operating spacecraft in orbits near the Sun. These difficulties remain valid today and future missions to the inner heliosphere and close to the Sun are only in the planning stages, as described later.

The Ulysses mission was conceived in the early 1970s, but was only launched in 1990. Its mission was to explore, for the first time, the regions over the poles of the Sun (Balogh *et al.*, 2001). Previous interplanetary spacecraft remained in the ecliptic plane, due to the Earth's orbital velocity inherited at launch. Ulysses was targeted towards Jupiter, for a close flyby, to be gravitationally deflected into a close-to-polar solar orbit, to study the heliospheric phenomena as a function of heliolatitude. Ulysses, illustrated in Fig. 7.83, is now in a solar orbit, its orbital plane has an 80° inclination with respect to the solar equator, with a perihelion of ∼1.4 AU (further than the Earth from the Sun) and an aphelion of 5.4 AU. Its orbital period is ∼6 years. The heliocentric range and heliolatitude of the spacecraft since launch is shown in Fig. 7.84, together with the monthly sunspot numbers. This shows that Ulysses has now explored heliospheric conditions at high solar latitudes over more than a complete solar cycle. The spacecraft carries a full particles and fields payload; although the instruments were developed and built around 1980, they all continue to operate successfully. The mission is due to be terminated in 2008. In the absence of an equivalent mission for the foreseeable future, the observations made in

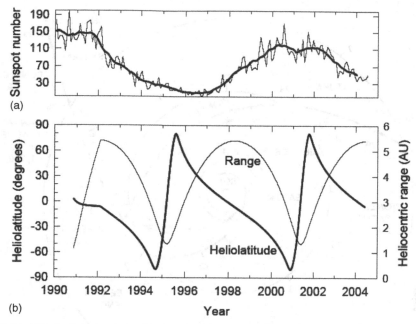

FIGURE 7.84. The heliolatitude and heliocentric range of the Ulysses spacecraft since its launch in 1990 (a). The monthly sunspot numbers (b) show the solar cycles that have been covered by the Ulysses observations.

the three-dimensional heliosphere by the Ulysses instruments will remain a high-value scientific investment for heliospheric research for a considerable time.

A particular advantage of Ulysses, apart from its unique orbit, is the close to 100% data recovery from the instruments. This has been achieved partly through periods of continuous telemetry recovery from the spacecraft by the use of three ground stations of NASA's Deep Space Network, and partly by the use of onboard tape recorders so that data recorded onboard can be transmitted at a later time. There is considerable scientific value in such a high-data recovery rate, as this is the only way to ensure that events and phenomena of great interest, often unpredictable, are not missed in the observations.

The remarkable achievement of Ulysses to cover almost two complete solar cycles of observations has been made possible by the longevity of both spacecraft and payload. The spacecraft has proved highly reliable, including its all-important tape recorders that enable the continuous observation programme to be implemented. In the past, such tape recorders had been considered to be the weak point of missions, in the light of their relatively high failure rate (all current and future missions use high-capacity solid sate memories for the purpose of recording and playing back the observations). The end of the Ulysses mission in 2008 will be due to the steady (and predictable) radioactive decay in its power source, the Radioisotope Thermoelectric Generator (RTG). At the same time, the instruments of the payload that were designed and built around 1980, continue to operate and provide data continuously.

The third class of heliospheric missions, those to the outer heliosphere, have been placed on escape orbits from the solar system. Two such missions, Pioneers 10 and 11 and Voyagers 1 and 2, both launched in the 1970s and each with two spacecrafts, have been undertaken by NASA up-to-date. A third mission, the Cassini Saturn Orbiter, launched in 1997, was placed into Saturn orbit in July 2004; this mission has collected

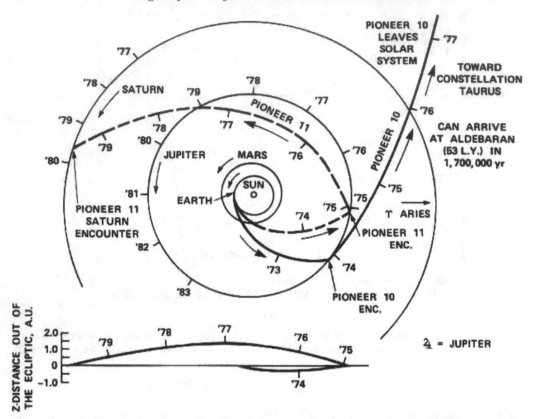

FIGURE 7.85. The trajectories of the Pioneer 10 and 11 spacecraft first encountering Jupiter, then Pioneer 11 travelling out to encounter Saturn at the opposite side of the Sun. In the lower left corner of the figure, the out-of-ecliptic trajectory of Pioneer 11 is shown, on its way to Saturn.

some limited amount of heliospheric data during its cruise to Saturn, but not all its fields and particles instruments operated during the cruise phase.

Two closely similar spacecrafts, Pioneers 10 and 11, launched in 1972 and 1973, respectively, were the first to reach the orbit of Jupiter and beyond, as illustrated in Fig. 7.85 (Ward, 1974). Following its close encounter with Jupiter, Pioneer 10 travelled to the outer reaches of the solar system and continued, intermittently, to provide data on the distant heliosphere. After encountering Jupiter, Pioneer 11 was placed on a trajectory to intercept Saturn. Its orbit towards Saturn took it about 16° above the ecliptic plane (as shown at the bottom of Fig. 7.85) where it was the first to confirm the limited heliolatitude extent of the interplanetary magnetic sector structure at the time of solar minimum activity. The two Pioneer spacecrafts travelled through the middle heliosphere, out to the orbit of Saturn and could observe the development and evolution of the CIRs, in particular the formation of leading and trailing shock waves that bound such regions. The two Pioneer spacecrafts remained in contact with the Earth up to the mid-1990s, when their power source, the RTG could no longer supply enough power to maintain the radio contact with NASA's Deep Space Network antennae.

The last two spacecraft of NASA's Mariner series, Voyager 1 and 2 were the first in that series to be sent to explore the outer solar system (Stone, 1977). Voyager 1 and 2 made studies of Jupiter and Saturn, their satellites and their magnetospheres as well as studies of the interplanetary medium. An option designed into the Voyager 2 trajectory, and ultimately exercised, was to direct it toward Uranus and Neptune to perform similar studies.

Although launched 16 days after Voyager 2, Voyager 1's trajectory was a faster path, arriving at Jupiter in March of 1979. Voyager 2 arrived about 4 months later in July 1979. Both spacecrafts were then directed on to Saturn with arrival times in November 1980 (Voyager 1) and August 1981 (Voyager 2). Voyager 2 was then diverted to the remaining gas giants, Uranus (January 1986) and Neptune (August 1989).

Data collected by Voyager 1 and 2 have not been confined to the periods surrounding encounters with the outer gas giants, with the various fields and particles experiments and the UV spectrometer collecting data nearly continuously during the interplanetary cruise phases of the mission. The two spacecrafts continue to operate as the recently renamed Voyager Interstellar Mission (VIM) searches for the boundaries of the heliosphere and exits the solar system. The VIM is an extension of the Voyager primary mission that was completed in 1989 with the close flyby of Neptune by the Voyager 2 spacecraft. At the start of the VIM, the two Voyager spacecrafts had been in flight for over 12 years having been launched in August (Voyager 2) and September (Voyager 1), 1977. Voyager 1 is escaping the solar system at a speed of about 3.5 AU per year, 35° out of the ecliptic plan to the north, in the direction of the Sun's (motion relative to nearby stars). Voyager 2 is also escaping the solar system at a speed of about 3.1 AU per year, 48° out of the ecliptic plane to the south. The ecliptic projection of the Pioneer and Voyager orbits is shown in Fig. 7.86 and the Voyager spacecraft and payload are illustrated in Fig. 7.87.

7.5.3. *Future Space Physics programmes and missions*

Space Physics is a mature branch of space sciences. After almost half a century since the first exploratory missions, the space environment is generally well known and many of its phenomena have been thoroughly studied. There remain some serious gaps in our understanding of the detailed phenomenology of space, in particular, we remain some way away from reliable predictions of the dynamics of space phenomena on both the large and the small scale. Several programmes have been devised and proposed with the goal to understand, in a predictive way, the dynamics of the magnetosphere's response to both steady state and transient inputs from the Sun. By analogy with terrestrial weather, the topic of magnetospheric conditions in response to the solar input has been called "space weather"; the international effort to forecast magnetospheric storms involves the monitoring of solar conditions, routine solar wind and interplanetary magnetic field observations, as well as continuous gathering near-Earth and ground-based data related to the state of the magnetosphere and the ionosphere. There are practical goals of forecasting and monitoring potentially damaging radiation conditions for scientific, military and commercial near-Earth spacecraft, as well as for the astronauts on the International Space Station.

NASA's current framework programme for the study and applications of Sun–Earth relations is the living with a star (LWS); there is also an international forum (ILWS) that includes other national and multinational space agencies for coordinating all relevant space research activities. The ILWS programme contains existing space missions, such as ACE, Wind, Cluster, Ulysses, SOHO, TRACE and others, as well as missions under development, such as STEREO and Themis. In addition, the programme coordinates the planning of new missions to add to, and extend existing capabilities in Space Physics over the next decade to 2015. An outline of some of the future elements of the ILWS programme will be described briefly below.

The STEREO mission of NASA consists of two, identically instrumented Sun-pointing spacecraft, one leading the Earth in its orbit, the other lagging behind as shown

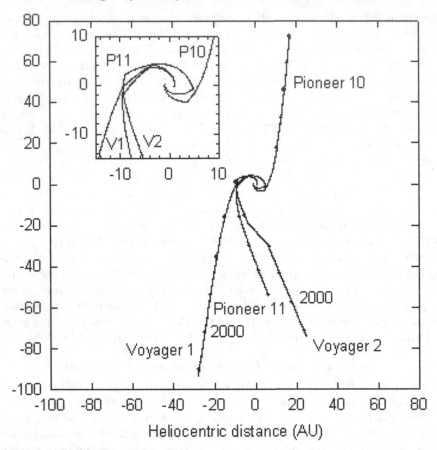

FIGURE 7.86. The ecliptic projection of the orbits of Pioneer 10 and 11 and Voyager 1 and 2. These spacecraft, launched in the 1970s, remain the only ones that have explored the outer heliosphere beyond the orbit of Saturn. The two Voyager spacecraft remain operational and it is expected that they might encounter the outer boundaries of the heliosphere in the next decade.

in Fig. 7.88. The objective of the mission is to provide two different viewing angles with respect to the Earth on solar eruptions and their propagation from the Sun to the Earth, to provide the images needed for a stereoscopic reconstruction of solar eruptions. Each spacecraft will carry a Cluster of solar telescopes (Coronagraphs, EUV Imager and Heliospheric Imager). Simultaneous telescopic images from the two spacecrafts will be combined with observations from ground-based telescopes and possibly, if these are available, from telescopes on low Earth orbit to obtain a full three-dimensional view of CME from the Sun to the Earth. A very complete set of *in situ* particles and fields instruments onboard the spacecraft will be used to provide the data for reconstructing the three-dimensional plasma structures of the CMEs in Earth orbit. Table 7.5 shows the suit of *in situ* instruments (Collective name: IMPACT) carried by each of the STEREO spacecraft.

Themis ("Time history of events and their macroscopic interactions during substorms") is managed by the University of California, Berkeley, and is one element of NASA's medium-class explorer (MIDEX) programme. The Themis mission consists of five identical small satellites with orbits designed to cover the key regions in the Earth's magnetotail. These spacecrafts are spin stabilized, with a spin period of 3 s and will be launched

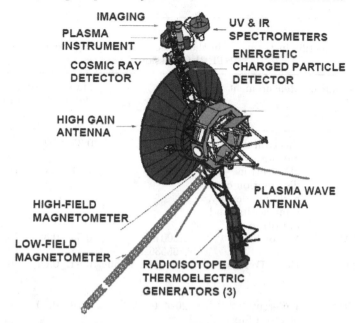

FIGURE 7.87. A schematic view of the Voyager 1 and 2 spacecraft with location of the *in situ* instruments.

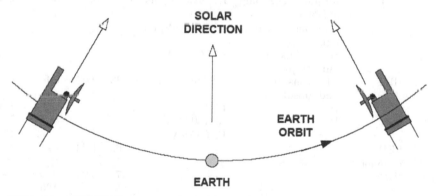

FIGURE 7.88. The STEREO concept of two identical spacecrafts, one leading the Earth in its orbit, the other lagging behind, to provide a stereoscopic view of CME from the Sun to the Earth.

simultaneously in 2007. The objective of the mission is to provide the coordinated, simultaneous measurements of plasma particles and fields to study the dynamics of the magnetotail, in particular much debated questions concerning the origin and development of magnetospheric substorms. The orbits of the five spacecrafts are equatorial; the three inner spacecrafts have orbital periods ~1 day, the outer two have orbital periods of 2 days and 4 days, respectively. The orbits (illustrated for two epochs around the year in Fig. 7.89) are designed to provide frequent near-radial alignments in the magnetospheric tail, near midnight local time, to provide simultaneous observations of processes occurring in the context of substorms at the critical distances. These observations, together with the extensive ground-based observations that can be made at the same time, are expected to resolve remaining controversies on the causal chain concerning the phenomenology of substorms. The configuration of the Themis probes is illustrated in Fig. 7.90.

TABLE 7.5. The scientific instruments in the particles and fields package (IMPACT) on the STEREO mission.

Experiment	Instrument	Measurement	Range	Mass (kg)	Power (W)	Data rate (bps)	Time resolution
SW	STE	Electron flux and anisotropy	2–100 keV	0.35	0.20	34	1 min
	SWEA	3D electron distribution core and halo density, temperature and anisotropy	~0–3 keV	1.41	1.10	41	1 min
MAG	MAG	Vector field	±500 nT, ±65536 nT	0.55	0.38	76	0.25 s
SEP	SIT	He to Fe ions	0.03–2 MeV/nuc	0.84	0.64	60	30 s
		^{3}He	0.15–0.25 MeV/nuc				30 s
	SEPT	Differential electron flux	20–400 keV	0.79	0.60	30	1 min
		Differential proton	20–7000 keV				1 min
		Anisotropies of e, p	As above				15 min
	LET	Ion mass 2–28 and anisotropy	1.5–40 MeV/nuc	0.51	0.18	80	1–15 min
		^{3}He ions flux and anisotropy	1.5–1.6 MeV/nuc				15 min
		H ions flux and anisotropy	1.5–3.5 MeV				1–15 min
	HET	Electrons flux and anisotropy	1–8 MeV	0.33	0.07	30	1–15 min
		H	13–100 MeV				1–15 min
		He	13–100 MeV				1–15 min
		^{3}He	15–60 MeV/nuc				15 min
	SEP common	- - -	- - -	1.69	1.44	- - -	- - -
Common	IDPU	- - -	- - -	1.78	3.22	33 or 126 (burst)	- - -

The payload of the Themis probes consists of the key instruments of Space Physics:

• Vector Fluxgate and Search Coil magnetometers (FGM, SCM) provide samples of waveforms up to 1024 vector/s.

• Vector Electric Field Instrument (EFI) provide samples of waveforms from DC up to 1024 vector/s.

• Electrostatic Analyser (ESA) measures ions and electrons of energy 5–30 keV over 4πgtr once per spin.

• Solid State Telescope (SST) measures ions and electrons of 20–1 MeV providing a coverage of $108° \times 360°$ once per spin.

Next are the Radiation Belt Mappers (RBM). Although the Earth's radiation belts (the Van Allen Belts) are generally well understood, the extensive use of Earth orbit for commercial and military satellites as well as for manned missions to the International

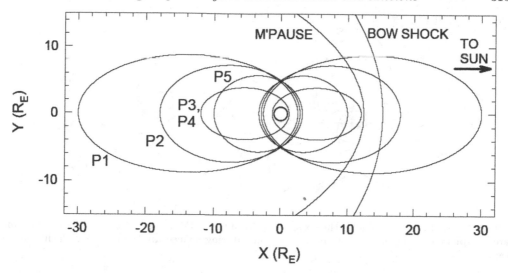

FIGURE 7.89. Orbits of the Themis probes projected into the equatorial plane. The probes are aligned in the direction of the magnetotail every 4 days when the apogees are in the midnight sector; similarly, alignments occur 6 months later in the direction of the Sun.

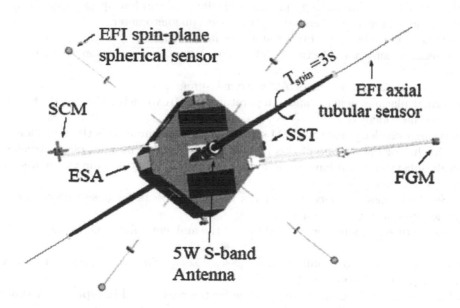

FIGURE 7.90. Sketch of the Themis spacecraft, showing the location of the payload instruments. (Used with permission of V. Angelopoulos, University of California, Berkeley.)

Space Station require a much better capability of modelling the details of the expected radiation environment than currently possible. This is particularly true of transient changes in the radiation environment during solar storm periods. The RBM mission concept (for possible launch in 2008) uses multiple, identical spacecraft on correlated orbits to make precise and simultaneous measurements in the radiation belts. The eventual objective is to make predictive models much more accurate than at present while also to provide "weather reports" from the radiation belts in real time. The orbits need careful selections,

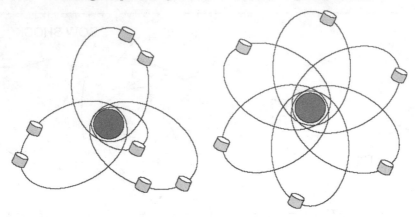

FIGURE 7.91. Possible orbits for the six spacecrafts of the RBM mission. Both orbit constellations are complemented by an inner radiation belt monitoring spacecraft (omitted for clarity on the right).

two possible orbit configurations are shown schematically in Fig. 7.91. The measurements to be made can be listed in order of priority for these spacecraft as follows:

- High-energy particles (1–20 MeV protons; 1–10 MeV electrons); mid-energy particles (30 keV–1 MeV); very-high-energy protons (20–500 MeV) on low-apogee spacecraft.
- Magnetic field measurements using a fluxgate magnetometer.
- Electric field probes (two orthogonal axes in spin plane).
- Low-energy ions and electrons (30 eV–30 keV), with the option to include mass analysis.
- Search coil magnetometer for AC wave measurements.
- Thermal plasmas density and temperature (<100 eV), with the option to include energy and mass analysis.

A particular challenge for the RBM spacecraft and instruments is the radiation environment itself. For the success of this mission, the design of the sensors, electronics and spacecraft subsystems will have to be more resistant to radiation damage than for most other space missions.

The Sentinels mission concept addresses the requirements for observations needed in interplanetary space to determine:

- the structure and long-term variability of the ambient solar wind in the inner heliosphere;
- the propagation and evolution of large-scale solar wind structures and solar transients in the inner heliosphere;
- what dynamic processes are responsible for the release and interplanetary evolution of solar transients that affect the Earth's space environment;
- the propagation of energetic particles through the inner heliosphere and how and where are released, accelerated.

The core of the Sentinels concept, the Inner Heliospheric Sentinels (IHS), is a quartet of spacecraft in heliocentric orbits as shown in Fig. 7.92. Launched together, the spacecraft use Venus Gravity assist to achieve their orbits. Careful phasing of the orbits leads to periods of scientifically useful conjunctions for observing the evolving structures of the solar wind from 0.5 to near 1 AU. The ILS spacecraft are relatively simple, spin-stabilized, and carry a basic complement of fields and particles instruments: solar wind plasma analysers (including composition measurements), magnetometers, energetic particle detectors and solar radio burst trackers.

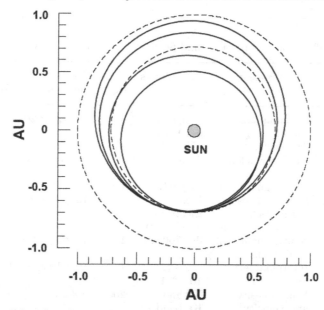

FIGURE 7.92. Possible orbits for the four Inner Heliospheric Sentinels mission. The orbits of the Earth and Venus are shown in dashed lines. The four orbits meet at the heliospheric distance of Venus (∼0.72 AU), the outer two spacecraft have an aphelion distance of 0.85 and 0.95 AU, respectively, the inner two spacecraft have perihelion distances of 0.5 and 0.6 AU, respectively.

The Sentinels concept has also envisaged the use of a solar observatory spacecraft at the far side of the Sun, at the same distance as the Earth. This spacecraft would be three-axis stabilized and would carry a full complement of remote sensing solar imagers and spectrographs. Its observations would provide a continuity of observations around the Sun, when also using a solar observatory on the side of the Earth, similar to SOHO.

In addition, the Sentinel concept also accommodates spacecraft monitoring the solar wind ahead of the Earth, in orbit around the L1 point, similar to the current ACE mission.

Plans for a new solar and heliospheric mission by ESA have been elaborated since the late 1990s. Although its launch date is now planned to be in or after 2013, the Solar Orbiter will be ESA's next such mission, following up Ulysses (launched in 1990, although conceived in the early 1970s) and SOHO (launched in 1995).

The scientific objectives of the Solar Orbiter are to determine *in situ* the properties and dynamics of plasma, fields and particles in the near-Sun heliosphere and to investigate the fine-scale structure and dynamics of the Sun's magnetized atmosphere, using close-up, high-resolution remote sensing.

Solar Orbiter will identify the links between activity on the Sun's surface and the resulting evolution of the corona and inner heliosphere. The mission's orbit will bring the spacecraft close to the Sun, to about 0.21 AU (closer than Helios reached in the 1970s) and will also reach heliolatitudes in excess of 30°, using Solar Electric Propulsion. The spacecraft will be Sun-pointing and will have a number of solar imagers, coronagraphs and spectrographs, as well as a comprehensive particles and field package as shown in Table 7.6. The sketch of the Solar Orbiter spacecraft is shown in Fig. 7.93.

For *in situ* measurements in the outer solar corona, a Solar Probe mission has been studied in detail by the space agencies, most recently by NASA. This mission, with a closest approach to the Sun at 4 RS, would carry typical *in situ* instrumentation for the first time into the solar atmosphere. However, the technical difficulties are formidable and,

TABLE 7.6. The proposed particles and fields instruments for ESA's Solar Orbiter.

Name	Measurement	Capabilities	Mass (kg)	Size (cm × cm × cm)	Power (W)	Telemetry (kb/s)
Solar Wind Plasma Analyser	Thermal ions and electrons	0–30 keV/Q; 0–10 keV	6	20 × 20 × 20	5	5
Radio and Plasma Wave Analyser	AC electric and magnetic fields	μV/m–V/m 0.1 nT–μT	10	15 × 20 × 30 (electronics box)	7.5	5
Magnetometer	DC magnetic field	to 500 Hz	1	10 × 10 × 10	1	0.2
Energetic Particle Detector	Solar and cosmic-ray particles	Ions and Electrons .01–10 MeV	4	10 × 20 × 20	3	1.8
Neutral Particle Detector	Neutral hydrogen and atoms	0.6–100 keV	1	10 × 10 × 20	2	0.3
Dust Detector	Interplanetary dust particles	Mass (g): 10^{-16}–10^{-6}	1	10 × 10 × 10	1	0.05
Radio Sounding	Wind density and velocity	X-band Ka-band	200g	5 × 5 × 5	3	0
Neutron detector	Solar neutrons	e > 1 MeV	2	10 × 10 × 10	1	0.15

FIGURE 7.93. The planned Solar Orbiter spacecraft of the European Space Agency planned for launch in 2013.

despite the scientific attraction of the mission, it is unclear whether it can be undertaken in the next 10 years or even beyond.

7.5.4. *Conclusions*

The brief survey of past, present and future Space Physics missions has illustrated some of their key features:

• The importance of the orbit selection. Given the maturity and capability of a package of *in situ* instruments, selecting the scientific objectives in specific regions of space involves a considerable effort to match the orbit, spacecraft and payload. The instruments that have been developed and refined over the past decades will perform the required measurements in all space environments, the key is to get the instruments in the required regions of space.

• Spacecraft carrying *in situ* instruments have become more complex, mostly through an effort to increase (or at least retain) the capabilities of earlier spacecraft, while reducing the mass and power required to operate its subsystems.

• The instruments themselves have become more complex and more capable. The increase in capability has been achieved through the use of improved design techniques, as in the case of plasma instruments that now use large and complex computer codes to determine the response of electrostatic analysers to incoming particles. Instruments also use complex onboard computer codes to optimize their data collection capability.

• The life cycle of Space Physics missions is generally very long. This means that the period between first proposal of a mission and its implementation and launch increased to typically a decade or more since the 1970s, when in was much shorter. The lifetime in orbit has also increased, thanks to a generally greater reliability of spacecraft and instruments; in most cases, missions are now capable of lasting 10 years or more. This also means that participating in a space physics mission is a significant fraction of researchers' careers.

• Space Physics missions need a preferably continuous data coverage, either through constant ground contact for continuous telemetry, or through the use of onboard recording and playback capabilities.

• Most instruments can generate data at rates that exceed by a large factor what can be transmitted through the telemetry. Increasing the telemetry rates is highly desirable; unfortunately, even current and future missions are seriously limited by the amount of data that they can transmit.

• In Space Physics observations, it is important to distinguish between spatial and temporal effects in the medium targeted by the missions. Space physics phenomena are naturally highly dynamic. Multi-point and coordinated observations in different regions is now the key consideration when devising new space physics missions.

REFERENCES

ACUÑA, M.H. *et al.* 1995 The global geospace science program and its investigations. *Space Sci. Rev.*, **71**, 5–21.

BALOGH, A., CARR, C.M. & ACUÑA, M.H. *et al.* 2001 The Cluster magnetic field investigation: overview of in-flight performance and initial results. *Ann. Geophys.*, **19**, 1207–1217.

BALOGH, A. *et al.* 1978 The low energy proton experiment on ISEE-C, *IEEE Trans. Geosci. Electron.*, **GE-16**, 176–180.

BALOGH, A., MARSDEN, R.G. & SMITH, E.J. 2001 The heliosphere and the Ulysses mission, in *The heliosphere near solar minimum: The Ulysses perspective*, pp. 1–41. Springer–Praxis Books in Astrophysics and Astronomy, London, Springer.

BAME, S.J., McCOMAS, D.J., YOUNG, D.T. & BELIAN, R.D. 1986 Diagnostics of space plasmas. *Rev. Sci. Instrum.*, **57**, 1711.

BAUMJOHANN, W. & TREUMANN, R. 1997 Basic Space Plasma Physics. Imperial College Press, London.

BOUGERET, J.-L., KAISER, M.L., KELLOGG, P.J. *et al.* 1995 "WAVES: The radio and plasma wave investigation on the WIND Spacecraft", *Space Sci. Rev.*, **71**, 5.

CARLSON, C.W., CURTIS, D.W., PASCHMANN, G. & MICHAEL, W. 1983 An instrument for rapidly measuring plasma distribution functions with high resolution. *Adv. Space Res.*, **2**, 67.

CRAVENS, T.E. 1997 Physics of Solar System Plasmas. Cambridge University Press.

DÉCRÉAU, P.M.E., FERGEAU, P., KRASNOSELSKIKH, V. & LE GUIRRIEC, E. *et al.* 2001 Early results from the Whisper instrument on Cluster: an overview, *Ann. Geophys.*, **19**, 1241–1258.

ESCOUBET, C.P., SCHMIDT, R. & GOLDSTEIN, M.L. 1977 Cluster – Science and mission overview. *Space Sci. Rev.*, **79**, 11–32.

GLOECKLER, J., GEISS, J., BALSIGER, H. *et al.* 1992 The solar wind ion composition spectrometer, *Astron. Astrophys. Suppl. Ser.*, **92**, 267.

GOMBOSI, T.I. 1998 Physics of the space environment. Cambridge University Press.

GURNETT, D.A. 1998 Principles of space plasma wave instrument design, in Measurement techniques in Space Plasmas – Fields, Pfaff, R.F., Borovsky, J.E., Young, D.T. (Eds.), pp. 121–136. Geophys. Monograph 103, American Geophysical Union.

HAMILTON, D.C., GLOECKLER, G., IPAVICH, F.M., LUNDGREN, R.A., SHELDON, R.B. & HOVESTADT, D. 1990 New high-resolution electrostatic ion mass analyser using time of flight. *Rev. Sci. Instrum.*, **61**, 3609.

KESSEL, R.L., JOHNSTONE, A.D., COATES, A.J. & GOWEN, R.A. 1989 Space plasma measurements with ion instruments, *Rev. Sci. Instrum.*, **60**, 3750.

KIVELSON, M.G. & RUSSELL, C.T. (EDITORS) 1995 Introduction to Space Physics. Cambridge University Press.

KOEHN, P.L., ZURBUCHEN, T.H. *et al.* 2002 Measuring the plasma environment at Mercury: The fast imaging plasma spectrometer. *Meteorit. Planet. Sci.*, **37**, 1173.

LAZARUS, A.J. & PAULARENA, K.I. 1998 A comparison of solar wind parameters from experiments on the IMP 8 and WIND spacecraft, in *Measurement techniques in Space Plasmas – Particles*, R.F. Pfaff, J.E. Borovsky, D.T. Young (Eds.), pp. 85–90. Geophys. Monograph 102, American Geophysical Union.

MARSDEN, R.G., HENRION, J., SANDERSON, T.R. *et al.* 1984 Calibration of a space-borne charged particle telescope using protons in the energy range 0.4 to 20 MeV. *Nucl. Instr. & Methods in Phys. Res.*, **221**, 619.

McCOMAS, D.J., NORDHOLT, J.E., BAME, S.J., BARRACLOUGH, B.L. & GOSLING, J.T. 1990 Linear electric field mass analysis: A technique for three-dimensional high mass resolution space plasma composition measurements. *Proc. Nat. Acad. Sci. U.S.A.*, **87**, 5925.

MÖBIUS, E., BOCHSLER, P., GHIELMETTI, A.G. & HAMILTON, D.C. 1990 High mass resolution isochronous time-of-flight spectrograph for three-dimensional space plasma measurements. *Rev. Sci. Instrum.*, **61**, 3104.

OGILVIE, K.W., DURNEY, A. & VON ROSENVINGE, T. 1978 Descriptions of experimental investigations and instruments for the ISEE spacecraft, *IEEE Transact. on Geosci. Electron*, **GE-16**, 151–153.

PASCHMANN, G., MELZNER, F., FRENZEL, R., VAITH, H. *et al.* 1997 The Electron Drift Instrument, in *The Cluster and Phoenix Missions*, Escoubet, C.P., Russell, C.T. & Schmidt, R. (Eds.), pp. 233–269. Kluwer Academic Publishers.

PEDERSEN, A., MOZER, F. & GUSTAFSSON, G. 1998 Electric field measurements in a tenuous plasma with spherical double probes, in *Measurement techniques in Space Plasmas – Fields*, Pfaff, R.F., Borovsky, J.E., Young, D.T. (Eds.), pp. 11–12. Geophys. Monograph 103, American Geophysical Union.

PFAFF, R.F., BOROVSKY, J.E. & YOUNG, D.T. (EDITORS) 1998a Measurement techniques in space plasmas: Fields. Geophysical Monograph 103, American Geophysical Union, Washington, D.C.

PFAFF, R.F., BOROVSKY, J.E. & YOUNG, D.T. (EDITORS) 1998b Measurement techniques in space plasmas: particles. Geophysical Monograph 10, American Geophysical Union, Washington, D.C.

RÈME, H., AOUSTIN, C., BOSQUED, J.M., DANDOURAS, I. *et al.* 2001 First multispacecraft ion measurements in and near the Earth's magnetosphere with the identical Cluster ion spectrometry (CIS) experiment. *Ann. Geophys.*, **19**, 1303.

SCUDDER, J., HUNSACKER, F., MILLER, G. *et al.* 1995 Hydra – A 3-dimensional electron and ion hot plasma instrument for the POLAR spacecraft of the GGS mission. *Space Sci. Rev.*, **71**, 459.

SMITH, C.W., ACUNA, M.H. *et al.* 1998 The ACE Magnetic Fields Experiment. *Space Sci. Rev.*

SMITH, E.J., CONNOR, B.C. & FOSTER, JR., G.T. 1975 Measuring the magnetic fields of Jupiter and the outer solar system. *IEEE Trans. Magn.*, **MAG-11**, 962.

STONE, E.C. 1977 The Voyager missions to the outer solar system. *Space Sci. Rev.*, **21**, 75.

STONE, E.C. *et al.* 1998 The Advanced Composition Explorer. *Space Sci. Rev.*, **86**, 1–22.

STONE, E.C., COHEN, C.M.S., COOK, W.R. *et al.* 1998 The Solar isotope spectrometer for the advanced composition explorer. *Space Sci. Rev.*, **86**, 355.

VASYLIUNAS, V.M. 1971 Deep space plasma measurements, in *Methods of Experimental Physics*, volume **9B**, Lovberg and Griem (Eds.), p. 49. Academic Press.

WARD, W.W. 1974 In celebration of the Pioneers – To Jupiter and beyond. *Technology Rev.*, **77**, 2–15.

WUEST, M. 1998 Time-of-flight ion composition measurement technique for space plasmas, in *Measurement techniques in Space Plasmas – Particles*, Pfaff, R.F., Borovsky, J.E., Young, D.T. (Eds.), pp. 141–156. Geophys. Monograph 102, American Geophysical Union.

ZURBUCHEN, T.H., GLOECKLER, G. *et al.* 1998 A low-weight plasma instrument to be used in the inner heliosphere, *SPIE Conf. Proc., Missions to the Sun II*, 217–224.

8. Planetary observations and landers

By A. CORADINI

Istituto di Astrofisica Spaziale e Fisica Cosmica, CNR, Roma, Italy

These lectures cover the principles of remote sensing instrumentation as commonly used for missions to solar system bodies. From the basic physical principles, an introduction to real instruments is provided. Particular attention is paid to the airborne visible infrared imaging spectrometer (AVIRIS) hyperspectral imager, Cassini-visual and infrared mapping spectrometer (VIMS) and the visible and infrared thermal imaging spectrometer (VIRTIS) series. We conclude with a review of the *in situ* science provided by landers, rovers and other surface elements.

8.1. Introduction

8.1.1. *Principles of remote sensing: the interaction of the radiation with matter*

Most of what we know about the planets comes from information that is obtained through the interaction between the planet's matter and light, or more generally, between its matter and radiation. Thus, it is appropriate to begin our study of the planets and how to observe them by examining the properties of light and the instruments that are used to detect it. In modern physics, light or electromagnetic (EM) radiation may be viewed in one of two complementary ways: as a wave in an abstract EM field or as a stream of photons. Although both are acceptable descriptions of light, for our purposes in introductory remote sensing, the wave description will be more useful.

The EM spectrum is the distribution of EM radiation according to energy, or according to frequency, or indeed wavelength since the two are in turn related (see Table 8.1).

Depending on the wavelength, the interaction of the radiation with matter changes: different wavelengths can be used to explore different properties of matter. If we think of an ideal situation, in which the matter can be simply represented by atoms and molecules, we can identify the type of interaction through its effect on the light. However, the average or bulk properties of EM radiation interacting with matter are systematized in a simple set of rules called *radiation laws*. These laws apply when a *blackbody radiator* can get close to the radiating body. Generally, blackbody conditions apply when the radiator has very weak interaction with the surrounding environment and can be considered to be in a state of equilibrium. The dark surface of planetary objects can be, as a first approximation, considered black bodies.

Interaction of light with atoms can generate atomic excitation and de-excitation: atoms can make transitions between the orbits allowed by quantum mechanics either by absorbing or emitting exactly the energy difference between the orbits. An atomic excitation is caused by absorption of a photon and an atomic de-excitation caused by emission of a photon, an electron transition in an atom. To raise an atom from a low energy level to a higher level, a photon must be absorbed, and its energy must match the energy difference between the initial and final atomic states. When an atom de-excites to a lower energy level, it must emit a sequence of one or more photons of the proper energy. Atomic transitions are characterized by photon energies that range from a few eV to the order of keV, depending on the type of atom and the particular transition involved. The frequency of the absorbed or emitted radiation is precisely in a way that the photon carries the energy difference between the two orbits. This energy may be calculated by dividing the product of the Planck constant and the speed of light (hc) by the wavelength of the light. Thus, an

TABLE 8.1. The EM spectrum.

Region	Wavelength (angstroms)	Wavelength (cm)	Frequency (Hz)	Energy (eV)
Radio	$>10^9$	>10	$<3 \times 10^9$	$<10^{-5}$
Microwave	10^9-10^6	$10-0.01$	$3 \times 10^9-3 \times 10^{12}$	$10^{-5}-0.01$
Infrared	10^6-7000	$0.01-7 \times 10^{-5}$	$3 \times 10^{12}-4.3 \times 10^{14}$	$0.01-2$
Visible	$7000-4000$	$7 \times 10^{-5}-4 \times 10^{-5}$	$4.3 \times 10^{14}-7.5 \times 10^{14}$	$2-3$
Ultraviolet	$4000-10$	$4 \times 10^{-5}-10^{-7}$	$7.5 \times 10^{14}-3 \times 10^{17}$	$3-10^3$
X-rays	$10-0.1$	$10^{-7}-10^{-9}$	$3 \times 10^{17}-3 \times 10^{19}$	10^3-10^5

atom can absorb or emit only certain discrete wavelengths (or equivalently, frequencies or energies). The photon energy due to an electron transition between an upper atomic level k (of energy E_k) and a lower level i is:

$$E = E_k - E_i = \frac{h}{\nu} = \frac{hc}{\sigma} = \frac{hc}{\lambda_v} \tag{8.1}$$

where ν is the frequency, σ, the wavenumber in vacuum and λ_v is the wavelength in vacuum. The most accurate spectroscopic measurements are determinations of transition frequencies, the unit being the Hertz ($1\,\text{Hz} = \text{cycle/s}$) or one of its multiples. A measurement of any one of the entities' frequency, wavenumber or wavelength (in a vacuum) is an equally accurate determination of the others since the speed of light is exactly defined. The most common wavelength units are the nanometre (nm), the Ångström ($1 = 10^{-1}\,\text{nm}$) and the micrometre (μm). The International System (SI) of wavenumber unit is the inverse metre, but in practice wavenumbers are usually expressed in inverse centimetres: $1\,\text{cm}^{-1} = 10^2\,\text{m}^{-1}$, equivalent to $2.99792458 \times 10^4\,\text{MHz}$.

In addition to frequency and wavenumber units, atomic energies are often expressed in electron volts (eV). One eV is the energy associated with each of the following quantities:

$$1\,\text{eV} = 2.4179884(7) \times 10^{14}\,\text{Hz}$$
$$1\,\text{eV} = 8065.5410(24)\,\text{cm}^{-1}$$
$$1\,\text{eV} = 1239.84244(37)\,\text{nm}$$
$$1\,\text{eV} = 11604.45(10)\,\text{K}$$
$$1\,\text{eV} = 1.6021773(5) \times 10^{-19}\,\text{J}$$

The atomic spectroscopy is simply based on this principle. The problem is then how to reveal these different light components, and in order to solve it, modern spectroscopy was born.

The different parts of the EM spectrum have very different effects upon interaction with matter. Each portion of the EM spectrum has quantum energies appropriate for the excitation of certain types of physical processes. The energy levels for all physical processes at the atomic and molecular levels are quantized, and if there are no available quantized energy levels with spacing which match the quantum energy of the incident radiation, then the material will be transparent to that radiation, and it will pass through it. For example, the quantum energy of microwave photons is in the range 0.00001–0.001 eV, which is in the range of energies separating the quantum states of molecular rotation and torsion.

8.1.2. *Exploring different phenomena*

8.1.2.1. *Radio waves*

Radio waves have wavelengths (in a vacuum) that are in excess of about 10 cm and have frequencies typically in the MHz region. The corresponding photon energies are about 10^{-5} eV or less. These are too small to be associated with electronic, molecular or nuclear transitions. However, the energies are in the right range for transitions between nuclear spin states (nuclear magnetic resonance) and are the basis for magnetic resonance imaging. Radio waves are used for subsurface planetary exploration and are generated by oscillating charges in antennas and electric circuits. Typical examples are the so-called Ground-Penetrating Radar (GPR).

8.1.2.2. *Microwaves*

The interaction of microwaves with matter other than metallic conductors will be to rotate molecules and produce heat as result of this molecular motion. Conductors will strongly absorb microwaves and any lower frequencies because they will cause electric currents which will heat the material. Most matter is largely transparent to microwaves. High-intensity microwaves, as in a microwave oven where they pass back and forth through the food millions of times, will heat the material by producing molecular rotations and torsions. Since the quantum energies are a million times lower than those of X-rays, they cannot produce ionization and the characteristic types of radiation damage associated with ionizing radiation. The wavelengths of microwaves are in roughly the 30-cm to 1-mm range, and have frequencies of the order of GHz. Microwaves correspond to photon energies typically between about 10^{-6} and 10^{-3} eV. These energies correspond to transitions between rotational states of molecules, as well as transitions between electron spin states (electron spin resonance). Microwaves are easily transmitted through clouds and the atmosphere, making them appropriate for radar systems, communication and satellite imaging. In planetary science, microwaves can be used to study planetary atmospheres through molecular spectroscopy, as well as to study thermal properties of different material through the determination of thermal inertia. The use of radars for planetary exploration developed through extensive experience in terrestrial remote sensing.

In this field, the project and design of different instruments, as well as the interpretation of their data, are favoured by the existence of a well-developed theory (i.e. the EM theory). Following Ulaby *et al.* (1983), the performance of radar is conveniently described by an equation known as the "radar equation", that relates the power received by the radar to the target parameter, as well as to the instrument parameters. The radar equation describes how EM waves propagate through a material. As it is usually used for radar passing through air, it neglects material losses. When a power P_t is transmitted by an antenna with gain G_t, the power per unit solid angle in the direction of the scatterer is $P_t G_t$, where the value of G_t in that direction is used. The received power is:

$$P_r = \frac{P_t G_t}{4\pi R^2} \sigma_{rt} \frac{A_{rs}}{4\pi R^2} \tag{8.2}$$

where σ_{rt} is radar cross section. The spreading loss $1/4\pi R^2$ is the reduction in power density associated with spreading of the power over a sphere of radius R surrounding the antenna. A_{rs} is the effective receiving area of the receiving antenna aperture at the polarization r. Note that the effective area A_{rs} is not the actual area of the incident beam intercepted by the scatterer, but rather the effective area; that is, it is that area of the incident beam from which all power would be removed if one assumed that the power going through all the rest of the beam continued uninterrupted. The actual value of A_{rs} depends on the effectiveness of the scatterer as a receiving antenna. Through this equation a relationship is established between the scattering characteristics of a target and the transmitted pulse power and pulse length, the effective area of the antenna, the average power returned to the receiver, the range, and the attenuation factor of the radar frequency.

In planetary science the target is made of solid or, less frequently, liquid particles. The radar equation describes how EM waves propagate through these different materials. In terrestrial remote sensing, this equation is usually used for radar through air, and, therefore, it neglects material losses.

For GPR, the radar equation describes how the propagating wave is modified or changed: by the transmitter antenna properties (gain, antenna pattern and frequency dependence), by coupling to the ground (efficiency, frequency dependence), by geometric spreading losses (such as the power being spread over the surface of a balloon being blown up around the antenna), by exponential material dissipation losses (including frequency-dependent dispersion from electrical conduction, dielectric or magnetic relaxation), by scattering (reflection, refraction, diffraction) from a change in properties (depending on contrast (Fresnel), orientation (Snell) and polarization (Stokes/Mueller)), and then in the direction of the receiving antenna, by exponential material losses, geometric spreading, coupling and the receiver antenna properties. Therefore, by analysing the returned echoes of the EM radiation penetrating under a planetary surface, and travelling through it, passing through layers of different materials, and being reflected from subsurface discontinuities, one can infer what is the local stratigraphy and the nature of the different layers (see e.g. Evans, 1969). The electrical and magnetic properties of rocks, soils and fluids (natural materials) control the speed of propagation of radar waves and their amplitudes. In most cases, the electrical properties are much more important than the magnetic properties. At radar frequencies, electrical properties are dominantly controlled by rock or soil density, and by the chemistry, state (liquid/gas/solid), distribution (pore space connectivity) and content of water.

Nearly all polarization of importance in Earth materials is the result of some interaction involving water (Franks, 1972). The dominant mechanisms of electrical conduction are ionic charge transport through water filling pore spaces in rocks and soils.

Below typical values of the dielectric constants for wet and dry soil, as well as water, are provided.

Material	Dielectric constant, ε (dimensionless)
Dry soil[a]	$2 - j\,0.05$
Wet soil[b]	$26 + j\,4$
Water[c]	$80 + j\,4$

[a]Virtually independent of temperature and frequency.
[b]Measured with silt loam soil at 1.4 GHz, 23°C, 40% volumetric moisture.
[c]Measured at 1 GHz at room temperature.

Therefore, using the radar equation as shown in Equation (8.2), the receiver power can be computed for various distances.

If we consider how important the search for water in the solar system is now, one can understand how powerful a GPR could be. In fact, the dielectric constant of water is significantly different from that of silicates, which are the most common material found on terrestrial planets.

8.1.2.3. *Infrared*

Electron transitions in molecules exhibit energy spacings that are similar to those in individual atoms; they are in the order of eV to keV. However, molecules are more complicated than atoms due to vibrational and rotational motions of the constituent atomic nuclei. A diatomic molecule consists of two atomic nuclei (each of mass M) and the various orbiting electrons. Even though this is the simplest type of molecule, it represents a complex multi-bodied system that has many degrees of freedom. Specifically, each of the two atomic nuclei has three degrees of freedom (corresponding to the three coordinate directions), and each electron in the molecule has the same. At face value, the problems connected with finding the energy states in even a light diatomic molecule appear almost insurmountable. There is a method by which molecular energy states can be determined, known as the Born–Oppenheimer approximation. The approximation consists of ignoring the motion of nuclei that are much more massive than electrons thus, they move much more slowly than the electrons in a molecule. As a first approximation, therefore, the nuclei can be regarded as having a fixed separation R in space. For each such separation, one then determines the possible values of the total electronic energy ϵ of the molecule; that is, there is an ϵ versus R curve associated with each possible electronic energy level. The quantum energy of infrared (IR) photons is in the range 0.001–1.7 eV which is in the range of energies separating the quantum states of molecular vibrations. IR is absorbed more strongly than microwaves, but less strongly than visible light. The result of IR absorption is heating of the material since it increases molecular vibrational activity. Indeed, IR radiation penetrates further than visible light. Due to basic atomic and molecular structure, the spectra associated with molecules typically involve IR wavelengths. Molecules are usually fragile, so molecular spectra are important mostly in objects that are relatively cool, such as planetary atmospheres, planetary surfaces and various interstellar regions.

Over the last several decades, remote spectroscopic studies have been very successfully used to explore the composition of various solid surfaces in the solar system. Building on these initial successes with mineral identification, recent efforts have focused on more detailed spectral inferences including assessing the composition and abundance of mafic minerals. Such efforts are complicated by the fact that most silicate-rich surfaces include lithological combinations of several mafic minerals which have overlapping absorption bands. One of the first experiments in this field was the Mariner-9 mission experiment where the spectroscopy of Mars was firstly observed. The IR spectroscopy experiment on the Mariner-9 mission was designed to provide information on atmospheric and surface properties by recording a major portion of the thermal emission spectrum of Mars. The original intent was to determine vertical temperature profiles, the surface temperature, and the atmospheric pressure at the surface, and to acquire information related to the surface composition. The experiment also was designed to search for minor atmospheric constituents, including H_2O vapour and isotopic components of CO_2 (Hanel *et al.*, 1972).

8.1.2.4. *Visible light*

The primary mechanism for the absorption of visible light photons is the elevation of electrons to higher energy levels. There are many available states, so visible light is absorbed strongly. Isolated atoms can absorb and emit packets of EM radiation having discrete energies dictated by the detailed atomic structure of the atoms. When the corresponding light is passed through a prism or spectrograph, it is separated spatially according to wavelength. The corresponding spectrum may exhibit a continuum, or may have superposed on the continuum bright lines (an *emission spectrum*) or dark lines (an *absorption spectrum*).

Thus, *emission spectra* are produced by thin gases in which the atoms do not experience many collisions (because of the low density). The emission lines correspond to photons of discrete energies that are emitted when excited atomic states in the gas make transitions back to lower-lying levels.

A *continuum spectrum* results when the gas pressures are higher. Generally, solids, liquids or dense gases emit light at all wavelengths when heated.

An *absorption spectrum* occurs when light passes through a cold, dilute gas and atoms in the gas absorb at characteristic frequencies; since the re-emitted light is unlikely to be emitted in the same direction as the absorbed photon, this gives rise to dark lines (absence of light) in the spectrum.

The near ultraviolet (UV) is absorbed very strongly in the surface layer, so that UV photons above the ionization energy of atoms and molecules can disrupt them.

8.1.2.5. *X-ray*

Since the quantum energies of X-ray photons are much too high to be absorbed in electron transitions between states for most atoms, they can interact with an electron only by knocking it completely out of the atom. That is, all X-rays are classified as ionizing radiation. This can occur by giving all of the energy to an electron (photoionization) or by giving part of the energy to the photon and the remainder to a lower energy photon (Compton scattering). At sufficiently high energies, the X-ray photon can create an electron positron pair.

One important practical consequence of the interaction of EM radiation with matter and of the detailed composition of our atmosphere is that only light in certain wavelength regions can penetrate the atmosphere well. These regions are called *atmospheric windows*. The dominant windows in the atmosphere are in the visible and radio frequency regions, while X-rays and UV are very strongly absorbed, and gamma-rays and IR are somewhat less strongly absorbed. We clearly see the argument for rising above the atmosphere with detectors on space-borne platforms in order to observe at wavelengths other than the visible and IR regions.

8.1.3. *Spectroscopy basic concepts*

Spectroscopy is the use of the absorption, emission or scattering of EM radiation by atoms or molecules (or atomic or molecular ions) to qualitatively or quantitatively study the atoms or molecules, or to study physical processes. The interaction of radiation with matter can cause redirection of the radiation and/or transitions between the energy levels of the atoms or molecules. A transition from a lower level to a higher level with transfer of energy from the radiation field to the atom or molecule is called *absorption*. A transition from a higher level to a lower level is called *emission* if energy is transferred to the radiation field, or non-radiative decay if no radiation is emitted. Redirection of light due to its interaction with matter is called *scattering*, and may or may not occur with transfer of energy; that is, the scattered radiation has a slightly different or the same wavelength (see e.g. Hapke, 1981; Hapke and Wells, 1981 and references therein):

• *Absorption*: When atoms or molecules absorb light, the incoming energy excites a quantized structure to a higher energy level. The type of excitation depends on the wavelength of the light. Electrons are promoted to higher orbitals by UV or visible light, vibrations are excited by IR light and rotations are excited by microwaves.

An absorption spectrum is the absorption of light as a function of wavelength. The spectrum of an atom or molecule depends on its energy level structure, and absorption spectra are useful for identifying compounds.

- *Emission*: Atoms or molecules that are excited to high energy levels can decay to lower levels by emitting radiation (emission or luminescence). For atoms excited by a high-temperature energy source this light emission is commonly called *atomic* or *optical* emission (see atomic emission spectroscopy), and for atoms excited with light it is called atomic fluorescence (see atomic fluorescence spectroscopy). For molecules it is called *fluorescence* if the transition is between states of the same spin, and *phosphorescence* if the transition occurs between states of different spin.

The emission intensity of an emitting substance is linearly proportional to analyse concentration at low concentrations, and is useful for the identification of emitting species.

- *Scattering*: When EM radiation passes through matter, most of the radiation continues in its original direction but a small fraction is scattered in other directions. Light that is scattered at the same wavelength as the incoming light is called *Rayleigh scattering*. Light that is scattered in transparent solids due to vibrations (phonons) is called *Brillouin scattering*.

- *Radiation penetration depth*: After having looked at the interaction of the radiation with matter, let us now consider the ability of radiation to penetrate materials. We know that, in addition to the energy of the radiation, the depth of penetration is also dependent on the density of the material being penetrated. We can describe the phenomena which occurs when energy interacts with matter as a function of energy of the penetration photon and as a function of density of the target. When the radiation is stopped and absorbed, it must transfer its energy to the material.

When a beam of photons passes through matter, a variety of interactions are possible, including scattering and absorption as well as undisturbed transmission. The process is random, with each photon independently having some small probability of interacting in any given way with each atom as it passes through the target. In the case of so-called "good geometry", any interaction between the incident quantum and the target removes that quantum from detection as part of the unscattered, unabsorbed radiation, the "main beam". One therefore measures and seeks to predict the intensity of the main beam as a function of the thickness of the target.

When we deal with high energy radiation, such as X-rays, the main process of interaction is ionization. We know that one of the factors affecting ionization is the *material type*. We also know that radiation is not able to easily penetrate dense materials. Photon energy is lost through collisions, but this process is a statistical one, so some photons may pass completely through the material with minimal or no interaction. Also, the depth of penetration for a given photon energy is dependent on material density (atomic structure). The more subatomic particles in a material (higher Z number), the greater the likelihood that interactions will occur and the radiation will lose its energy. On the other hand, if the radiation wavelength is known, the nature of this interaction can help in deciphering the nature of the observed material.

The absorption of radiation starts as soon as the radiation enters a material. The process is progressive and continues as the radiation penetrates deeper into the material. Additional energy is absorbed through the various processes of ionization. At some point in the material, there is a level at which the radiation intensity becomes one-half at the surface of the material. This depth is known as the Half-Value Layer (HVL) for that material. Each material has its own specific HVL thickness. Not only is the HVL material-dependent, but it is also energy-dependent. This means that for a given material, if the radiation energy changes, the point at which the intensity decreases to half its original value will also change.

If we raise the energy of the radiation interacting with the same material, the HVL will occur deeper in that material. X-rays and gamma-rays with shorter wavelengths will have more energy that must be absorbed and, therefore, more energy will reach deeper into the material or through the material. Conversely, if we lower the radiation energy, the HVL will occur at shallower levels. In the case of safety shielding, on the other hand, one must consider scattered and secondary radiation as well as the main beam.

If the detector is a Geiger–Muller tube, or any similar device that does not permit distinguishing these other radiations from the main beam, and if the geometry permits significant numbers of scattered or secondary quanta to reach the detector, then the prediction of the counting rate as a function of absorber thickness is complicated. The simpler theory that is appropriate in the case of good geometry will be developed first.

We consider a monochromatic beam (e.g. a beam composed of photons all of which have the same energy, and therefore the same frequency, wavelength and, if visible, colour) passing through successive layers of thickness Δx, and measure the intensity, $I(x)$, of the beam of unscattered, unabsorbed photons remaining at any particular depth, x:

$$I(x + \Delta x) = I(x) - \mu \Delta x I(x)$$

is characteristic of the material and photon energy. The relationship must be like this: because all the photons are identical, a specific fraction of the incoming photons will be scattered or absorbed in each layer, and clearly the thicker the layer, the more will interact. The increased absorption with increased thickness will be linear (only in the limit of thin layers, for which $\mu \Delta x \ll 1$). In essence the atoms at the front of the layer must not shield the atoms at the back of the layer, and there must be no double scattering. Therefore, the above equation leads to the following:

$$\frac{I(x + \Delta x) - I(x)}{\Delta x} = -\mu I(x) \qquad (8.3)$$

in other words, the rate of change of the intensity is proportional to the intensity. Hence, the calculation:

$$I(x) = I(0) \, e^{-\mu x} \qquad (8.4)$$

or

$$T(x) = e^{-\mu x} \qquad (8.5)$$

for monochromatic photons, with definitions of symbols as below, following Cember (1969):

- $I(0) =$ incident intensity;
- $x =$ absorber thickness;
- $I(x) =$ intensity after passing through an absorber of thickness x;
- $T(x) =$ transmission probability of an absorber of thickness x;
- $\mu =$ absorption coefficient.

The transmission probability is defined in such a way that a transparent object has $T = 1$ and an opaque object has $T = 0$.

There are four standard ways to specify the absorber thickness and hence four corresponding ways to specify the absorption coefficient:

- If the thickness is measured in centimetre, the absorption coefficient is measured in inverse centimetre.

- If the thickness is measured in grams/square centimetre, the absorption coefficient is measured in square centimetres/gram.
- If the thickness is measured in atoms/square centimetre, the absorption coefficient is measured in square centimetres/atom.
- If the thickness is measured in atoms/barn, the absorption coefficient is measured in barns/atom, and is usually called the atomic "cross section" and given the symbol σ:

$$1 \text{ barn} = 10^{28} \text{ m}^2 \tag{8.6}$$

1 barn = 1 square Fermi = 1 square femtometre, the approximate physical cross-section size of a small nucleus. The mass density, in grams/cubic centimetre, the atomic weight, in grams/mole and Avogadro's number, in atoms/mole, can be used for conversion in these four methods of measurement.

Another common way to describe the passage of radiation through matter is in terms of the mean distance travelled through the absorber, the "mean penetration depth", D. In exact analogy to radioactive decay, we can write:

$$D = \frac{1}{\mu} \tag{8.7}$$

Since D is the mean penetration depth, we expect that many photons will be absorbed or scattered before reaching that depth, and that many will penetrate significantly farther than D.

8.1.4. *From the principles to the experiment*

This introduction has been rather drawn out, for a reason: all imaging systems map some spectroscopic property of the imaged object onto a detector. In a large fraction of imaging modalities, the imaged object directly or indirectly emits, reflects, transmits or scatters EM radiation. Such is the case, for example, in radar and remote sensing systems, radio astronomy, optical microscopy, X ray and positron emission tomography, fluorescence imaging, magnetic resonance imaging and telescope-based systems, to name but a few areas. In the following sections we will discuss the properties of the sensors, which are able to detect specific phenomena. We will also study how to build up a new remote sensing sensor starting from specific scientific requirements.

8.2. Radiation detection

The first step that should be taken when designing an experiment, particularly a planetary one, is to establish a clear definition of scientific requirements. Once these have been correctly defined, the next step is to define the technique to be used. In remote sensing experiments, we use the spectral range, which clearly indicates the physical problems that should be addressed. In planetary science, especially at the beginning of this kind of study, the definition of specific requirements was difficult due to the lack of knowledge of the specific object. For this reason, at the beginning of planetary exploration, an attempt was made to put together integrated payloads able to cover specific aspects, such as the surface of the planet, its composition, its interior, its magnetic field and its atmosphere. Advantage should be taken of the reduced distance from the planet in order to expand on our knowledge. For this reason, when planning a new mission to a planet never before observed, the most important instrument to develop is the camera.

Images are clearly needed in order to have a preliminary vision of the planetary object. For the first lunar explorations, for example, cameras with simple film were used. Since that time, other kinds of detectors have been used in order to build up the image. Cameras are now complex instruments with many functions including the possibility to create colour and stereo images, and in some cases, to build under-sampled spectra able to help in the characterization of surfaces composition.

Early ground-based and spacecraft visible wavelength imaging relied upon photographical techniques, often supplemented by the use of broadband colour filters. However, photographical film is a highly non-linear detector that is extremely difficult to calibrate, so alternatives were sought which allowed for the derivation of more quantitative information from images. A major advance in spacecraft imaging after the film systems of lunar orbiter was the vidicon, first successfully flown on the Mariner-4 fly-by of Mars in 1965 (Leighton *et al.*, 1967).

An imaging instrument uses optics such as lenses or mirrors to project an image onto a detector, where it is converted to digital data. Natural colour imaging requires three exposures of the same target through different colour filters, typically selected from a filter wheel. Earth-based processing combines data from the three black and white images, reconstructing the original colour by utilizing the three values for each picture element (pixel). In the past, the detector that creates the image was a vacuum tube resembling a small Cathode-Ray Tube (CRT), called a *vidicon*. In a vidicon, an electron beam sweeps across a phosphorous coating on the glass where the image is focused, and its electrical potential varies slightly in proportion to the levels of light it encounters. This varying potential becomes the basis of the digital video signal produced. Viking, Voyager and Mariner spacecraft used vidicon-based imaging systems. A vidicon requires a fairly bright image to detect. Vidicon imaging systems work in a two-step process: first, the photoconductive surface is exposed, much like film in a conventional camera, after an electron beam passes over the back of the photoconductor, priming it with a negative charge. Photons incident upon the photoconductor surface reduce the negative charge in proportion to scene brightness. After this picture-forming step, the resultant image on the photoconductor is read out by a scanning electron beam. The areas of the image with a less negative charge will draw more electrons for the beam, and this variation in current is read out as the video signal current. After readout, the image is digitized and sent back to Earth. Early vidicon images (Mariners -4, -6 and -7) were crude and had low photometric accuracy by today's standards. However, they provided a dramatic first look at Mars. Improved vidicon cameras were extremely reliable with a photometric accuracy generally better than 10 (Mariner-10, Viking, Voyager). The photoconductive surface was sensitive to visible light in a range of about 350–650 nm. To obtain multispectral data, a filter wheel was employed that allowed for several broadband filters typically placed between 400 and 600 nm. Colour filters led to the mapping of gross colour heterogeneities on a scale of 1–10 km for Mercury, Mars and selected outer planet satellites (e.g. Soderblom *et al.*, 1978; Rava and Hapke, 1987). The first of two Mariner C probes, later renamed Mariner-3 and -4, was ready for launch in early November 1964. The probe lifted off on 5 November in the afternoon, was briefly put into a parking orbit and finally launched towards Mars by the second ignition of the Agena stage. The probe was not working since the solar panels remained in their stowed position and thus were not generating any power. On 28 November, Mariner-4 left its mother planet and headed for a Type I interplanetary transfer orbit and a 246,378 km fly-by of Mars. This was required to minimize the chances of the unsterilized Agena rocket with Mars. On 5 December, the probe corrected its trajectory and lowered its minimum distance from Mars to 9600 km inside an optimal fly-by window dictated by many constraints (including: Mars occultation, Sun and Canopus visibility, angular distance between Canopus and the two tiny Martian Moons). The probe reached the minimum distance of 9846 km from the planet at 1.01 and this was followed, 30 min later, by 54 min of Mars disk occultation. The experiment showed what the expected data based on terrestrial observation had predicted, namely that the atmosphere was rarefied, with a ground pressure of 80 hPa and a ground temperature close to 0°C. Mariner-4 showed the atmosphere to be extremely thin, having a ground pressure, depending on the composition (not to be measured by the probe), of between 4.1 and 7 hPa, and a ground temperature close to −100°C. Twenty-one complete images were recorded, in addition to 21 rows of the 22nd, covering less than a hundredth of the Martian surface starting at 37° North, 173° East, to end in the night hemisphere at 50° South, 255° East. The first historical image showed 365 km of the 17,600-km-long Martian limb. It was the first surprise of the mission, for it showed a bright halo parallel to the limb. Starting from the third image, once contrast was stretched to compensate for the very high Sun (near the zenith), some object clearly related to Martian surface morphology started appearing, but only from image seven on did they clearly show what they were: hundreds of craters ranging in size up to 35 km. These first seven images covered the bright areas of Amazonis and Zephyria, but the next six images covered two dark maria: Mare Cimmerium and Mare Sirenum. Here the probe took its most famous picture, number 11, showing a huge 150 km diameter crater flanked by many smaller craters. Due to the geometry and illumination of the picture this was also the one with the best resolution, showing details as small as 1.4 km. The crater, located at 33° South, 197° East, was later named Mariner crater to honour the probe which discovered it. A total of 300 craters were found in the probe's images, with diameters ranging from 5 km, close to the camera resolution limit, to 120 km. None of the infamous "canals" were captured in the images, although the maps of Schiaparelli and Lovell showed a few of them in the imaged areas. The data relayed by Mariner-4 took the scientific community by surprise, for they showed a planet with a much thinner atmosphere than previously thought and looking quite similar to the Moon. The first crater counts were performed, and the idea of using crater counting as a method of relative dating was suggested (Chapman *et al.*, 1969). In the same paper it was also observed that craters are degraded by erosion, indicating that, despite the fact that the atmosphere is very tenuous, it has a strong influence on the evolution of Martian geomorphology.

The last two spacecraft of National Aeronautics and Space Administration's (NASA) Mariner series, Voyagers 1 and 2 were the first in that series to be sent to explore the outer solar system. Preceded by the Pioneer 10 and 11 missions, Voyagers 1 and 2 were to make studies of Jupiter and Saturn, their satellites, and their magnetospheres as well as studies of the interplanetary medium. An option was designed into the Voyager 2 trajectory, and finally directed towards Uranus and Neptune to perform similar studies. Initially, NASA intended the twin robotic probes to visit just Jupiter and Saturn, wrapping up their tour by 1981. Instead, their missions were extended, with Voyager 2 going to Neptune and Uranus. Between 1979 and 1989 the two probes studied 48 Moons and four planets – more of the solar system than any other spacecraft. The Voyager journey of discovery continues. After travelling through space for more than 26 years, Voyager 1 is approaching a new milestone. On 5 November 2003, the spacecraft reached 90 Astronomical Units (AU) from the Sun.

The main discoveries made by Voyager were done by the Imaging Science Subsystem (ISS). The Voyager ISS is a modified version of the slow-scan vidicon camera designs that were used in the earlier Mariner flights. The system consists of two cameras: a high-resolution Narrow-Angle (NA) camera and a lower resolution, more sensitive Wide-Angle (WA) camera. Unlike the other on-board instruments, operation of the cameras is not autonomous, but is controlled by an imaging parameter table residing in one of the spacecraft computers, the flight data subsystem (FDS) (Snyder, 1979). The Voyager experience was also extremely important since it showed, that, provided that the instruments survive and there is the possibility to re-programme their main electronics, new operative modes can be introduced and new compression algorithms can be implemented. Telemetry enhancement techniques were used by Voyager-2 to reduce telemetry transmission rates by over 50% compared to those used at Saturn, with negligible loss in information return. An new image data compressor algorithm for noiseless coding techniques was implemented (Urban, 1987). In that way, it was possible to observe effectively Uranus and Neptune as well.

The Mariner-10 television system centred around two vidicon cameras, each equipped with an eight-position filter wheel. The vidicons were attached to telescopes mounted on a scan platform that allowed movement in vertical and horizontal directions for precise targeting on the planetary surfaces. These folded optics (Cassegrain) telescopes were required to provide narrow-angle, high-resolution photography. An auxiliary optical system mounted on each camera allowed the acquisition of a wide-angle, lower-quality image. Changing to the wide-angle photography was done by moving a mirror on the filter wheel to a position in the optical path of the auxiliary system. In addition to wide-angle capability, the filter wheels included blue bandpass filters, UV polarizing filters, minus UV high-pass filters, clear apertures, UV bandpass filters, defocusing lenses for calibration and yellow bandpass filters. A shutter blade controlled the exposure of the 9.8 by 12.3-mm image face of the vidicon for an interval that could be varied from 3 ms to 12 s. The light image formed on the photosensitive surface of the vidicon produced an electrostatic charge proportional to the relative brightness of points within the image. During vidicon readout, an electron beam scanned the back side of the vidicon and neutralized part of the charge so as to produce electric current variations proportional to the point charge being scanned at the time. These analogue signals produced from the vidicon readout process were electronically digitized as 832 discrete dots or picture elements (pixels) per scan line, and presented to the flight data system in the form of 8-bit elements for transmission. Each TV frame-one picture consisted of 700 of these vidicon-scan lines. All timing and control signals, such as frame start, line start/stop, frame erase, shutter open/close and filter wheel step, were provided by the systems on board the spacecraft.

Most of the scientific findings about Mercury comes from the Mariner-10 spacecraft which was launched on 3 November 1973. It flew past the planet on 29 March 1974 at a distance of 705 km from the surface. On 21 September 1974 it flew past Mercury for the second time and on 16 March 1975 for the third time. During these visits, over 2700 pictures were taken, covering 45% of Mercury's surface. Up until this time, scientists did not suspect that Mercury would have a magnetic field. Until Mariner-10, little was known about Mercury because of the difficulty in observing it from Earth telescopes. At maximum elongation it is only 28° from the Sun as seen from Earth. Due to this, it can only be viewed during daylight hours or just prior to sunrise or after sunset. When observed at dawn or dusk, Mercury is so low on the horizon that the light must pass through 10 times the amount of Earth's atmosphere than it would if Mercury was directly overhead.

8.2.1. *Solid state detectors*

The imaging systems of former Mars missions, as well as those of other planetary missions, had only limited spectral capabilities and were limited in spectral range due to spectral sensitivity cut-offs of the vidicon cameras used. Meanwhile a new sensor technology, called Charge-Coupled Devices (CCDs), was developed which would allow imaging from the UV to the near-IR region. Such measurements would give important information on chemical/mineralogical composition and weathering state of the Martian soil and about meteorological phenomena (clouds, dust, frost) in the form of both visual impression and spectral information. This was already stated in the 80s by different scientists (see e.g. Neukum, 1982 and references therein). CCDs are an integral part also of the modern telescope and a basic tool of modern astronomy. A review of the basic semiconductor physics, methods of construction and astronomical uses can be found in Mackay (1986) and McLean (1997). Filippenko (1992) outlines many of the advantages of CCDs, which have been summarized by Bode (1995). The most significant of these are: (i) they have relatively high detective Quantum Efficiency (QE); the number of electrons emitted per photon incident upon a photodetector, with respect to a photomultiplier tubes, enable fainter objects to be observed; (ii) the effective entrance aperture is defined by the image itself and can therefore be matched more flexibly to seeing conditions, thus decreasing the sky contribution optimally in a given exposure and (iii) CCD data are extremely well adapted to computer manipulation due to their digitally quantized and discrete pixelated nature, which provides information in a natural form for computers.

The heart of the modern cameras are the detectors. The first modern cameras were based on CCDs, which are actually optical transducers. In fact, optical radiation detectors are transducers: they transform optical radiant energy at high frequencies via a square-law process into some other form of energy that is more readily measurable. This normally takes the form of electrical signals, since the technology of electrical measurement is well developed. We have already stated that for planetary science it is extremely important to collect images, and modern spacecraft use CCDs. CCDs are also the best-known solid state detectors.

In order to understand how they work, it is better to recall the physical principles on the basis of which they have been developed. Despite their working range, solid state detectors, such as CCD or IR focal plane array (IRFPA) (described further on in this chapter), are all based on the same principle: the electrical properties of semiconductors are strongly affected by the absorption of light, or more specifically of photons. Therefore, by carefully selecting different kinds of materials one can "build up" the appropriate sensor. In this respect, semiconductors are completely different from metals that have high electrical conductivity, being only marginally affected by the interaction with photons. In addition, by adding a small quantity of impurities, semiconductors can strongly modify their electrical properties. For this reason over the past 10 years semiconductors have become the real basis of any electronic sensor.

Elemental semiconductors are silicon and germanium. Introducing impurities into the semiconductor materials (doping process) can control their conductivity. Doping processes produce two groups of semiconductors: the negative charge conductor (n-type) and the positive charge conductor (p-type). The combination of these two semiconductor types generates devices with special electrical properties which allows electrical signals (diode, transistor, silicon-controlled rectifier (SCR), etc.) to be controlled. Different elements are used to obtain a negative or positive charge:

• *n-type*: In n-type doping, phosphorus or arsenic is added to the silicon in small quantities. Phosphorus and arsenic each have five outer electrons, so they are out of place when they get into the silicon lattice. The fifth electron has nothing to bond to, so it is free to move around. It takes only a very small quantity of the impurity to create enough free electrons to allow an electric current to flow through the silicon. An n-type silicon is a good conductor (electrons have a negative charge, hence the name n-type).

• *p-type*: In p-type doping, boron or gallium is the dopant. Boron and gallium each have only three outer electrons. When mixed into the silicon lattice, they form "holes" in the lattice where a silicon electron has nothing to bond to. The absence of an electron creates the effect of a positive charge, hence the name p-type. Holes can conduct current: a hole readily accepts an electron from a neighbour, moving the hole over a space. A p-type silicon is a good conductor.

TABLE 8.2. Relevant energy ranges for quantum detectors.

E_g (eV)	λ_c (μm)
0.04	31
0.1	12.4
0.4	3.1
1	1.24
4	0.31

- *Photoelectric or quantum detectors*: optical radiant energy is transformed directly into an electrical signal. Two primary detection mechanisms can be identified.
- *Thermal detectors*: optical radiant power is first transformed into heat and then into an electrical signal.

Although photoelectric or quantum detectors were developed more recently than thermal detectors, we will describe them first seeing as they are extremely important for planetary science.

8.2.2. *Quantum photon detectors*

Photons with sufficient energy can directly excite electrons (or holes) from a bound state (valence band) to a free state (conduction band). In useful detectors, there is an energy gap between the valence band and the conduction band. Gap energy E_g lies approximately between 0.04 and 4 eV for useful photodetectors. In Table 8.2 the energies associated with different wavelengths are shown.

In pure materials the E_g is a well-defined energy gap. Absorbed photon has sufficient energy to elevate the electron from a bound state in valence band into a free state in conduction band. Electrons are now available for emission or for conduction under an applied electric field. The efficiency of this process depends on the material, on which the structure of the gap depends, and on the thermal state. One can define the intrinsic carrier concentration as due to thermal activity:

$$n_i^2 = \text{constant} \times T^3 e^{-E_g/kT} \qquad (8.8)$$

The constant is about 2×10^{31} cm^{-6} K^{-3} for most materials. kT is in electron–volts (eV), and is 0.025 85 at 300 K. Since n_i must be low in order to have sufficient carriers available for photon excitation, low-E_g detectors designed for long-wavelength operation must be cooled. These detectors are suitable for use as a standard, depending primarily on properties of the responsive QE η. Early linear arrays started as long, narrow single detectors, then slotted to make linear arrays for punched paper and card readers. $(n+1)$ wires for n elements could read simultaneously with n readout amps. These were later replaced with planar process arrays. Two-dimensional (2D) arrays were not even attempted because of the difficulty in wiring (it takes $xy+1$ wires), so 2D arrays were developed. After planar monolithic and Metal Oxide Semiconductor (MOS) technologies evolved, better results were had with arrays, leading to monolithic 2D arrays.

A CCD is usually a large-scale integrated circuit having a 2D array of hundreds of thousands, or millions, of charge-isolated wells, each representing a pixel. Light falling on a well is absorbed by a photoconductive substrate such as silicon and releases a quantity of electrons proportional to the intensity of light. The CCD detects and stores accumulated electrical charge representing the light level on each well over time. These charges are read out for conversion to digital data. CCDs are much more sensitive to light of a wider spectrum than vidicon tubes, they are less massive, require less energy, and they interface more easily with digital circuitry. It is typical for CCDs to be able to detect single photons. A charge-transfer device has an MOS structure that is composed of many independent pixels where the charge is stored in such a way that the charge pattern corresponds to the irradiation pattern. These devices can be linear or 2D. According to the method used to detect the charge pattern, two types of charge-transfer devices can be distinguished: CCDs and Charge-Injection Devices (CIDs). In a CCD the signal charge is transferred to the edge of the array for readout.

Alternatively, multiplexing can be used. The charge packets are transferred in discrete time increments by the controlled movement of potential wells. In a linear CCD the charge is moved in a stepwise fashion from element to element and is detected at the end of the line. A 2D array CCD consists of a 2D assembly of interconnected linear CCDs. As the charge from wells located far from the output must undergo many hundreds of transfers, the Charge-Transfer Efficiency (CTE) is of concern. The full-frame array has a single photosensitive array for photon collection, charge integration and charge transport. It is read out a line at a time and incident radiation must be blocked during the readout process. A frame-transfer array is composed of two arrays in series, the image and storage arrays. The storage array is covered with an opaque mask. After the image array is irradiated, the entire exposed electronic image is rapidly shifted to the storage array for readout. While the masked storage array is read out, the image array may acquire a charge for the next image. Direct X-ray and broad wavelength-band imaging and detection can be performed by a thinned CCD irradiated from the side opposite the electrodes.

8.2.3. *Pixel binning*

The mechanism by which charge transfer from a CCD is performed lends itself to on-chip pixel binning. To quickly review charge transfer in a CCD, the following diagram (Fig. 8.1) is shown. As light (photons) falls on the surface of the CCD, charge (electrons) accumulates in each pixel. The number of electrons that can accumulate in each pixel is referred to as well depth. For the KAF-0400 and KAF-1600, this is 85,000 electrons. Some CCDs, such as the SITe 502AB have well depths exceeding 350,000 electrons. Once the exposure is complete, this charge must be transferred to a single output and digitized. This is accomplished in two steps. First, an entire row is transferred in the vertical direction to the horizontal register. Second, charge is transferred horizontally in this register to the output amplifier.

Most CCDs have the ability to clock multiple pixel charges in both the horizontal and vertical direction into a single larger charge or "super pixel". This super pixel represents the area of all the individual pixels contributing to the charge. This is referred to as binning. Binning of 1×1 means that the individual pixel is used as is, whereas a binning of 2×2 means that an area of four adjacent pixels have been combined into one larger pixel, and so on. In this instance the sensitivity to light has been increased by four times (the four pixel contributions), but the resolution of the image has been cut in half. Figure 8.2 illustrates the effect.

Binning can be a very useful tool, which can be used to effectively increase the pixel size while also increasing sensitivity. It is also a good method for focusing, because image acquisition speeds up greatly while giving greater sensitivity to lower out-of-focus light levels. A Kodak KAF-0400 having 796×512, $9\,\mu$m pixels can appear to have 398×256, $18\,\mu$m pixels with a binning of 2×2. With 3×3 binning, the sensor appears to have

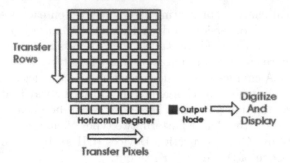

Transfer Rows

Digitize
And
Display

■ Output Node

Horizontal Register

Transfer Pixels

FIGURE 8.1. Charge transfer in CCDs.

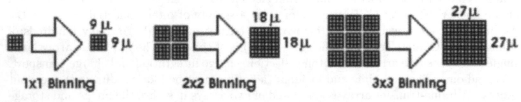

1x1 Binning   2x2 Binning   3x3 Binning

FIGURE 8.2. Pixel binning.

265×170 pixels with a size of $27\,\mu m$ and nine times the sensitivity! At 3×3 binning, a KAF-1600 will appear as 512×341 pixels each $27\,\mu m$ in size.

Depending on the focal length and optics used, the optimum pixel size for the type of observing can change. The ability to do at least 3×3 binning will provide greater flexibility. Binning also has an impact on well depth: While each pixel on a Kodak KAF-0400 has a maximum well depth of 85,000 electrons, the output node which collects binned charges has a capacity of greater than 140,000 electrons. This allows a greater charge to be collected for each super pixel than was possible with unbinned pixels.

8.2.4. *Quantum efficiency*

The QE of a sensor describes its response to different wavelengths of light. Standard front-illuminated sensors, for example, are more sensitive to green, red and IR wavelengths (in the 500–800 nm range) than they are to blue wavelengths (400–500 nm). This is most noticeable in tricolour imaging with colour filters, where exposures taken with the blue filter need to be much longer than the exposures taken with the green and red filters, in order to achieve proper colour balance. The different response to wavelengths also accounts for the fact that unfiltered CCD exposures of spiral galaxies will typically suppress star-formation regions (blue light) in the arms, and accentuate emission nebulae and dust lanes (red and IR light). Back-illuminated CCDs have exceptional QE compared to front-illuminated CCDs, due to the way they are constructed. To make a back-illuminated CCD is simple: a front-illuminated CCD is thinned to only $15\,\mu m$ thick, and it is then mounted upside down on a rigid substrate. The incoming light now has a clear way to the pixel wells without those annoying gate structures blocking the view. Note that CCDs with anti-blooming have about $1/2$ the QE of those without anti-blooming. CCDs with back illumination (such as the SITe SI003AB and SI502AB) can boost QE to over 85%. The new Blue Plus CCDs from Kodak reach a peak of about 65% and are about 30% at 400 nm.

8.2.5. *System gain*

System gain, also known as inverse system gain, is a way of expressing how many electrons of charge are represented by each count (analogue digital unit, ADU). A gain of 2.5 electrons/ADU indicates that each count or grey level represents 2.5 electrons. This implies that the total well depth (85,000 electrons) of a Kodak KAF-0400 pixel could be represented in $85,000/2.5 = 34,000$ counts.

As long as the total well depth of a sensor can be represented, a lower gain is better to minimize the noise contribution from the electronics and give better resolution. Gains which are unnecessarily high can result in more digitization noise, while gains which are too low will minimize noise at the expense of well depth. For example, a gain of 1.0 would certainly minimize the electronics contribution to noise, but would only allow $65,536/1.0 = 65,536$ electrons of the 85,000 to be digitized. System gains are designed as a balance between digitization counts, digitization noise and total well depth.

8.2.6. *Digitization*

Digitization, also referred to as analogue-to-digital conversion, is the process by which charge from the CCD is translated into a binary form used by the computer. The term binary refers to the base 2 number system used. A 12-bit camera system will output 2 raised to the 12th power or 4096 levels, a 14-bit system will output 2 raised to the 14th power or 16,384 levels and a 16-bit camera will output 2 raised to the 16th power or 65,536 levels. The objection may be raised that some monitors will only display 256 grey shades, and that 65,536 levels seems a bit excessive, but image processing software can contain a variety of image "scaling" algorithms to deal with this. They include histogram equalization, gaussian, linear, minimum–maximum, etc. These algorithms decide how to best "fit" the raw data into the available grey shades on the computer. With high dynamic range systems, such as we produce, there are many different ways to view a single image to bring out the desired detail in the area of interest.

8.2.7. *Readout noise*

Readout noise is specified both for the CCD sensor and the total system. Noise can be thought of in two ways. First, there is not perfect repeatability each time charge is dumped out of the CCD and digitized. Conversions of the same pixel with the same charge will not always yield exactly the same result from the A/D. The second aspect of noise is the injection of unwanted random signals by the sensor and electronics which end up being digitized along with the pixel charge. In addition, every analogue-to-digital conversion circuit will show a distribution or spread about an ideal conversion value. In either case the result is a certain "uncertainty" which is referred to as noise, specified in electrons (e^-).

Galileo's solid state imaging (SSI) instrument, which pioneered the technology, contains a CCD with an 800×800 pixel array. The optics for Galileo's SSI, inherited from Voyager, consist of a Cassegrain telescope with a 176.5-mm aperture and a fixed focal ratio of $f/8.5$. Since the SSI's wavelength range extends from the visible into the near-IR, the experimenters are able to map variations in the satellites' colour and reflectivity that show differences in the composition of surface materials.

Not all CCD imagers have 2D arrays, for example the Mars Orbiter Camera (MOC) on the Mars Global Surveyor (MGS) spacecraft has a detector made of a single line of CCD sensors. A 2D image is built up as the image of the Martian surface moves across the detector as the spacecraft moves in orbit, much as a page moves across the detector in a fax machine.

8.2.8. *IR detectors*

High-quality imaging at near-IR (NIR) and mid-IR wavelengths has recently become practical because of advances in IR-sensitive arrays. Usually the detection system is based on solid sate devices. When a photon strikes a semiconductor, it can promote an electron from the valence band (filled orbitals) to the conduction band (unfilled orbitals) creating an electron($-$)–hole($+$) pair. The concentration of these electron–hole pairs is dependent on the amount of light striking the semiconductor, making the semiconductor suitable as an optical detector. There are two ways to monitor the concentration of electron–hole pairs. In photodiodes, a voltage bias is present and the concentration of light-induced electron–hole pairs determines the current through the semiconductor. Photovoltaic detectors contain a p–n junction, that causes the electron–hole pairs to separate to produce a voltage that can be measured. Specifically, arrays constructed from indium and antimony (InSb) substrates have had spectacular success in achieving high Signal-to-Noise Ratio (SNR), and high dynamic range for telescopic and spacecraft imaging applications. Other IR-sensitive substrates, including silicon–arsenic (SiAs), germanium (Ge) and indium–gallium–arsenic (InGaAs), have also been used with good results. A particular advantage of many of these arrays is their ability to operate effectively with only modest cooling requirements. Ge and InGaAs arrays, for example, can operate effectively even at temperatures as high as 250–270 K. InSb arrays can produce good data when cooled to liquid nitrogen temperatures (77 K), although they perform even better at liquid helium temperatures (4 K). Military applications dominate the requirements today, especially for cooled IRFPAs. In addition to the many military applications for IR systems such as target acquisition, search and track, and missile seeker guidance, there is also great potential for IR systems on the commercial market. This forces industry to find a way to produce IRFPAs in low volume and to turn out new designs quickly.

8.2.8.1. *HgCdTe*

Hg1-xCdxTe is unique because the cut-off wavelength is tunable. By mixing a semiconductor (CdTe) with an energy gap of 1.61 eV and a semi-metal (HgTe) which has a negative energy gap of -0.3 eV (the conduction band and valence band form an indirect gap) one can tune this alloy's energy gap by the mole fraction ratio to any desirable energy gap. Hg1-xCdxTe is unique because it is intrinsic. This results in two important characteristics:

• Cooling requirements for equal cut-off wavelengths to extrinsic are less stringent (operates at a higher temperature). This is because HgCdTe has no impurity levels in the perfect lattice configuration and, therefore, the cooling is only required to keep thermal (kT) carrier generation below the energy gap. For extrinsic there are impurity bands in the forbidden gap which can capture kT generated electrons. Therefore, the temperature must be lower to avoid excitation to these Si:X or Ge:X levels.

• The absorptivity of HgCdTe is better. The number of atoms capable of capturing a photon is larger, thus a large capture cross section. For the same total absorption, the length of HgCdTe is approximately 1/50th of excited silicon.

8.2.8.2. *Thermal IR detectors*

Thermal detectors work by means of a two-step process: first the detector absorbs radiation, producing temperature rise T; afterwards the temperature is measured using one of several means, such as change of resistance, thermoelectric voltage/current, change of detector charge/capacitance or measure temperature rise directly (calorimetry). Thermoelectric effect first used for optical radiation measurements (solar radiometry) by Nobili and Melloni in 1835, the first quantitative detector. Later versions used for early

TABLE 8.3. Thermal and photon sensors comparison.

	Thermal	Photon (Quantum)
Point	Bolometer	Photodiode
	Thermocouple	Photoconductor
	Pyroelectric	Photomultiplier
Area	Pyroelectric array	Eye
	Evaporagraph	CCD array
	Microbolometer array	Film

IR spectroscopy. Signals can be increased by placing several junction pairs in series, thermally connecting alternate junctions to a blackened-receiving element exposed to incoming optical radiation and thermally connecting other junctions to heat sink. Devices using multiple thermocouple junctions are called *thermopiles*; most practical thermoelectric sensors have multiple junctions. Early thermopiles were fabricated by soldering or welding fine wires (usually Bi–Sb). Recent thermopiles are made by vacuum evaporating alternating layers of Bi and Sb onto a substrate.

8.2.8.3. *Bolometers*

In 1878, Langley invented the bolometer, a radiant-heat detector that is sensitive to differences in temperature of one hundred-thousandth of a degree Celsius (0.00001°C). Composed of two thin strips of blackened ($e = 0.97$) platinum, a Wheatstone bridge, a battery and a galvanometer. This instrument enabled Langley to study solar irradiance in the far IR region and to measure the intensity of solar radiation at various wavelengths. Bolometers are used individually for variety of applications, but recently they were largely replaced by pyroelectric detectors. In the recent years, planetary scientists started using microbolometer arrays fabricated from silicon, that are revolutionizing low-cost thermal imaging. The operational characteristics which are making microbolometers interesting are the uniform wavelength response (which depends on crystal absorption or external blackening), high sensitivity without cooling, small dimensions and the possibility to be used without an external bias source. They behave like a capacitor with a variable dielectric and can be used in either current- or voltage-mode. Their negative characteristics are that they are slow when operated at maximum responsivity. They can be used for high-speed applications only if sufficient input power is available. In Table 8.3 the comparison between thermal and photon detectors is shown.

Finally, in Table 8.4, the main characteristics of different sensors are compared.

8.3. Planetary remote sensing techniques

In the previous sections, we have seen that several kind of sensors can be used to explore planetary surfaces at several wavelengths, obtaining different, and usually complementary, information. If we would like to obtain images of the surface first, then we have to concentrate our efforts on obtaining high-quality images. However, images are merely the first step in the interpretation of the nature of a planet, and usually these images only give information about the structure of the surface or about the atmosphere surrounding it. We usually need more than a simple description: in some cases, for

TABLE 8.4. Sensor comparison.

Description	Collector to base voltage (V)	Maximum collector current (mA)	Maximum collector dark current at 25°C (mA)	Minimum light current (mA)	Maximum power dissipation at 25°C (mW)	Typical response time (μs)	Package
	$V_{(BR)CBO}$	I_c	I_D	L_L	P_D		
NPN, Si, Vis–IR	30 (VCEO)	40	100@10V VCE	1.0	150	6	T046 Flat Lens
NPN, Si, Vis–IR	80	40	20@5V VCE	12	200	2	T046 Dome Lens
NPN Photo-light det.	30 (VCEO)	–	100@10V VCE	100 μA@5V	150	–	–
NPN, Si, Photo Darlington	60 VCE	–	100@10V VCE	5@5V	150	–	–
NPN, Si, Darlington Vis–NIR	50	250	100	12	250	151	T018 Dome Lens
NPN, Si, Vis–IR	50	50	200@30V VCE	10@3V	150	2	T018 Dome Lens
NPN, Si, Vis	25	20	500@10V VCE	5@3V	50	1.5	–
NPN, Si, Phototransistor	20 (VCEO)	30	1 μA@10V VCE	1 Typ	100	10 max	–
NPN, Si, Darlington Vis–IR	20 (VCEO)	30	500@10V VCE	0.2	100	100	–
NPN, Si, Darlington, IR, Vis cut-off, Narr.Accept.	35 (VCEO)	50	10 μA@10V VCE	1.5	75	80	–
NPN, Si, Darlington, IR, Vis cut-off	35 (VCEO)	50	10 μA@10V VCE	0.2	75	400	–

example, we may need to associate the knowledge of the composition of the different geological units present on the planet surface with the investigation of that surface, which is the basis of planetary geology. In other words, we need to associate mineralogical and elemental composition to the geomorphology of a planetary surface. When possible, we would also like to find out about the nature of the planetary subsurface, as well as of the planet's interior. Geomorphological information can result from broadband visible orbital imaging in the case of planets with thin to non-existent atmospheres. For planets with atmospheres, the opacity of the atmosphere as a function of wavelength must be taken into account; in certain cases, such as for Venus, the only choice for orbital geomorphological studies is microwave (radar) imaging. Surface mineralogy can be investigated using spectroscopy from the visible to IR, since many minerals exhibit diagnostic absorption and emission features at these wavelengths. If we would like to associate this with mineralogy and geochemistry, we use different techniques which can explore the nuclei directly, such as X-ray and gamma-ray spectroscopy.

8.3.1. *Spectroscopy*

Spectroscopy is the study of electromagnetic (EM) spectra – the wavelength composition of light – due to atomic and molecular interactions. For many years, spectroscopy has been an important tool in the study of physics, and it is now equally important in astronomical, biological, chemical and other analytical investigations. Much of physics is concerned with spectroscopy: the probing of material with EM waves of specific frequency and wavelength, and the study of the response of the material to such probes. By understanding the nature of the scattering of such waves, we can deduce very precise and detailed properties of the material under study. Therefore, we can broaden the definition of spectroscopy to include electron and other matter waves (such as gamma-rays) as applied to nuclear and particle physics. Almost all astronomy relies on spectroscopy for species identification, as well as for the measurement of the distances of the galaxies through the red-shift. In planetary sciences, essentially three types of spectroscopic observations are used: reflectance spectra, thermal emission spectra and X-ray and gamma-ray spectra. Each offers specific advantages for answering certain types of compositional or mineralogical questions. Planetary atmospheres as well as surfaces can be studied through spectroscopy. When photons hit a surface, they are absorbed in minerals by several processes, thus the variety of absorption processes and their wavelength dependence allows us to derive information about the chemistry of a mineral from its reflected or emitted light (Sunshine and Pieters, 1998).

Spectrometers are instruments which are able to diffract light. Once we have selected the heart of the experiment, namely the detector, as was extensively explained in Section 8.2, we have to concentrate on the dispersing element. Obviously this dispersing element should be optimized to disperse the light efficiently in the selected spectral range. *Prisms and gratings* are the diffracting elements most used. In optics, a prism is a device used to reflect light or to break it up (to disperse it) into its constituent spectral colours, traditionally built in the shape of a right prism with triangular base. The main effect of a prism is to *deviate* a beam of light. Due to the dispersion in the refractivity of transparent materials, the deviation is slightly different for light of different colours. This slight difference in the deviation, only 1% or 2%, is the dispersive power of a prism. The prisms used for dispersing light sometimes have only two adjacent faces polished; the third side and the bases are often left in a rough-ground state, and may be painted black to absorb unwanted reflections. The angle between the two polished faces is called the *refracting* angle. If the prism is to be used for its dispersion, the refracting angle is usually about 60°. Refracting prisms are sometimes used to produce a very small angular deviation. These have a small refracting angle, and are often called *wedges* (although they do not have a sharp edge, which would be fragile and easily broken). Sometimes a very thick wedge is used to produce a small angular deviation by refraction, and the dispersion is merely a nuisance. For small angles, the deviation of a glass prism is about 1/3 of the prism angle, so a wedge angle of only a degree and a half would suffice to mimic the refraction of the standard atmosphere. Prisms can also be used to break up light into its constituent spectral colours because the refractive index depends on frequency; the white light entering the prism is a mixture of different frequencies, each of which gets bent slightly differently. Blue light is slowed down more than red light and will therefore be bent more than red light. Prism can be made off different materials, such as optical glass, quartz, glass ceramics, borosilicate glass, CaF_2, MgF_2, germanium or filter glass, depending on the wavelength range of the light to be diffracted. When selecting a prism is important not only is the material used important, but also the accuracy in the dispersion. For space application, small size is also a bonus. Table 8.5 shows the characteristics of a prism produced by a private company, Berliner Glas.

A diffraction grating is a collection of reflecting (or transmitting) elements separated by a distance comparable to the wavelength of light under study. It may be thought of as a collection of diffracting elements, such as a pattern of transparent slits (or apertures) in an opaque screen, or a collection of reflecting grooves on a substrate. The multiple grooves of the grating can be considered as individual slit-shaped sources, and each of

TABLE 8.5. Prism characteristics (Berliner Glas) for optical applications.

Types of prisms	Size (mm)	Accuracy (angular s)
Reflecting prisms		
Pentaprism	10–60	Up to 1
90° prism	5–100	Up to 3
Triple prism	10–70	Up to 1
Dove prism	10–80	Up to 3
45–60° prism	10–70	Up to 3
Roof prism	10–60	Up to 1
Rhomboid prism	10–70	Up to 3
Dispersing prisms		
Equilateral dispersing prism	10–70	Up to 3
Beam splitting prism		
Beam splitting cube	5–100	Up to 5

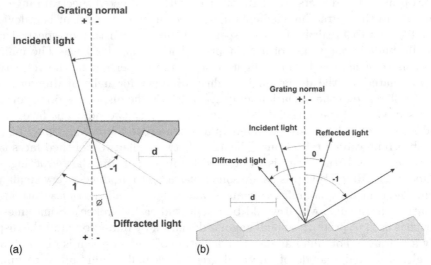

(a)   (b)

FIGURE 8.3. The typical path of a plane diffraction grating. (a) A transmission grating: the incident and diffracted rays lies on opposite sides of the grating. (b) A reflection grating: the incident and diffracted rays lie on the same side of the grating.

them diffracts the incident light, which combine to form a diffracted wavefront. The total reflected light consists of a coherent superposition of each individual contribution. Diffraction gratings function by the principle of interference. Constructive interference occurs only in a unique set of directions (discrete angles), where all of the "partial" waves from the grooves are in phase for a given groove spacing. In all other directions the partial wave contributions are out of phase and cancel by destructive interference. A *reflection grating* consists of a grating superimposed on a reflective surface, whereas a *transmission grating* consists of a grating superimposed on a transparent surface. An EM wave incident on a grating will, upon diffraction, have its electric field amplitude, or phase, or both, modified in a predictable manner (see Figure 8.3).

The diffraction grating is of considerable importance in spectroscopy, due to its ability to separate (disperse) polychromatic light into its constituent monochromatic components. In recent years, the spectroscopic quality of diffraction gratings has greatly improved. The grating equation is:

$$m\lambda = d(\sin\beta + \sin\alpha) \tag{8.9}$$

Here m is the (integral) diffraction order (usually $|m| = 2$), λ the wavelength, d the groove spacing, α and β are the angle of diffraction (measured from the normal). The efficiency behaviour of transmission gratings is simpler than that for reflection gratings, since no metals are present to introduce complicated EM effects and since the angles of diffraction are usually small. Thus, polarization effects are virtually absent. The so-called "blazed gratings" are often used in the applications. The peak efficiency of a blazed (triangular-groove) transmission grating occurs when the refraction of the incident beam though the mini-prism that constitutes a groove lies in the same direction as the diffraction given by the grating equation. Unlike reflection gratings, the groove angle is much larger than the blaze angle for a transmission grating, since the phase retardation doubles upon reflection but is multiplied by $n-1$ for a transmission grating, where n is the refractive index of the grating medium.

Generally, most of the incident light will be diffracted either into the zero order or positive first order and very little light will go into higher orders.

The quality of the science that we are able to perform depends not only on the knowledge of the process that we achieved, but also on our technical ability to build up adequate experiments. The extremely high precision required of a modern diffraction grating put high demands on the mechanical dimensions of diamond tools, ruling engines and optical recording hardware, as well as their environmental conditions, which should be controlled to the very limit of what is physically possible. If these standards in the technical accuracy fail to be reached, the produced gratings have little technical or scientific value. A personal experience of this author demonstrates this important concept: the heart of the author's experiment Visible and IR Thermal Imaging Spectrometer (VIRTIS), which was launched on the Rosetta mission and which will be present on Venus Express and the "Discovery" mission *Dawn*, was manufactured in such a way that it was possible to diffract the light from the visual part of the spectrum to the IR, which has resulted in a very high-performing experiment, with savings in both mass and power.

Having covered the detector and the dispersive element, we will now go on to discuss the manufacturing of a spectrometer. The discussion of different kinds of spectrometers is beyond the purpose of this chapter; however, in the next section, we will discuss some of the imaging spectrometers that are among the most powerful instruments developed for planet exploration.

Spectrometers in the visual and IR are able to detect the light reflected from planetary surfaces or planetary atmospheres or rings. For this reason, the reflectance spectroscopy principles will briefly be described here. Figure 8.4 shows the classical layout of a prism spectrometer.

Suppose that we have a light L illuminating the entrance slit S_1 located in the focal plane of collimator lens L_1. The parallel beam from L_1 passes through the prism P where it is deviated (refracted) by an angle $\theta(\lambda)$ depending on the wavelength λ. The focusing lens (camera lens) L_2 forms an image $S_2(\lambda)$ of the entrance slit S_1. The position $l(\lambda)$ of this image along the imaging plane B is therefore a function of the wavelength, and multiple images $S_2(\lambda_1, \lambda_2, \ldots, \lambda_n)$ of slit S_1 are consequently formed. The linear dispersion of the spectrograph, $dl/d\lambda$, depends on the spectral dispersion $dn/d\lambda$ of the prism material and the focal length of lens L_2.

THE CLASSICAL PRISM SPECTROGRAPH

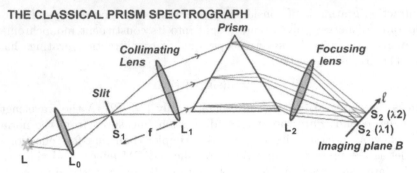

FIGURE 8.4. The typical optical layout for the "classical" prism spectroscope (after Dalton, 2004).

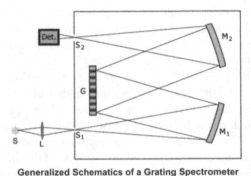

Generalized Schematics of a Grating Spectrometer

FIGURE 8.5. Layout of a very simple grating spectrometer (after Dalton, 2004).

When a reflecting diffraction grating is used to separate spectral lines $S_2(\lambda)$ the two lenses L_1 and L_2 are usually replaced by two mirrors M_1 and M_2 to image the entrance slit S_1 onto the imaging plane B defined above. A general spectrometer schematic using a grating is shown in Fig. 8.5. Obviously this a very simplified situation: for remote sensing observations, real spectrometers have to be combined with telescopes that are able to collect from the source the necessary amount of light to be diffracted.

The grating and prism are sometimes combined, to create a new optical element, called a grism. The traditional grism is constructed from a prism that has had a diffraction grating applied on one surface. The objective of such a design is to use the prism wedge angle to select the desired "in-line" or "zero-deviation" wavelength that passes through on an axis. The grating on the surface of the prism provides much of the dispersion for the spectrometer. A grism can also be used in a "constant-dispersion" design which provides an almost linear spatial scale across the spectrum. The basic optic concepts that are needed to build instruments of this kind, and the basic technicalities that are necessary to build a real spectrometer, can be found in several books, here we suggest the fundamental work by Lena *et al.* (1998), as well as that of Kitchin (1998). The schematic figures used are from the web site of Steve Dearden (March 2004).

Following Clark (1999), there are four general parameters that describe the capability of a spectrometer: (1) spectral range, (2) spectral bandwidth, (3) spectral resolution and (4) SNR. Spectral range is important to cover enough diagnostic spectral absorptions to recognize the mineral. Spectral bandwidth is the *width* of an individual spectral channel in the spectrometer. Spectral resolution is the ability to distinguish two close features: it

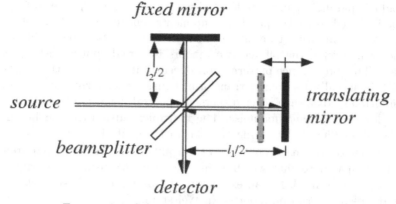

FIGURE 8.6. Schematics of a Fourier spectrometer.

depends on the spectral bandwidth and on the optical characteristics of the spectrometer. The higher the spectral resolution, the narrower the absorption features the spectrometer can measure and separate. Bandwidths and resolutions greater than 25 nm are needed to resolve important mineral absorption features (*ibid.*). SNR are kept high in order to distinguish small spectral signatures from the noisy background. In appropriate observing conditions, a good spectrometer can easily reach SNR of about 100. Obviously, there are several ways to improve *a posteriori* the SNR, for example by adding up spectra from the same region; however, a good SNR should be defined *a priori* within the instrumental requirements.

Depending on the physical problem to be investigated, it is possible that a higher spectral resolution, SNR, or spatial resolution may be required. For the study of a planetary surfaces, a high spatial resolution, in the order of milliradiants, is necessary in order to detect spectral features of localized areas, but usually moderate spectral resolution not greater than 200 is used.

If a higher spectral resolution is to be reached, as is the case when there is a planetary atmosphere, spectrometers based on interference principles should be used. These spectrometers can have very high spectral resolution, but cannot be used to make imaging. We will now describe a few examples of them. A Michelson interferometer works as follows: it consists of two mirrors located at right angles to each other and oriented perpendicularly, with a beamsplitter placed at the vertex of the right angle and oriented at a 45° angle relative to the two mirrors. Radiation incident on the beamsplitter from one of the two "ports" is then divided into two parts, each of which propagates down one of the two arms and is reflected off one of the mirrors. The two beams are then recombined and transmitted out the other port (Fig. 8.6). When the position of one mirror is continuously varied along the axis of the corresponding arm, an interference pattern is swept out as the two phase-shifted beams interfere with each other.

For example, an interferometer constructed using a half-silvered mirror inclined at a 45° angle to the incoming beam. Half the light is reflected perpendicularly and bounces off a beamsplitter, and half passes through and is reflected from a second beamsplitter. The light passing through the mirror must also pass through an inclined compensator plate to compensate for the fact that the other ray passes through the mirror glass three times instead of once. The path length difference for emerging parallel rays from a point source is:

$$\Delta r \simeq 2L \cos \vartheta \qquad (8.10)$$

As this light is parallel, it must be focused with a lens. A net phase shift of π radians must also be included, since the parallel components reflects off the front of the first and second mirrors. By smoothly translating one mirror, the optical path difference $x = 2L$ between the beams reflecting off the two mirrors is varied continuously, producing an interferogram. The interference pattern is produced that "contains" the spectrum of the source, which is actually its Fourier transform. Fourier transform spectrometers have a so-called "multiplex advantage" over dispersive spectral detection techniques for signal, but a multiplex *disadvantage* for noise. The multiplex advantage can be summarized, following Weisstein (2004), by saying that an instrument simultaneously measuring a signal over a range of frequencies obtains a $t^{\frac{1}{2}}$ advantage in the time t required to obtain a given SNR compared to that which would be necessary using dispersive methods if the noise is detector-limited. For an exhaustive discussion on this subject we suggested reading the detailed description reported in Weisstein.

8.3.2. *Reflectance spectroscopy*

Once we have built a spectrometer, we have to use it correctly. By this is meant that we must perform an *extensive campaign of calibration,* that will involve the use of different reference sources and, where possible, of known reference natural samples. To do that we have to achieve a clear knowledge of the physical processes of the phenomena affecting natural samples. Only in this way we will be able to interpret correctly remotely sensed data, discriminating the instrumental effects from the real features that are present in the samples. For this reason, we will now briefly discuss the main physical effects that are related to different mineralogical assemblies, and which consequently are diagnostic of it.

When reading a spectral curve for its information content, the most useful feature is the wavelength location and depth of the diagnostic absorption bands. Several atomic mechanisms, in fact, are responsible for the absorption of radiant energy that is then expressed by these bands. These include crystal-field disturbance, charge-transfer absorption, conduction band shifts, colour centre absorption and vibrational processes. For this reason reflectance spectroscopy provides diagnostic information on the mineralogy and degree of crystallinity of the uppermost microns of a planetary surface. In fact, the active layers, namely the layers that are involved in the absorption/reflection process, are usually in the order of a few times the considered wavelength. This technique involves measuring the spectrum of sunlight reflected from a planetary surface, and is thus restricted to the wavelength range where the Sun's flux is highest and where the amount of energy reflected from the object is greater than the amount that is thermally emitted (the typical wavelength range is from 0.3 to 3.0 μm at the distance of the Earth, 0.3 and 5.0 μm at the distance of Saturn). Many materials, including primary and secondary minerals, exhibit electronic spectral features and vibrational overtone bands at these solar wavelengths.

8.3.2.1. *Reflectance spectra*

Reveal absorption features that are characteristic of certain minerals and ices and/or indicate the presence of certain cations. For example, the mineral pyroxene, a common component of basaltic rocks on the Earth, can be detected remotely by the measurement of diagnostic absorption features near 1.0 and 2.0 μm. The typical spectra of pyroxene are shown in Fig. 8.7. The presence of Fe^{2+} causes the two absorption bands, near 1 and 2 μm, to deepen and shift notably towards lower wavelengths.

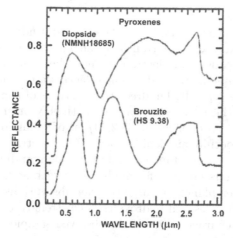

FIGURE 8.7. Spectra of different pyroxene. The influence of iron is evident in this spectral plot, through parts of the visible–NIR and short-wave–IR ranges of two pyroxenes. Diopside ($CaMgSi_2O_6$) contains almost no iron. Bronzite ($[Mg, Fe]SiO_3$) has Fe but no Ca. The presence of Fe^{2+} causes two absorption bands, near 1 and $2\,\mu m$, to deepen and shift notably towards lower wavelengths.

Pyroxene spectra have long been observed on the Martian surface (Adams and Mc-Cord, 1969). More recently, their presence was confirmed at high spectral resolution by the infrared imaging spectrometer (ISM) on board the Phobos-2 spacecraft in 1989. Mustard et al. (1997) has de-convolved ISM spectra to identify the components of the spectra which comprise the pyroxene signal and determine the 1 and $2\,\mu m$ band centre locations. Steutel et al. (2003) have developed a quantitative model for determining pyroxene Ca/Fe/Mg composition from the centre position of the 1 and $2\,\mu m$ features in laboratory pyroxene spectra. They have applied our model to the identified band centres of the ISM pyroxene spectra to more precisely estimate the compositions of pyroxenes on the Martian surface. This shows that when high-quality spectra are available, the quantity of information hidden within them can still be understood even after several years have passed, and not necessarily by the same group of researchers who developed the instruments and made the first interpretations.

8.3.2.2. *Electronic processes*

Involve absorption of photons with specific energies (hence, wavelengths) that cause an electron jump from a lower energy state to an electron shell at a higher energy state. If the electron returns to a lower state (lower energy shell), it may emit a photon. For shared electrons between atoms, the energy jump spreads over a range of values, producing "energy bands". A common electronic movement results from crystal field disturbance. A crystal field describes the effects of perturbing the "d" orbital shells of a transition metal (Fe, Cr, Ni, Co, Ti, V, etc.), distributed within the atomic lattice of a crystal. This metal cation interacts with the electric field imposed by surrounding anions (negatively charged) or dipolar groups (ligands). The charge symmetry is thus distorted. Variations in the abundances of Fe and Ca in the pyroxene can also be inferred based on subtle shifts in the positions of these bands. Reflectance spectroscopy is currently the most useful technique for remotely measuring the mineralogy of planetary surfaces, and specific minerals have been identified on the Moon, Mars and a number of asteroids. Additional detailed information on reflectance spectroscopy theory and techniques can be found in Burns (1993) and Gaffey et al. (1993).

8.3.2.3. *Charge transfers*

The mineralogy of a natural sample can be inferred studying the absorption bands caused by charge transfer. Charge transfers can cause absorption bands when the absorption of a photon causes an electron to move between ions. The transition can also occur between the same metal in different valence states, such as between Fe^{2+} and Fe^{3+}. Their strengths are typically hundreds to thousands of times stronger than crystal field transitions. Usually these transitions appear in the UV or in the visible. Morris *et al.* (1985) studied the spectral properties (0.35–2.20 μm) of submicron powders of hematite, maghemite, magnetite, goethite and lepidocrocite. The spectral properties of the iron oxides and oxyhydroxides are important not only for understanding the basic physics and chemistry of the compounds but also for applications such as the remote sensing of the Earth and Mars and other planets. Charge-transfer absorptions are also the main cause of the red colour of iron oxides and hydroxides. The strength of the different signatures depends on the physical characteristics of the observed samples: powdered materials, as an example, are characterized by a decrease in the absorption band intensity. This could be due to the increased surface to volume ratio at small grain size results in a greater proportion of grain boundaries where crystal field effects are different (Morris *et al.*, 1985; Clark, 1999).

8.3.2.4. *Vibrational processes*

The bonds in a molecule or crystal lattice are able to vibrate around an equilibrium position. The *frequency* of vibration depends on the strength of the bond (atomic or lattice) and their masses. The fundamental modes of vibration are $3N - 6$, for a molecule made by N atoms. We can observe also additional vibrations are called overtones when they involve multiples of a single fundamental mode or their combinations. A vibrational absorption will be seen in the IR spectrum only if the molecule responsible shows a dipole moment. Vibrations from two or more modes can occur at the same frequency, and because they cannot be distinguished, are said to be degenerate. An isolated molecule with degenerate modes may show the modes at slightly different frequencies in a crystal because of the non-symmetric influences of the crystal field. One can measure the absorption spectra, and determine what must be the molecular structure. Details of such calculations are in various texts on spectroscopy (e.g. Hollas, 2004). Background material, particularly with reference to minerals, can also be found in a series of excellent papers by Hunt and Salisbury (1970).

8.3.2.5. *Water and hydroxyl*

Water and OH (hydroxyl) produce particularly diagnostic absorptions in minerals. The water molecule (H_2O) has $N = 3$, so there are $3N - 6 = 3$ fundamental vibrations. In the isolated molecule (vapour phase) they occur at 2.738 μm (v_1, symmetric OH stretch), 6.270 μm (v_2, H–O–H bend) and 2.663 μm (v_3, asymmetric OH stretch). In liquid water, the frequencies shift due to hydrogen bonding: $v_1 = 3.106$ μm, $v_2 = 6.079$ μm and $v_3 = 2.903$ μm.

The overtones of water are seen in reflectance spectra of H_2O-bearing minerals. The detection of liquid water on other planets was not yet done ... however ice was detected on several planets from Mars to the outer regions of the solar system, including Uranus and Neptune. Water vapour was also seen in Mars atmosphere.

8.3.2.6. *Carbonates*

Carbonates also show diagnostic vibrational absorption bands. The absorptions are due to the planar CO_3^{-2} ion. There are four vibrational modes in the free CO_3^{-2} ion: the symmetric stretch, v_1: $1063 \, \text{cm}^{-1}$ ($9.407 \, \mu\text{m}$); the out-of-plane bend, v_2: $879 \, \text{cm}^{-1}$ ($11.4 \, \mu\text{m}$); the asymmetric stretch, v_3: $1415 \, \text{cm}^{-1}$ ($7.067 \, \mu\text{m}$) and the in-plane bend, v_4: $680 \, \text{cm}^{-1}$ ($14.7 \, \mu\text{m}$). The v_1 band is not IR active in minerals. There are actually six modes in the CO_3^{-2} ion, but two are degenerated with the v_3 and v_4 modes. In carbonate minerals, the v_3 and v_4 bands often appear as a doublet. The doubling has been explained in terms of the lifting of the degeneracy (e.g. see White, 1974) due to mineral structure and anion site.

To have a detailed analysis of the diagnostic signature of different rock-forming material, it is suggested to read carefully the text of Clark (1999) that represents a fundamental text to understand planetary spectroscopy.

8.3.3. *Thermal emission spectroscopy*

Thermal emission spectroscopy also provides diagnostic information on the mineralogy of planetary surfaces, as well as additional information on surface thermophysical properties like temperature and thermal inertia. Most of the major rock-forming minerals exhibit their fundamental molecular vibration spectral features at mid-IR wavelengths, typically from 3.0 to $25.0 \, \mu\text{m}$, where thermal radiation emitted from planetary surfaces at temperatures from 200 to 400 K dominates over reflected sunlight.

Unlike reflectance spectra, thermal IR spectra can exhibit features in both emission and absorption, depending on the nature of the planetary environment. Thermal IR spectra are more difficult to interpret, but they also allow the potential, through radiative-transfer modelling, to infer additional information about a planetary surface, such as emissivity, particle size and degree of compaction, and the subsurface temperature profile (e.g. Hapke, 2001). In this book are collected fundaments of reflectance and emittance as quantitative tools to measure the properties of surfaces and materials. It is intended primarily for use in the interpretation of remote observations of the surfaces of the Earth and other planets, and it will also be useful to chemists, physicists, geologists, engineers and others who deal with particulate media. In this book are also discussed the propagation and absorption of light in continuous media, reflection by smooth surfaces, scattering by spheres and irregular particles, reflectances and emissivities of particulate media, reflectance and emittance spectroscopy and the polarization of light scattered by particulate media. Knowledge of all these subject are essential in understanding and interpreting remote sensing data.

Thermal emission spectra have been used, for example, to obtain remote compositional information on variations in terrestrial basaltic lava flows and to constrain the thermal inertia and rock abundance of the Martian surface (e.g. Christensen, 1986; Kahle, 1994). Christensen established a new technique by mapping through the Viking IR Thematic Mapper (TM) the spatial distribution of rocks exposed on the Planet's surface concluding that only a 6% areal coverage rock abundance. He concluded also that most of the surface was covered by dust, using in order to do that, the thermal inertia maps. More details on the theory and application of thermal emission spectroscopy can be found in, for example, Salisbury *et al.* (1991), Hanel *et al.* (1992) and Salisbury *et al.* (1992).

The spectroscopic techniques discussed above have typically been constrained by instrumentation to obtain compositional or mineralogical information only for specific, possibly small places on planetary surfaces. The imaging techniques discussed in the previous chapters have traditionally been used primarily for obtaining morphological information, at the expense of more detailed compositional or mineralogical data. On the other hand, elegant spectroscopic techniques were developed in order to obtain characteristics of planetary atmospheres and identifying the nature of their constituents. However, we have to stress out that, when looking a planet from orbit, if an atmosphere is present, it is very difficult to discriminate between the surface and atmospheric signal. To do that, different techniques are used: as an example a recent and important development in remote sensing is the combination of imaging and spectroscopic techniques to allow for the determination of compositional information and the mapping of this information across a planetary surface at high resolution, that will be discussed in the next section. Here we will describe instead a spectrometer that does not have imaging capability, as IR interfero spectrometer (IRIS) that flew on Mariner-9, Planetary Fourier Spectrometer (PFS), that is observing the surface of Mars on Mars Express Mission, and a spectrometer that has limited spectral resolution, but very good image quality such as thermal emission spectrometer (TES), since they were both essential for Martian exploration.

8.3.4. *Infrared spectrometers*

Four spacecraft spectrometers have returned spectra of Mars of the thermal emission wavelength regions: the 1969 Mariner-7 IR spectrometer (IRS), (1.9–14.4 μm); the 1971 Mariner-9 IRIS (5–50 μm); the Global Surveyor TES (6–50 μm) and, more recently on Mars Express PFS (wavelength range from 1.2 to 45 μm) (Kirkland *et al.* 1999). The first three were mounted on NASA spacecraft, while the fourth one is still collecting data on Mars Express Mission, that is an European Space Agency (ESA) mission.

IRS returned high-quality spectra of Mars covering the wavelength region from 1.8 to 14 μm. IRS measured with the highest SNR of the three NASA spectrometers IRS spectra (1.8–14.4 μm) provide unique information of the 3–7 μm region. This region contains overtone features, which is important because finely particulate minerals commonly have strong diagnostic overtone features, but very weak fundamental features (which occur at longer wavelengths). At the time, the importance of IRS data was probably not totally understood but, recently, the spectra acquired by the IRS in 1969, have been recovered and re-calibrated, and several new discoveries were obtained (Kirkland *et al.*, 1999, 2000). Absorptions detected at 2.4, 3, 11.25 and 12.5 μm provide strong spectral evidence for the presence of a hydrous weathering product on the Martian surface, interpreted to be goethite. The 11.25 and 12.5 μm bands do not correlate with atmospheric path length, neither do they correlate with 12.6 μm atmospheric CO_2 band depth, or the 9 μm atmospheric dust band depth. This is a clear example of how good data remain a rich patrimony of the scientific community, for a much longer time that expected by the same people who developed the original experiment.

The IRS experiment on the Mariner-9 mission, named IRIS, was designed to provide information on atmospheric and surface properties by recording a major portion of the thermal emission spectrum of Mars (Hanel *et al.*, 1972). IRIS is a Michelson infrared interferometer spectrometer (IRIS-M) that records the spectral interval from 200 cm^{-1} (50 μm) to about 2000 cm^{-1} (5 μm) with a nominal spectral resolution of 2.4 cm^{-1}. The original intent was to determine vertical temperature profiles, the surface temperature and the atmospheric pressure at the surface, and to acquire information related to the surface composition. The experiment also was designed to search for minor atmospheric constituents, including H_2O vapour and isotopic components of CO_2. The most striking result of the experiment was the strong effect of the atmospheric dust on the emission spectra. The authors (*ibid.*) stressed out that the entire spectrum, with the exception of the strongly absorbing part of the 667-cm^{-1} (15-μm) CO_2 molecular band, apparently is influenced to varying degrees by the opacity of the dust in the atmosphere. Recently also IRIS data were re-calibrated (Zasova *et al.*, 1999). An instrumental effect that was unidentified before was discovered. This effect, although small enough so that does not affect essentially previous results, made more difficult the data interpretation. The newly calibrated IRIS data allowed to identify albite in the dust present in the Martian atmosphere (Maturilli *et al.*, 2002).

More recently two new experiment were developed TES on MGS and PFS on Mars Express. TES has three subsections that measure incoming IR and visible energy: a Michelson interferometer, a broadband radiance sensor and a solar reflectance sensor. The interferometer is at the heart of the thermal IR spectrometer portion of TES. It covers the wavelength range from 6 to $50\,\mu m$ (~ 1650–$200\,cm^{-1}$) with 5 and $10\,cm^{-1}$ spectral resolution. The broadband sensor measures radiance in a single band from 5.5 to $100\,\mu m$. To measure the brightness of reflected solar energy, a third sensor views the Planet in the 0.3–2.7-μm region. Each sensor has six detectors arranged in a 3 by 2 array, each with a Field Of View (FOV) of 8.3 mrad. This translates to a spatial resolution of 3 km from the MGS orbit. The TES instrument is quite small ($24 \times 35 \times 40\,cm$) and weighs 32 lb (14.4 kg). TES is powered by the spacecraft's solar cells using only 14.5 W.

TES was particularly successful in different fields, such as surface mineralogy, polar processes, atmospheric processes and thermophysical properties of the surface. Following Bandfield *et al.* (2000), three main results shall be stressed out: two are positive results, in terms of identification of large-scale characteristics of Martian surface, and one is negative. They are: first, the large-scale identification of the mineralogy of large volcanic regions on Mars, whose mineralogy varies from basaltic, composed of plagioclase, feldspar, clinopyroxene, olivine, plus or minus sheet silicates, to andesitic, dominated by plagioclase feldspar and high-silica volcanic glass (Fig. 8.8) (Bandfield *et al.*, 2000). The basalts occur primarily in the ancient, southern hemisphere highlands, and the andesites occur primarily in the younger northern plains. Second important result is the identification of aqueous mineralization occurred in limited regions under ambient or hydrothermal conditions. In Fig. 8.9 is shown the concentration of hematite measured by the MGS TES instrument. The abundance of hematite is shown in grey, with increasing brightness indicating increasing hematite abundance. These units provide evidence for the long-term stability of liquid water near the surface of Mars (Anderson *et al.*, 2003). The third one is also important and is a negative one, in the sense is that no evidence for carbonates has been found. All these discoveries have great implications for the Mars evolution. TES has been also extremely important in identifying the appropriate landing site for the

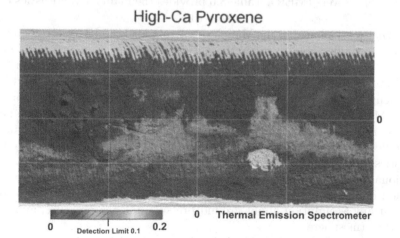

FIGURE 8.8. Concentration map of high-calcium pyroxene at 4 pixels per degree (ppd). High concentrations correspond to basaltic surfaces. Mineral concentration maps are available in digital and image formats at 1, 2 and 4 ppd (after Bandfield *et al.*, 2000).

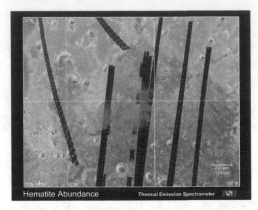

FIGURE 8.9. Concentration of hematite measured by the MGS TES instrument. The abundance of hematite is shown in shaded grey in the black strips, with increasing darkness indicating increasing hematite abundance.

TABLE 8.6. Characteristics of the NASA spectrometers.

	1969 IRS	1971 IRIS-D	1996 TES
Wavelength range (μm)	1.8–14.4	5–50	∼6–50
Spectral resolution	$10\,cm^{-1}$	$1.2\,cm^{-1}$	10 or $20\,cm^{-1}$
At $10\,\mu$m	(1%)	(0.12%)	(1 or 2%)
Measurements per spectrum	1340	1500	143
Spatial resolution (km)	130–500	125–1000	3
SNR at $2.2\,\mu$ma	190	–	–
SNR at $6\,\mu$ma	253	18	33
SNR at $10\,\mu$ma	711	100	345
SNR at $25\,\mu$ma	–	400	388

Mars Exploration Rovers (MER). Table 8.6 provides the main characteristics of the TES instrument.

The PFS that was included in the Mars Express mission is an IR spectrometer optimized for atmospheric studies able to cover the wavelength range from 1.2 to $45\,\mu$m divided in two channels with a boundary at $5\,\mu$m. The spectral resolution is $2\,cm^{-1}$. The instrument FOV is about $2°$ for the Short-Wavelength Channel (SWC) and $4°$ for the Long-Wavelength Channel (LWC) which corresponds to a spatial resolution of 10 and 20 km when Mars is observed from a height of 300 km (nominal height of the pericentre). Each channel is based on a double pendulum interferometer concept. PFS studies the IR radiation reflected and emitted by the planet Mars. In Table 8.7 are reported the main characteristics of PFS, and in Fig. 8.10 is shown the optical scheme of PFS.

The rational of this kind of experiment is multifold: the aim to solve the problem of the inversion of the 15-μm CO_2 band to obtain the vertical temperature profile in all cases without any assumption about ground pressure and temperature, and about dust content of the atmosphere.

The PFS is presently on board the ESA mission named Mars Express orbiting Mars. On 10 January 2004, it was for the first time switched on to measure IR radiation from the planet. The first spectra measured with unprecedented accuracy CO_2, CO,

TABLE 8.7. Main characteristics of the PFS.

	SW	LW
Spectral range (μm)	1.2–5.0	5.0–45
Spectral range (cm^{-1})	2000–8000	222–2000
Spectral resolution (cm^{-1})	2	2
FOV (rad)	0.02	0.044
Nebulosity (NEB) (W cm^{-2} sr^{-1})	5×10^{-9}	4×10^{-8}
Measurement cycle duration (s)	10	10
Detector type	Photoconductor	Pyroelectric
Material	PbSe	LiTaO$_3$
Interferometer type	Double pendulum	
Reflecting elements	Cubic corner reflectors	
Beamsplitter	CaF$_2$	CsI

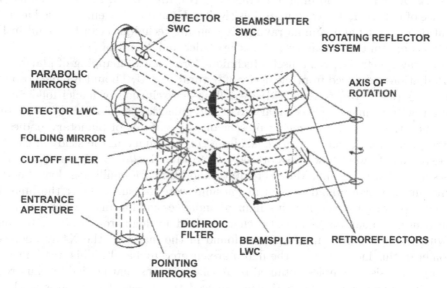

FIGURE 8.10. PFS optical scheme.

H_2O, clearly identified with several bands. In particular the 15- and 4.3-μm CO_2 bands were measured with unprecedented accuracy and resolution, allowing accurate temperature profiles retrieval, the first step towards a global study of the Martian atmosphere (Formisano *et al.*, 2004). Recently, the discovery by means of PFS has been extensively discussed in several conferences. The implications of Methane discoveries can be important since its presence in the Martian atmosphere was never firmly suggested before, and could be suggestive of active volcanic phenomena on the planet, or of life activity (Formisano *et al.*, in preparation).

PFS has given unique data necessary to improve our knowledge not only of the atmosphere properties but also about mineralogical composition of the surface and the surface–atmosphere interaction.

PFS is a complex experiment and extremely performing experiment. However, it does not have the spatial resolution to make accurate mineralogical mapping of the surface.

For that purpose Mars Express hosts other optimized experiments, that we will describe in the next section. This is a clear example of how difficult could be, in a real case, to discriminate atmospheric effects from surface effects. For this reason usually the planetary missions have complex payloads with different sensitivities and different spectral ranges in order to discriminate complex and interacting phenomena.

8.3.5. *X-ray and gamma-ray spectra*

X-ray and gamma-ray spectra provide diagnostic information on the abundances of specific elements in the outermost layers of a planetary surface. These spectroscopic techniques take advantage of the fact that high energy galactic cosmic-ray particles and lower energy solar X-rays and charged cosmic-ray particles penetrate up to several centimetres into the surface and excite X-ray radiation that is characteristic of specific elements. This secondary radiation is then emitted from the surface and can be detected using both scintillator and semiconductor detectors at gamma-ray energies and gas-filled proportional counters at X-ray energies. These techniques can provide quantitative information on the abundances of many rock-forming minerals (e.g. Na, Mg, Al, Si, P, S) as well as on the abundance of natural radioactive materials (K, Th, U) and hydrogen. Detailed additional information on X-ray and gamma-ray remote sensing techniques can be found in Evans *et al.* (1993) and in the fundamental book by Adler and Trombka (1980).

Gamma-ray spectroscopy is a useful technique for chemical mapping of planetary surfaces. Radiations produced from the decay of natural (e.g. U, Th and K) and cosmogenic radionuclides (e.g. ^{22}Na, ^{26}Al, ^{153}Gd) as well as de-excitation of nuclei following their interaction with primary and secondary solar and galactic cosmic-rays are characteristic of the concentration of the elements present on the surfaces of planetary bodies without atmosphere. Particularly effective has been this technique in the study of the Moon. These experiments were flown on both *Apollo 15* and *Apollo 16*. Generally, over mare regions such as Mare Crisium, Mare Serenitatis and Mare Tranquillitatis, low abundances of aluminum and high abundances of magnesium were measured. Over the lunar highlands, the opposite pattern (high aluminum abundances and low magnesium abundances) were measured. Lunar maria are also characterized by high iron content. In combination with the low aluminum abundance found in the maria by the X-ray fluorescence spectrometer, this indicates that the maria are covered by basalt. This was known from laboratory analysis of samples obtained by Apollo missions that landed in mare regions, but the orbital geochemistry experiments allowed this knowledge to be extended to a larger segment of the Moon.

While lunar gamma-ray spectroscopy was first demonstrated with Apollo gamma-ray measurements, the full value of combined gamma-ray and neutron spectroscopy was shown for the first time with the Lunar Prospector Gamma-Ray and Neutron Spectrometers (LP-GRS and LP-NS).

Another example of the use of gamma-ray spectroscopy is the near mission spectrometer. This experiment combines an X-ray spectrometer with a gamma-ray one. The near X-ray/Gamma-Ray Spectrometer (XGRS) has been used to determine compositional properties of the asteroid. The experiment consisted of two instruments, the X-ray fluorescence spectrometer and the GRS. In addition, there are two X-ray solar monitors to determine the incident solar X-ray spectrum. The scientific objectives of this instrument were to provide a compositional map of the surface of asteroid 433 Eros to establish constraints on the origin and geological history of the asteroid, and to determine what relationship, if any, Eros has to meteorite types collected on Earth. In Fig. 8.11 is shown the gamma-ray spectrum from the surface of Eros. The spectrum has been obtained

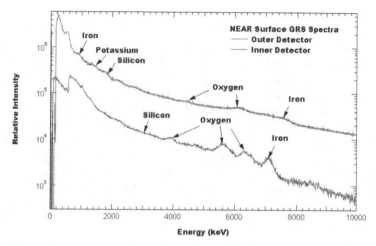

FIGURE 8.11. Gamma-ray spectrum from the surface of Eros. These data – the first ever collected on the surface of an asteroid – result from 7 days of measurements following near Shoemaker's historic landing on 12 February.

combining result from 7 days of measurements following near Shoemaker's area (Evans *et al.*, 2001).

The near GRS consists of a prime detector and shield detector mounted on the instrument deck. The prime detector is a 2.5 × 7.5 cm NaI (Tl) scintillator with an FOV of about 60° and 8.5% Full Width at Half Maximum (FWHM) minimum resolution at 662 keV. The shield detector is an 8.9 × 14 cm bismuth germanate (BGO) scintillator cup with 15% FWHM minimum resolution at 662 keV. The GRS can measure an energy range from 0.3 to 10 MeV in 10-keV channels. Gamma-rays are produced by naturally occurring radioactive elements such as K, Th and U, and by the interaction of cosmic-ray protons and energetic particles associate with solar flares with other elements near the surface such as Fe, Si, O, Mg and H. They can be detected from material from the surface down to about 10 cm in depth (Trombka *et al.*, 2001).

The X-ray fluorescence spectrometer detectors are three gas proportional counters collimated to 5° FOV to determine X-ray line emissions from the asteroid. Balanced filters on two detectors (Al on one and Mg on the other) are used to separate Mg, Al and Si lines, and the Fe, S, Ti and Ca lines can be resolved. The detectors have a 25 cm^2 active aperture area and a 25 μm beryllium window with a beryllium liner and window support (Trombka *et al.*, 2001).

One of the most important success of this technique was obtained during the Mars Odyssey mission by the GRS (Saunders and Meyer, 2001). This instrument, in fact is particularly effective in revealing the presence of H, and consequently, also of the H generated by cosmic-rays interaction with water. The gamma-ray detector is a large (1.2 kg) high-purity germanium (Ge) crystal. The crystal is held at a voltage of approximately 3000 V. Little or no current flows (less than 1 nano-Amp) unless a high-energy ionizing photon or charged particle strikes it. The electric charge from such a strike is amplified, measured and digitally converted into one of 16,384 (214) channels, or bins. After a specified number of seconds, a histogram is produced, which shows the distribution of events (number of strikes) as a function of energy (channel number). This histogram is one gamma-ray spectrum.

The spatial resolution of the instrument is about 300 km and a region of these dimensions receives about 6 h of accumulation time near the equator at the end of the mission and much longer accumulation times near the poles (about a factor of 5 more). Elements which need longer accumulation times can be determined with degraded spatial resolution by summing spectra over larger regions of the Planet. For example, oxygen, silicon, chlorine, potassium and iron can be determined in a 300 km spot, but nickel and chromium can only be determined by summing the data over very large regions, for example all of the highlands or all of the lowlands. According to Jeff Taylor (2001), the concentration of hydrogen is so large that it must be in the form of ice. The amount of ice in the upper meter or so begins to rise at about −60° latitude and continues to increase towards the South Pole. Detailed analysis of the data indicates that the ice-rich layer resides beneath a hydrogen-poor upper layer. The thickness of the upper layer decreases from about 75 cm at −42° to about 20 cm at −77°. The amount of ice in the lower layer is between 20 and 50 wt% (weight per cent), with a best estimate of 35 wt%. As ice is much less dense than mineral grains, this translates to more ice than rock by volume. So the conclusion is that Mars is rich in dirty ice.

8.3.6. *Spectroscopy: conclusions*

After this description of these experiments we can try to draw some preliminary conclusion: to study a planet that is characterized by many different and complex phenomena we have to face many different aspect: first we have to try to have a fist knowledge of it surface or atmosphere, as in the case of Mars, that is now the target of an extremely complex exploration through the efforts of different agencies and several groups of scientists. Spectroscopy is one of the best tool in order to acquire a quantitative knowledge of the compositional aspect of a planet, from the atmosphere (UV to far-IR) to its surface (Vis–NIR X-gamma) and also to its subsurface (gamma, neutrons).

8.4. From spectral imaging to imaging spectroscopy

We have seen, in the previous lectures how many information are synthetically collected in images, and, independently how relevant in terms of understanding the evolution of a planet are the surface and atmosphere spectra: for this reason, in the recent years it has been tried to combine spectroscopy and imaging. In fact the localization of materials with different composition on the surface of the planet allow to interpret these data in a geological context. At the same time, also atmospheric physics takes great advantage from the identification of spatial distribution of different molecules. Significant advance in sensor technology was achieved subdividing spectral ranges of radiation into bands (intervals of continuous wavelengths), allowing sensors in several bands to form multi-spectral images.

Reflectance imaging spectroscopy measurements are currently based on two different techniques. The first technique is to combine an imaging instrument with a modest number (10–20) of discrete narrow-band filters placed at key wavelengths for the detection of specific minerals and/or ices. The result is a high spatial resolution dataset at high enough spectral sampling and resolution to allow for the mapping of various spectral units. The tradeoff for this method is that not all of the wavelengths can be obtained simultaneously, so spacecraft motion is used in order to cover with the same filters the same surface (see Fig. 8.12).

The second technique is to combine a spectrometer with a 2D array to allow for the spectra of different spatial locations to be obtained simultaneously. The result is a high spectral resolution data set at high enough spatial resolution to again allow for the mapping of surface units. The tradeoff for this method is that only one axis of spatial

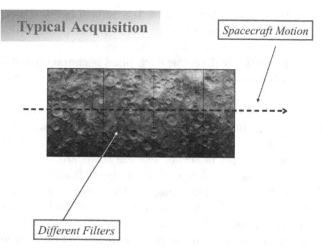

Typical Acquisition

Spacecraft Motion

Different Filters

FIGURE 8.12. Typical acquisition when using several filters. The same area on the surface is seen first by the first filter (dark grey in the figure) and then by the other. In each point of the figure a spectrum in the four filters can be generated.

information can be obtained at a time, so in order to build an image, the second axis of spatial information must be obtained by scanning the instrument and/or moving the spacecraft. The former approach has been used very successfully by ground-based and Hubble Space Telescope observations of the Moon, Mars and asteroids, and on the Galileo and Clementine spacecraft missions. The latter technique has also been successfully used by ground-based observers and by instruments on the Phobos-2 and Galileo, Cassini and Rosetta missions. The choice of technique depends primarily on the importance of obtaining high spectral resolution in respect to the spatial one.

For cases where lower resolution is adequate, the discrete filter technique offers excellent image quality; for cases where detailed compositional or mineralogical information is required, then spatial image quality must be sacrificed for higher spectral resolution.

The simplest way to obtain images is to use 2D sensors, that are equivalent to a photographical acquisition. The previously described cameras technique (see Section 8.2) belongs to this category. When the radiation coming from the entire scene is instantaneously measured at once the systems are called framing systems. The most important characteristic of these sensors is the FOV that define the size of the scene: this in turn depends on the apertures and optics of the system. As noted by Norwood and Lansing (1983), the earliest frame sensors grew out of television industry and provided the basis for vidicon that were adapted for imaging the Moon and Mars.

If the scene is sensed point by point (generating "picture elements" or *pixels*) in a small FOV (Instantaneous FOV, IFOV) along successive lines over a typical time (usually named *integration time*), we have a scanning system. Most non-camera sensors operating from moving platforms image the scene by scanning. This concept is extremely useful to understand the basic of imaging spectroscopy. The optical setup for imaging sensors can be image plane or optical plane focused (depending on where the photon rays are converged by a lens). The sensor can be either a one dimensional (1D) of a 2D array. It is usually called *swath*, the area imaged on the surface of the planet by the sensor. Imaging swaths for a space-borne sensors generally are between tens and hundreds of kilometres wide.

Each line acquired (swath) is subdivided into a sequence of individual spatial elements that represent a corresponding usually square or rectangular area on the scene surface being imaged (or in the target when we look the atmosphere). Thus, along any line is an array of contiguous small elements, from each of which emanates radiation. These elements are sensed along the line. In the sensor, each cell is associated with a pixel that is connected to a microelectronic detector; each pixel is characterized for a brief time by some single value of radiation (e.g. reflectance) converted by the photoelectric effect into electrons.

The IFOV is defined as the solid angle extending from a detector to the area on the ground it measures at any instant, and is a function of the optics of the sensor, the sampling rate of the signal, the dimensions of any optical guides, the size of the detector and the altitude above the target or scene. The electrons are removed successively, pixel by pixel, to form the varying signal that defines the spatial variation of radiance from the progressively sampled scene. The image is then built up from these variations – each assigned to its pixel as a discrete value called the digital number (a DN, made by converting the analogue signal-to-digital values of whole numbers over a finite range). Using these DN values, a "picture" of the scene is recreated on another support.

The term scanning can be applied both to movement of the entire sensor and, in its more common meaning, to the process by which one or more components in the detection system either move the light gathering, scene viewing apparatus or the light or radiation detectors are read one by one to produce the signal. Two broad categories of most scanners are defined by the terms "optical-mechanical" and "optical-electronic", distinguished by the former containing an essential mechanical component (e.g. a moving mirror) that participates in scanning the scene and the latter by having the sensed radiation move directly through the optics onto the linear or array detectors. There is a wide range of mechanical scanners: a typical one could be a flat-scan mirror located in front of the collecting aperture (Norwood and Lansing, 1983).

The remote sensors can be classified on the basis of the way in which they gather their data in respect to the motion of the satellite around the planet. In fact, depending on the optical system and on the orbit could be convenient perform cross-track or along track scanning. In this respect planetary sensors are not different from those developed for Earth remote sensing. They usually monitor the path over an area out to the sides of the path; the already mentioned swath width. The width is determined by that part of the scene encompassed by the telescope's full angular FOV which actually is sensed by a detector array – this is normally narrower than the entire scene's width from which light is admitted through the external aperture (usually, a telescope).

The *cross-track* mode normally uses a rotating (spinning) or oscillating mirror to sweep the scene along a line traversing the ground that is very long (km) but also very narrow (m), or more commonly a series of adjacent lines (see for an exhaustive discussion Remote Sensing Tutorial (RST)†. This is sometimes referred to as the *whisk-broom* mode from the vision of sweeping a table side to side by a small handheld broom. The essential components of an instrument like that are: (1) a light gathering telescope that defines the scene dimensions at any moment; (2) appropriate optics, within the light path train; (3) a mirror; (4) a device (spectroscope, spectral diffraction grating, band filters) to break the incoming radiation into spectral intervals; (5) a means to direct the light so dispersed onto a battery or bank of detectors; (6) an electronic means to sample the photoelectric effect at each detector and to then reset to a base state to receive the next incoming light packet, resulting in a signal that relates to changes in light values coming from the ground or target and (7) a recording component that either reads the signal as an

†http://rst.gsfc.nasa.gov/Homepage/Homepage.html, A Remote Sensing Tutorial (RST) by Nicholas M. Short

analogue (displayable as an intensity-varying plot (curve) over time or converts the signal to DN.

The *along-track* mode does not have a mirror looking off at varying angles. Instead a 2D matrix of CCD is used. In this mode, the pixels that will generate the image correspond to the pixel on the CCD or on the line array. As the satellite advances along the track, at any given moment radiation from each ground cell area along the ground line is received simultaneously at the sensor and the collection of photons from every cell impinges in the proper geometric relation to its ground position on every individual detector in the linear array equivalent to that position. This type of scanning is also referred to as *push-broom scanning* (*ibid.*).

Several detector systems, can be used in remote sensing, being CCD the most common. The list includes photoemissive, photodiode, photovoltaic and thermal detectors, like those that we have extensively described. This approach to sensing EM radiation was developed in the 1970s, which led to the Push-broom Scanner, which uses CCDs as the detector.

There are also a few important characteristic of instruments of this kind, that is better recall here. They are:

- *Spatial resolution*: Ability to separate closely spaced objects on an image or photograph. Resolution is commonly expressed as the most closely spaced line-pairs per unit distance that can be distinguished.
- *Spectral resolution*: The ability to resolve two contiguous spectral features.
- *Resolving power*: A measure of the ability of individual components and of remote sensing systems, to separate closely spaced targets.
- *Dwell Time*: Time of residence of an image in the IFOV of the experiment.

8.4.1. *Hyperspectral imaging*

Another major advance, now coming into its own as a powerful and versatile means for continuous sampling of broad intervals of the spectrum, is hyperspectral imaging. The multispectral sensors are usually characterized by broad bands in which spectral radiation is integrated within the sampled areas to cover ranges, such as $0.1\,\mu m$. In hyperspectral data, that interval narrows to $10\,nm$. Thus, we can subdivide the interval between 0.38 and $2.55\,\mu m$ into 217 intervals, each approximately $10\,nm$ ($0.01\,\mu m$) in width, as in the case of airborne visible/IR imaging spectrometer (AVIRIS; see below). We can use detectors for the visual component of the reflected light, as silicon microchips; we can also use other detectors for short-wave IR (SWIR, between 1.0 and $2.5\,\mu m$) as indium–antimony (In–Sb) alloy or mercury–cadmium–tellurium (Hg–Cd–Te) for longer wavelengths (1–$5\,\mu m$), provided that the detector are cooled down. If a radiance value is obtained for each such interval, and then plotted as intensity versus wavelength, the result is a sufficient number of points through which we can draw a meaningful spectral curve (see Fig. 8.13). The typical data product of an hyperspectral imager are the so-called image cubes. A hyperspectral image can be thought of as a single spatial image taken at several spectral bands or wavelengths. By stacking these 2D images in order of their wavelength, we obtain an image cube that varies spatially in 2D and spectrally in the third, practically the product can be shown as a cube. Mathematically we have to deal with three-index matrices, instead of the two-index ones to which we are used when dealing with images. Every element has a unique spectral signature determined by the way that material reflects, absorbs and emits EM radiation. Due to of this property, the spectrum of a scene can provide clues and information about the objects in that scene and the chemical properties of those objects. Hyperspectral images can be very large, consisting of hundreds of frames of data (Fig. 8.14). Effective *compression* algorithms

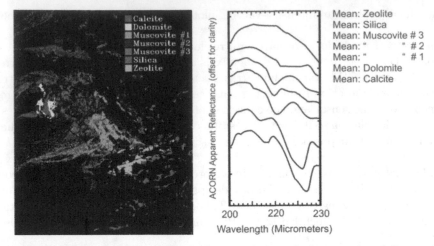

FIGURE 8.13. AVIRIS hyperspectral imager. It is a unique optical sensor that delivers calibrated images of the upwelling spectral radiance in 224 contiguous spectral channels (bands) with wavelengths from 400 to 2500 nm. AVIRIS has been flown on two aircraft platforms: a NASA ER-2 jet and the Twin Otter turboprop.

FIGURE 8.14. The AVIRIS image cube. The hyperspectral data cube are displayed with false-colour three-band image on the front face of the cube and pseudo-coloured edges of bands shown on the bottom and side of the cube. It is constructed from AVIRIS data from the area of cuprite, Nevada (courtesy NASA).

are needed to be able to store the images or to transmit the images with reasonable bandwidth and time constraints. These algorithms are specially important when data shall be transmitted from very far, as from distant planets.

The Jet Propulsion Laboratory (JPL) has produced two hyperspectral sensors, one known as airborne imaging spectrometer (AIS), first flown in 1982, and the other known as AVIRIS, which continues to operate since 1987. AVIRIS consists of four spectrometers with a total of 224 individual CCD detectors (channels), each with a spectral resolution of 10 nm and a spatial resolution of 20 m. Dispersion of the spectrum against this detector array is accomplished with a diffraction grating. The total interval reaches from 380 to 2500 nm (about the same broad interval covered by the Landsat TM with just seven bands). It builds an image, push-broom like, by a succession of lines, each containing 664 pixels. From a high-altitude aircraft platform such as NASA's ER-2 (a modified U-2), a typical swath width is 11 km.

AVIRIS has shown, since its first observation campaign to be extremely effective both in the identification of different kind of mineral and atmospheric bands. However, it has been immediately clear that this kind of hyperspectral sensors need an accurate and systematic ground calibration (see e.g. Vane *et al.*, 1988). Moreover, hyperspectral images may be corrupted with additive noise and have lost samples due to equipment tolerances. Therefore, methods and algorithms are needed to remove the noise and recover or estimate the lost samples without destroying the information contained in the image.

In particular AVIRIS has been extensively used to detect hydrothermal activity. As an example, the AIS and the AVIRIS were used to map the alteration mineralogy of a hydrothermal system in the northern Grapevine Mountains of Nevada and California, using detailed spectral characteristics. Areas of quartz–sericite–pyrite alteration were identified based on the presence of sericite (fine-grained muscovite) spectral features near 2.2 μm. Areas of argillic alteration were defined based on the presence of montmorillonite (Kruse and Taranik, 1988). The experience developed with AVIRIS has been extremely precious also to develop ability of recognizing distinctive signatures of rock-forming minerals in other planes, for example on Mars. On Mars, in fact, identifying and mapping areas of hydrothermal alteration is important in order to understand the water cycle on that planet (Guinness *et al.*, 2003). The authors (Guinness *et al.*, 2003) used the summit of Mauna Kea, Hawaii, that provides an environment that exhibits hydro-thermal alteration within a volcanic terrain, to define an analogous to Mars. Mapped alteration minerals on Mauna Kea include kaolinite, montmorillonite, saponite, hematite and jarosite (*ibid.*), that will be hopefully identified on Mars during the next OMEGA (Observatoire pour la Mineralogie, l'Eau, le Glace e l'Activité) Mars Express mission.

8.4.1.1. *ISM and OMEGA*

In planetary science, the first imaging spectrometer flown was ISM, that was built in France and that flown on the mission Phobos. ISM instrument is an infrared imaging spectrometer developed by French CNRS laboratories with the support of CNES, the French space agency. It was part of the payload of the Phobos-2 probe with the Soviet Russian for Camera and UV-visible Spectrometer of the Mission (KRFM) spectropho-tometer. Partly complementary to KRFM goals, its major scientific objectives were to provide a mineralogical mapping of the low-latitude zone of Mars for a better under-standing of the history of the planet and to return significant data on the time and space variability of the atmosphere. For Phobos it was supposed to give information about the mineralogical composition of the surface with high spatial resolution for assessing the mean composition and the level of heterogeneity of this small body.

The ISM instrument is a scanning imaging spectrometer that covers the spectral range 0.76–3.16 μm. For each pixel, 128 spectral measurements are acquired simultaneously. A 2D image of the surface was obtained by rotating the entrance mirror to scan in the cross-track direction for the image samples and the forward motion of the spacecraft provides the image lines. The spectral dispersion is obtained by using a grating, whose the first and second orders are exploited. These two orders are separated by a beam-splitter and filters, and measured by four groups of 32 cooled PbS detectors, designated first and second order odd and even (see Fig. 8.15).

ISM opened a new world in Martian science: in fact, before the acquisition of ISM data it was assumed that the surface of Mars, being covered by a large amount of dust, should have been mineralogically homogeneous. The ISM data, despite their limited number

Optical scheme of ISM

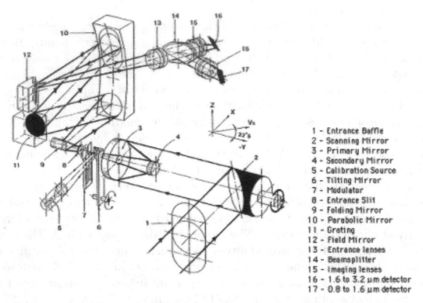

1 – Entrance Baffle
2 – Scanning Mirror
3 – Primary Mirror
4 – Secondary Mirror
5 – Calibration Source
6 – Tilting Mirror
7 – Modulator
8 – Entrance Slit
9 – Folding Mirror
10 – Parabolic Mirror
11 – Grating
12 – Field Mirror
13 – Entrance lenses
14 – Beamsplitter
15 – Imaging lenses
16 – 1.6 to 3.2 µm detector
17 – 0.8 to 1.6 µm detector

FIGURE 8.15. Optical scheme of the ISM instrument. The main characteristics of the instrument can be summarized as follows: telescope $f/4$; 100 mm focal length; 25 mm aperture; FOV 20 (by a 2400 steps scanner); imaging method whiskbroom (full aperture, object-space scanner); $f/4$; grating disperser.

and their pioneering quality, clearly showed that the Mars surface is characterized by a noticeable mineralogical variability. Observations of Mars with the ISM on board the Phobos-2 spacecraft provided more than 40,000 spectra in the wavelength range 0.7–3.2 μm with a spatial resolution of about 20 km. These spectra bear features of both atmospheric and surface origin (Titov *et al.*, 1994). The first major difficulty was actually to calibrate the data, and the second one was to discriminate the different contribution to the spectra, coming from the surface, from the atmosphere and, in non-negligible way, from the aerosol present in the atmosphere. A valuable effort was made to calibrate ISM using telescopic and laboratory data (Mustard and Bell, 1994). Six regions of variable albedo and geological setting were identified where ISM and 1988 opposition telescopic coverage either overlapped physically or sampled the same surface geological unit. The sizes and positions of the regions measured telescopically were compared with ISM pixels falling within these spots and were averaged to produce a spatially convolved spectrum that simulates what would have been seen telescopically. In this way, it was possible to obtain a valuable calibration of ISM.

The discrimination between the atmospheric contribution and the surface one was possible by modelling the Martian atmosphere (Castronuovo and Ulivieri, 1995). The spectra between 1.6 and 3.2 μm show evidence for scattering (probably by dust particles) and for minor gaseous constituents (H_2O and CO). The imaging capabilities of the ISM spectrometer allowed the study of spatial variability of these atmospheric constituents. The observations of the Martian volcanoes whose vertical extension is higher than the atmospheric scale height give spectra corresponding to different altitudes of the Martian surface, which are used to study the vertical mixing ratio of H_2O and CO (Combes *et al.*, 1991; Encrenaz *et al.*, 1991). Thanks to ISM it was possible also to apply to Mars a technique of inversion of the depth of CO_2 band to determine the relative height of different area. The measurement of the CO_2 column density on Mars, from the 2-μm band, in the Phobos/ISM spectra allows the determination of the altitude of the surface of Mars, with a horizontal spatial resolution of 20 km (Bibring *et al.*, 1991).

The recent re-examination of ISM data allowed to anticipate the OMEGA data showing that hematite, where present, can be seen in the spectra in the 0.4- to 2.5-μm range, the silicate-clay host is spectrally active beyond 3 μm and can be identified from this domain; and that, phyllosilicates such as montmorillonite or smectite may be abundant components of the Martian soils, although the domain below 3 μm lacks the characteristic features of the most usual terrestrial-clay minerals (Erard and Calvin, 1997). Automatic mapping of mineralogically different provinces using multivariate classification procedure was also successfully attempted (Cerroni and Coradini, 1995).

OMEGA is a second generation instrument based on the ISM instrument. OMEGA is a visible and near-IR mapping spectrometer, operating in the spectral range (0.35–5.1 μm). Combining imagery and spectrometry, OMEGA is designed to provide the mineralogical and molecular composition of the surface and atmosphere of Mars through the spectral analysis of the diffused solar light and surface thermal emission. OMEGA was originally developed for the Mars-96 mission by IAS and DESPA in France, IFSI in Italy and IKI in Moscow. OMEGA spare unit has been selected to fly on board the Mars Express ESA mission, launched in 2003. The main differences with ISM performances include extended spectral range at both shorter and longer wavelengths (0.5–5.2 μm), and increased angular and spectral resolutions. The instrument is made up of two co-aligned channels (visible and NIR), each using a dedicated telescope and fore optics.

The visible channel works in push-broom mode, using a Thomson CCD matrix 384 × 288 pixels (spatial × spectral). The total FOV is 8.8°. In baseline modes, pixels are binned either 3 × 3 or 3 × 2, providing a spatial sampling (IFOV) of 4.1 arcmin and a spectral sampling of 7 and 4 nm, respectively: the beam is spread through an holographical grating over the spectral range 0.5–1.1 μm, over 96 or 144 "spectels".

The NIR channel works in whisk-broom mode, with an IFOV of 4.1 arcmin. The incoming radiation is divided in two subchannels, each using a grating spectrometer working in the first order. A cross-track-scanning mechanism is used to build swaths of 16, 32, 64 or 128 IFOV width, depending on the spacecraft altitude.

For the IR channel is extremely important the temperature control. In fact, particularly in the range 2.5–5.0 μm, the sensor "senses" also the IR photons due to the temperature background of the spectrometer. Therefore, this shall be cooled in order to reduce the background noise. The detector shall be cooled as well in order to reduce the dark currently and therefore to increase the SNR. The optical head of OMEGA is cooled by a radiator, that looks at the deep space (3 K). The IR focal plane of OMEGA is cooled down by means of an active cooler. This is the first time that an active cooling system is used in space.

The lesson learned by the ISM spectrometer, pushed the OMEGA team to pay special attention to OMEGA laboratory calibration. OMEGA has been extensively calibrated in laboratory at the Orsay Institute.

OMEGA has already obtained valuable results, that were recently published (Bibring *et al.*, 2004). The main results obtained can be summarized as follows: OMEGA has contributed to make the inventory of water and carbon dioxide reservoirs on Mars. In fact has been firmly demonstrated that not only the northern cap is mainly composed of water ice, but also the southern one (see Fig. 8.16). Bibring *et al.* (2004) presented the first direct identification and mapping of both carbon dioxide and water ice in the Martian high southern latitudes, at a resolution of 2 km, during the local summer, when the extent of the polar ice is at its minimum. They also observed that this south polar cap

FIGURE 8.16. OMEGA observed the southern polar cap of Mars on 18 January 2004, as seen on all three bands. The right one represents the visible image, the middle one the CO_2 (carbon dioxide) ice and the left one the H_2O (water) ice.

contains perennial water ice in extended areas: as a small admixture to carbon dioxide in the bright regions; associated with dust, without carbon dioxide, at the edges of this bright cap; and, unexpectedly, in large areas tens of kilometres away from the bright cap (*ibid.*). New results are expected by OMEGA since it will map the mineral composition of the surface at 100-m resolution for the next 4 years!

8.4.1.2. *Cassini visual and infrared mapping spectrometer*

The Cassini visual and infrared mapping spectrometer (VIMS) is an imaging spectrometer covering the wavelength range 0.3–5.2 μm in 352 spectral channels, with a nominal instantaneous FOV of 0.5 mrad. VIMS consists of two integrated bore-sighted slit-grating spectrometers with separate reflecting telescopes (Brown *et al.*, 2003). These two spectral channels cover the ranges 0.350–1.050 μm (visual channel) and 0.8–5 μm (IR channel) with nominal spectral sampling of 7.3 nm (96 bands) and 16.6 nm (256 bands), respectively. The visual channel has a 2D CCD array detector and the IR channel has a linear InSb array detector. Both channels effect 2D imaging while measuring all spectral channels simultaneously using a scanning mirror in one (visual) or two (IR) dimensions. Thus, the nominal data set is an image cubes with two spatial and one spectral dimensions (Coradini *et al.*, 2004; Mc Cord *et al.*, 2004).

The data returned by VIMS consist of instrument response data numbers for each spectral channel at each spatial pixel. In that, VIMS does not differ from the previously described detectors. Each spectral channel is assigned a wavelength determined from the wavelength calibration, and the data numbers are multiplied by a radiometric response function for each pixel. The radiometric calibration is derived from measurements made on the ground before launch and in-flight of Venus, the Moon, Galilean satellites and several stars. The result is I/F (reflectance for the specific geometry) as a function of wavelength for the solar-illuminated part of the FOV when no other radiation is present. Sometimes, such as for Io, thermal radiation by the target adds to the reflected solar radiation at some wavelengths, making it difficult or impossible to derive I/F values.

VIMS-V employs a Shafer telescope, an Offner spectrometer, where the dispersive element is a holographical diffraction grating, and a 256×512 CCD focal plane array (FPA) where dispersed spectral images are collected. This kind of grating was used for

(a)   (b)

FIGURE 8.17. VIMS-IR (a) and visual (b) channels. The visible channel (VIMS-V) is a 4.5 cm telescope that deflects its beam through slit, and then through a diffraction grating. The slit determines the FOV, allowing in only light along a line. The diffraction grating is a grooved mirror such that light reflecting from each groove interferes with the light coming from other grooves in a way that causes light to be dispersed according to wavelength. The light is focused on a CCD array of 256×512 elements. The IR part of VIMS, VIMS-IR, is similar but only has a 1D InSb (Indium-Antimonide, a chemical compound) detector. Thus it can only take the spectrum of one point at a time. In the amount of time it takes the optical half to make one complete exposure of the spectrum of a line of points, the IR half has to take the spectrum of 64 different points one at a time. To do this it requires a larger telescope with which to collect light, 23 cm, than did the optical assembly.

the first time in VIMS-V and was specially manufactured by Zeiss. The instrument is housed in two separate boxes (see Fig. 8.17): the visible channel optical head (VCOH) and the visible channel electronics (VCE). To obtain 2D images of either low- or high spatial resolution each channel of VIMS uses a scanning mirror: at each position of the scanning mirror the spatial elements within the slit are spectrally dispersed by the grating in 352 spectral bands (96 for -V and 256 for -IR).

The resulting data set is again an image cube. While VIMS-V operates in push-broom mode, VIMS-IR uses a whisk-broom mode. In that respect VIMS is similar to OMEGA. However the VIMS visual channel, VIMS-V, has a wider spectral coverage and a higher spectral resolution. Each x–y plane in the data cube is a monochromatic spatial image (where x is the fast-scan direction and y is the slow-scan direction in the classic definition of a raster scan with $x = 0$ and $y = 0$ being the upper left of the raster); for a fixed spatial position within an image cube, the corresponding vector normal to this position is the spectrum at the given spatial location (i.e. a "skewer" through the cube in the z direction). The plane x–z corresponds to the CCD plane and is named "frame". Both channels are cooled down by passive radiators: for VIMS-V a small radiator surmount the optical head, while for the IR channel a two stages radiator has been used. The two channel seen in Fig. 8.17 have independent acquisition electronics (proximity electronics). In the proximity electronics data are digitized and then transmitted to the main electronics. The main electronics assembly drives and reads the data from both VIMS-V and VIMS-IR. The data from the two halves of the instrument are fused together, partially reduced, and compressed before being relayed to the main Cassini computer for storage and later transmission to Earth.

VIMS has immediately shown its potential value when the first Moon images were acquired. Moon data (Bellucci *et al.*, 2002) were also used to improve the ground-based calibration, particularly for the visual channel (Coradini *et al.*, 2004). The first scientific results were acquired a few months later when Venus was observed. In this case, the first detection and profile characterization of thermal emission from the surface of Venus at 0.85 and 0.90 μm was performed. It was observed in the spectra acquired by VIMS during its Venus fly-by. The strength and shape of these two newly observed nightside emissions agree with theoretical predictions based on the strength of the strong emission observed at 1.01 μm. These emissions, together with previously reported surface emission features at 1.01, 1.10 and 1.18 μm, potentially provide a new technique for remotely mapping the mineralogical composition of the Venusian's surface (Baines *et al.*, 2000).

A new important step in the VIMS story was attained during the first Jupiter fly-by (Mc Cord *et al.*, 2001). VIMS returned many both targeted and serendipitous observations of the Galilean Satellites during the Cassini mission fly-by of Jupiter in December 2000 and January 2001 (Mc Cord *et al.*, 2001; Coradini *et al.*, 2004). These observations consist of reflectance and emission spectra between 0.35 and 5 μm with 12-nm spectral resolution of the integral disk of the satellites at a variety of phase angles and central longitudes. The extreme distance of the fly-by resulted in spatial resolution of less than a pixel for all satellites. Selected observations have been calibrated and analysed. The objectives are to search for new spectral features, validate and expand on earlier analyses and improve the calibrations of VIMS. Among the most important results it should be mentioned (Brown *et al.*, 2003):

- The detections of methane fluorescence on Jupiter.
- The confirmation of the existence of jovian NH_3 absorption at 0.93 μm.
- A surprisingly high opposition surge on Europa.
- The first visual–NIR spectra of Himalya.
- The first Jupiter's optically thin ring system spectrum (Brown *et al.*, 2002).
- The first NIR observations of the rings over an extensive range of phase angles (0–120).
- The study of the auroras showing similarities in the centre-to-limb profiles of H^+ and CH_4 emissions indicating that the H^{+3} ionospheric density is solar-controlled outside of the auroral regions.

Himalya has a slightly reddish spectrum, an apparent absorption near 3 μm, and a geometric albedo of 0.06 ± 0.01 at 2.2 μm (assuming an 85-km radius). If the 3-μm feature in Himalya's spectrum is eventually confirmed, it would be suggestive of the presence of water in some form, either free, bound or incorporated in layer-lattice silicates.

A mean ring-particle radius of 10 μm is found to be consistent with Mie-scattering models fit to VIMS NIR observations acquired over 0–120 phase angle (Brown *et al.*, 2003).

The most important data of VIMS will be acquired in the next 4 years, during the nominal Cassini mission. However, we had already an example of the validity of these data. After the insertion in orbit of Cassini, the 1 July 2004, VIMS acquired immediately data of Phoebe. The spatially resolved spectra of Phoebe indicate a low surface albedo, from <1 to 6 absorption features due to materials which occur with variable abundances and/or grain sizes in different locations on the body. The VIMS observations of Phoebe revealed the presence of several kind of ices (Clark *et al.*, 2004). These include: water ice, bound water and trapped CO_2. A broad 1-μm feature is interpreted to be due to Fe^{2+}-bearing minerals. Water ice is observed with absorptions at 3.1, 2, 1.5, 1.25 and 1.04 μm. Absorptions located in the 3.3 and 1.7 μm region indicate the presence of organic molecules (Fig. 8.18). The complex chemistry of Phoebe seem to indicate that it incorporate materials formed farther out in the solar system, where it is cold enough for them to remain stable. Since VIMS data refer to only a few microns of the surface, it

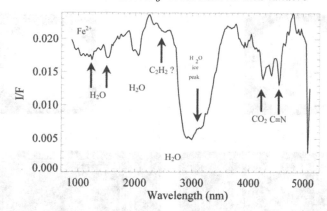

FIGURE 8.18. Average Phoebe spectra. The data were acquired on 11 June, 20:54:49 UTC; the IR image dimensions: 48 S 48 L and IR integration time: 380 ms; the Sun distance was 9.0121 AU. Water ice features occurring at about 1.25, 1.5 and 3.0 mm can be easily recognized, as well as the Fresnel's reflectance peak of this ice at around 3.1 mm; Fe^{2+} can be responsible for some features towards the shortest IR wavelengths (before 1.2 mm); CO_2 feature takes place at 4.25 mm, with an asymmetric shape which is common for "trapped" ice. Anyway, many other features remain to be identified. The data were elaborated by Tosi and Coradini (2004 private communication).

is difficult to say if the observed material represents the bulk composition of Phoebe or is simply a veneer covering the surface. However, the Phoebe orbit, indicating that it could easily be a captured object and its peculiar composition point in the direction that Phoebe is a body coming from the outer regions of the solar system.

The most interesting results were acquired during the 2 Titan fly-bys performed in July and October 2004. The Cassini VIMS obtained spectra in the 0.35–5.1 μm range that include narrow windows in the methane spectrum near 1.6, 2.0, 2.8 and 5.0 μm where the surface might have been observed with spatial resolution up to about 100×200 km during the Saturn orbit insertion phase on 30 June 2004 (Mc Cord *et al.*, 2004b). Cassini's VIMS pierced the haze that surround Titan. VIMS, through the atmospheric windows of methane, that is mostly responsible of the absorption of the Titan atmosphere, reveals an exotic surface bearing a variety of materials in the south and a circular feature that may be a crater in the north (see Fig. 8.19). At some wavelengths, we see dark regions of relatively pure water ice and brighter regions with a much higher amount of non-ice materials (such as simple hydrocarbons). This is different from what we expected. The general idea is that now, the ocean on Titan surface is not present. The source of methane can be under the icy-organic crust that apparently covers Titan.

8.4.1.3. *The VIRTIS family*

VIRTIS is a sophisticated imaging spectrometer that combines three unique data channels in one compact instrument. Two of the data channels are committed to spectral mapping and are housed in the Mapper (**-M**) optical subsystem. The third channel is devoted solely to spectroscopy and is housed in the High-resolution (**-H**) optical subsystem. The experiment was built in order to study the complex world of comets to be part of the ESA Cornerstone mission Rosetta.

The nature of the solid compounds of the comets (silicates, oxides, salts, organics and ices) is still largely unknown. These chemical compounds can be identified by IR spectroscopy using high spatial resolution imaging to map the heterogeneous parts of a nucleus and high spectral resolution spectroscopy to determine the composition unambiguously. The visual and IR spectrum of the comet coma is characterized by a number

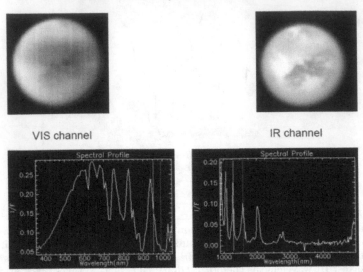

FIGURE 8.19. Image of Titan collected during the second fly-by. Hyperspectral data cube taken from a distance of 2.23×10^5 km from Titan with a solar phase angle of $13°$. Image acquired 26/10/2004 at 04:15Z; Visible channel exposure time 5.12 s IR channel exposure time 80 ms; RGB images were built up by using the spectral channel indicated in the enclosed average spectra; several other elaborations will be released by using other spectral channels in order to investigate the different aspect of Titan's structure and composition. In this case, the image were processed to highlight the surface morphology seen through the methane windows of the satellite dense atmosphere. In fact, as in the case of the Earth, the atmosphere becomes transparent where absorption bands are absent or faint: this allows to retrieve the surface signal. The images were elaborated in INAF: Istituto Nazionale di Astrofisica by the Cassini VIMS team members (A. Coradini, F. Capaccioni and G. Filacchione).

of components comprising both gas emission bands and a dust continuum. Ground-based visual spectroscopy has detected various atomic, radical and ionic species formed through photo-dissociation by solar UV radiation of the so-called "parent" molecules which are sublimated from the nucleus and possibly from a halo of volatile grains. Previous IR comet observations have shown that various hydrocarbons show emission bands between 3 and 4 μm.

A multispectral imager – covering the range from the near UV (0.25 μm) to the NIR (5.0 μm) and having moderate to high spectral resolution and imaging capabilities – is an appropriate instrument for the determination of the comet global (size, shape, albedo, etc.), and local (mineralogical features, topography, roughness, dust and gas production rates, etc.) properties.

The VIRTIS instrument is one of the most complex imaging spectrometer ever built. It has also a great flexibility and for this reason it has been integrated into three different missions: the already mentioned Rosetta, the ESA Venus Express mission, that will study the Venus atmosphere and the NASA Discovery mission Dawn, that will study the main belt asteroids Vesta and Ceres. This is a great success for the group that has proposed VIRTIS, that can be considered a standard for this decade. One of the most important reasons for this success is the flexibility of the experiment, that has several operations modes, that allow to collect data in many different operating conditions. The other reason is the resiliency – is needed to survive in different operating conditions – requested by the Rosetta mission that will operate from about 3 AU (Aphelion of Churyumov–Gerasimenko) from the Sun up to 1 AU (Perihelion of Churyumov–Gerasimenko). For

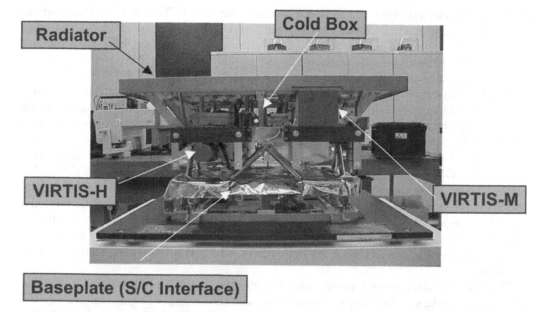

Radiator

Cold Box

VIRTIS-H

VIRTIS-M

Baseplate (S/C Interface)

FIGURE 8.20. Schematic view of VIRTIS.

this reason, VIRTIS with minor modification, has been included in the Venus Express payload.

The instrument is divided into four separate modules: the optics module – which houses the two -M and -H optical heads and the stirling cycle cryocoolers used to cool the IR detectors to 70°K–, the two proximity electronics modules (PEM) required to drive the two optical heads, the main electronics module – which contains the data handling and support unit (DHSU), for the data storage and processing, the power supply and control electronics of the cryocoolers and the power supply for the overall instrument. A detailed description of the experiment is given in Coradini *et al.* (1998, 1999), in what follows we will briefly summarize the main features of the instrument along with a description of the most recent activities carried out.

PEM description. Each optical head shall require specific electronics to drive the CCD, the two IRFPAs, the covers, the thermal control; the PEMs are two small boxes interfaced directly to the spacecraft and placed in the vicinity of the optics module to minimize interference noise.

Optics Module Description. The -M imaging spectrometer and the -H echelle spectrometer optical heads are located inside the optics module (see Fig. 8.20), which in turn is divided into two regions thermally insulated from each other by means of Multilayer Insulation (MLI): the Cold Box and the Pallet.

The Pallet is mechanically and thermally connected to the spacecraft; inside the Pallet are located the two stirling cycle cryocoolers. The heat produced by the cryocoolers compressors (a maximum of 24 W in closed-loop mode) is dissipated to the spacecraft.

The Cold Box contains the two optical heads and its main function is to act as a thermal buffer between the optical heads, working at 130 K, and the external environment (the S/C temperature ranges from 250 to 320 K). The Cold Box is mechanically connected to the Pallet through 8 Titanium rods, whose number and size were selected to minimize conductive heat loads and to avoid distortion upon cooling from room temperature. The structural part of the Cold Box is a ledge which is supported by the 8 titanium rods;

on the ledge the two optical heads are mechanically fixed. Thermal insulation of the Cold Box is improved by means of MLI, while thermal dissipation from the Cold Box is achieved by means of a two-stage passive radiator: the first stage keep the Cold Box temperature in the range 120–140 K, while the second stage is splitted in two parts, one for each optical head, and allows to reach the required 130 K.

Another important component of the instrument are the two covers; they provide a double function: protection against dust contamination, internal calibration by means of an internally reflecting surface finish. They use a step motor and their operation is controlled by the PEMs.

VIRTIS-M description. The VIRTIS-M optical head perfectly matches a Shafer telescope to an Offner grating spectrometer to disperse a line image across two FPAs. The Shafer telescope produces an anastigmatic image, while Coma is eliminated by putting the aperture stop near the centre of curvature of the primary mirror and thus making the telescope monocentric. The result is a telescope system that relies only on spherical mirrors yet remains diffraction limited over an appreciable spectrum and field: at $\pm1.8°$ the spot diameters are less than $6\,\mu$m in diameter, which is seven times smaller than the slit width.

The Offner grating spectrometer allows to cover the visible and IR ranges by realizing, on a single grating substrate, two concentric separate regions having different groove densities: the central one, approximately covering 30% of the grating area is devoted to the visible spectrum, while the external region is used for the IR range. The IR region has a larger area as the reflected IR solar irradiance is quite low and is not adequately compensated by the IR emissions of the cold comet.

The visible region of the grating is laminar with rectangular grooves profile, and the groove density is 268 grooves/mm. Moreover, to compensate for the low solar energy and low CCD QE in the UV and NIR regions, two different groove depths have been used to modify the spectral efficiency of the grating. The resulting efficiency improves the instrument's dynamic range by increasing the SNR at the extreme wavelengths and preventing saturation in the central wavelengths.

Since the IR channel does not require as high a resolution as the visible channel, the lower Modulation Transfer Function (MTF) caused by the visible zone's obscuration of the IR pupil is acceptable; the groove density is 54 grooves/mm. In any case, the spot diagrams for all visible and IR wavelengths at all field positions are within the dimension of a $40\,\mu$m pixel. For the IR zones, a blazed-groove profile is used that results in a peak efficiency at $5\,\mu$m to compensate for the low signal levels expected at this wavelength.

VIRTIS-H description. In -H the light is collected by an off-axis parabola and then collimated by another off-axis parabola before entering a cross-dispersing prism made of lithium fluoride. After exiting the prism the light is diffracted by a flat reflection grating which disperses in a direction perpendicular to the prism dispersion. The prism allows the low groove density grating, which is the echelle element of the spectrometer, to achieve very high spectral resolution by separating orders 9 through 13 across a 2D detector array: the spectral resolution varies in each order between 1200 and 3500.

Since the -H is not an imaging channel, it is only required to achieve good optical performance at the zero field position. The focal length of the objective is set by the required IFOV and the number of pixels allowed for summing. While the telescope is $F/1.6$, the objective is $F/1.67$ and requires five pixels to be summed in the spatial direction to achieve a $1\,\text{mrad}^2$ IFOV ($5 \times 0.45\,\text{mrad} \times 0.45\,\text{mrad}$).

Main electronics module description. The main electronics is physically separated from the optics module. It houses the power supply for all the experiment, the cooler electronics, the spacecraft interface electronics, for telemetry and telecommanding, the interfaces

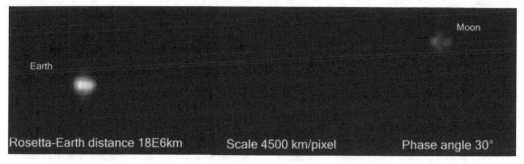

FIGURE 8.21. Earth and Moon seen by VIRTIS during the Rosetta commissioning. The Earth is only partially illuminated and the non-illuminated part is seen in red.

with the optics module subsystems, and the DHSU which is the electronics for the data handling, processing and for the instrument control. The data processing and the data handling activities into the DHSU are to be performed using the on-line philosophy. The data will be processed and transferred to the spacecraft in real time. The mass memory (solid state recorder, SSR) of the spacecraft will be used to store or buffer a large data volume. The main electronics contains no additional hardware component for data processing and compression. All data processing will be performed by software.

The Rosetta mission with its 10 years of cruise and with the comet harsh environment poses a number of constraints on the materials and components; all the mechanisms have to undergo specific testing activity to guarantee proper functioning. In particular, the stirling cycle cryocoolers are among the most critical components of our experiment and, although already used extensively in space, have never been qualified for the environmental conditions expected for the Rosetta mission. A dedicated test programme approved by the ESA, has then been carried out at Officine Galileo (Florence) on the choosen RICOR K508 integral rotary cryocoolers to verify the coolers performances after thermal and vibration tests.

Visible and IRFPAs. The CCD in VIRTIS-M has been selected as the Thomson TH7896; this is a frame-transfer device with multi-pinned-phase (MPP) to reduce the dark current at room temperature, even if it will be operated at low temperature (160 K) and the main concern is not the dark current (which decreases drastically with the temperature), but the reduction in CTE which is relevant below the 155 K. A chrome–gold thin film on the front surface of the CCD is used to improve the capture cross section of the long-wavelength photons, while lumogen coating is used to enhance the efficiency in the UV range.

VIRTIS has been already tested during the extensive commissioning campaign developed by ESA for the Rosetta mission. It has already acquired images of the Earth and the Moon, allowing to test the good instrument performances (see Fig. 8.21).

8.5. *In situ* science from Apollo to MER

The planetary exploration usually follows a typical trend: first a planet is visited remotely, through one or more than one fly-by, then, when the current technology allows to do it, the insertion in orbit is attempted. The first phase is usually exploratory, therefore the planet images are the most important data. Cameras of different kind have shown to us new words, like the outer planets satellites, whose surface and geological characteristics were totally unknown. Usually, the new data collected during a fly-by are

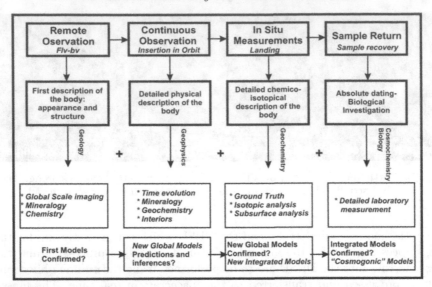

FIGURE 8.22. The built up of a knowledge of a complex system such as the solar system evolves through different steps, and each step represents the basis for the evolution of the understanding of each object observed as well as of the general scheme of the evolution of the solar system. The combination of global measurements, as those collected by far, with specific local observations is absolutely needed to confirm the accuracy of the theories developed by the scientific community.

the basis to plan new missions, stimulating the organized curiosity of the scientific community. Therefore, more complex missions, usually with orbit insertion, are the obliged next step. In this case, the new artificial satellite inserted in orbit around the planet has a typical remote sensing scientific payload. In general orbiters are characterized by a complex suite of different experiments, including cameras, spectrometers at different wavelengths, laser or radar altimeters, magnetometers, dust detectors and plasma analyser. The dream of planetary scientists is to understand the main physical processes that have determined the present surface appearance and interior structure of planets and all the complex systems that sometimes accompany them, such as planetary satellites, rings, radiation and particle belts.

Unfortunately, remote sensing measurement are often ambiguous, and therefore the need to confirm what has been remotely observed, has pushed the scientific community to plan the landing on the planet surface. It is clear that a landing mission offers a better knowledge of the complex reality of a planetary body, but, it asks for a higher level the resiliency and reliability of the overall mission. The further improvement is the sample return. Once we were able to collect samples and send them back toward the Earth, but, unfortunately, this ability is now apparently lost (see Fig. 8.22). The only extraterrestrial body for which the complete cycle depicted in Fig. 8.22 is the Moon. We hope that, thanks to the new measurements obtained on Mars, a similar path could be followed also for Mars and other planets.

8.5.1. *The inner solar system*

8.5.1.1. *Apollo program*

The most important enterprise never done on another planet was the so-called "Apollo program". The Apollo program included a large number of test missions without crew and 12 crewed missions: three Earth orbiting missions (Apollo 7, 9 and Apollo-Soyuz), two lunar orbiting missions (Apollo 8 and 10), a lunar swing-by (Apollo 13), and six Moon landing missions (Apollo 11, 12, 14, 15, 16 and 17) (see for details Brooks *et al.*, 1979).

Two astronauts from each of these six missions landed on the Moon, and they remain the only humans to have walked on another solar system body. The observations made from orbit during the Apollo program allowed to reach a noticeable knowledge of the lunar geology and to establish the basis of planetary geology (see e.g. El-Baz, 1974; El-Baz and Wilhelms, 1975; Burnett, 1975; Taylor, 1975).

Apollo 11 was the culmination of the Apollo programme however six more increasingly complex missions were flown to the Moon prior to the end of Apollo. The typical Apollo Mission included a Command and Service Module (CSM), made of two distinct units: the Command Module (CM), which housed the crew, spacecraft operations systems and re-entry equipment, and the service module (SM) which carried most of the consumables (oxygen, water, helium, fuel cells and fuel) and the main propulsion system. The total length of the two modules attached was 11.0 m with a maximum diameter of 3.9 m. The landing module was called CSM. It carried two astronauts. Included on the LM was the Early Apollo Scientific Experiment Package (EASEP). Similar packages were also present on the further lunar missions, and the results were published in several papers and several technical reports (Apollo Lunar Surface Experiment Package (ALSEP) Termination Report, NASA Reference Publication 1036, April, 1979). Apollo 11 carried photographical equipment and materials to obtain photographs of the transposition, docking, lunar module (LM) ejection maneuver and the LM rendezvous sequence from both the command and LM, as well as of the lunar ground track and of the landing site from the low point of the LM's flight path: it was also able to record the operational activities of the crew, obtain long-distance Earth and lunar terrain photographs with 70-mm still cameras, and make a photographical recording of the activities of the two astronauts who landed on the Moon. The camera equipment carried by Apollo 11 consisted of one 70-mm Hasselblad electric camera, two Hasselblad 70-mm lunar surface superwide-angle cameras, one Hasselblad El data camera, two 16-mm Maurer data acquisition cameras and one 35-mm lunar surface stereoscopic camera. The photographs taken were more then 1000 frames of 70-mm format, about 60,000 frames of 16-mm photography, and 17 stereoscopic pairs. It is thanks to these equipments that we have a punctual documentation of the astronauts' activities (see Fig. 8.23).

Apollo 13 was almost a failure, and fortunately not a slaughter due to the explosion of an oxygen tank in the SM. Astronauts survived, but the mission was not completed. Apollo 12 and 14–17 were successful and provided much of the data on which our current scientific understanding of the Moon is built. Here we want to underline, that even if under evaluated in comparison to its great political value, the scientific value of the lunar manned exploration was far from being negligible, also non-considering the tremendous advantage of having brought samples from another celestial body to the Earth. All the Apollo missions had scientific packages, including Apollo 11, in which the following experiments were present:

- The *soil mechanics investigation* studied the properties of the lunar soil (Costes *et al.*, 1970; Scott, 1975).
- The *solar wind composition experiment* collected samples of the solar wind for analysis on Earth (Geiss *et al.*, 1970).

Other experiments were deployed by the crew and then monitored from Earth by radio telemetry after the crew departed. This group of experiments was termed the EASEP. It was less extensive than the experiments performed on later missions, both because of time restrictions on the Extra-Vehicular Activity (EVA) and because of limitations on the payload mass carried on the first landing attempt.

FIGURE 8.23. The Apollo 11 Command Module "Columbia" carried astronauts Neil Armstrong, Edwin Aldrin and Michael Collins on their journey to the Moon and back on 16–24 July 1969. During the Apollo 11 mission the lunar excursion module (LEM).

- The *Passive Seismic Experiment* (PSE) detected lunar "moonquakes" and provided information about the internal structure of the Moon. PSE was designed to detect vibrations and tilting of the lunar surface and measure changes in gravity at the instrument location. The vibrations are due to internal seismic sources (moonquakes) and external (meteoroids and impacts from the spent Saturn IVB (Rocket stage) (S-IVB) and LM ascent stages). The primary objective of the experiment was to use these data to determine the internal structure, physical state and tectonic activity of the Moon. The secondary objectives were to determine the number and mass of meteoroids that strike the Moon and record tidal deformations of the lunar surface. The PSE unit was constructed principally of beryllium and had a mass of 11.5 kg, including the electronics module and thermal insulation (Nakamura *et al.*, 1982).
- The *laser ranging retroreflector* measured very precisely the distance between the Earth and Moon. The laser ranging retroreflector experiment was deployed on Apollo 11, 14 and 15. It consists of a series of corner-cube reflectors, which are a special type of mirror with the property of always reflecting an incoming light beam back in the direction it came from (see Fig. 8.24).
- The *lunar dust detector* studied the effects of lunar dust on the operation of the experiment package.

During Apollo 12 to the previous suite were added the lunar surface magnetometer, that measured the strength of the Moon's magnetic field, the cold cathode gauge that measured the abundance of gases in the lunar atmosphere, and the suprathermal ion detector experiment that studied the lunar ionosphere and the solar wind spectrometer measured the composition of the solar wind.

During the following missions more ambitious experiment were included, such as the heat flow experiment. The rate at which a planet loses its internal heat to space is an important control on the level of tectonics (faulting and folding of the planet's surface due to internal deformation) and volcanic activity on the planet. This loss of internal heat was measured by the *heat flow experiment* on Apollo 15 and 17. This experiment was also attempted on Apollo 16, but failed due to a broken cable connection.

FIGURE 8.24. The laser beam reflected from the Moon is observed with the telescope, providing a measurement of the round-trip distance between Earth and the Moon. This is the only Apollo experiment that is still returning data from the Moon. Many of these measurements have been made by McDonald Observatory in Texas. From 1969 to 1985, they were made on a part-time basis using the McDonald Observatory 107-in. telescope. Since 1985, these observations have been made using a dedicated 30-in. telescope (LPI Photo: webmaster@lpi.usra.edu).

The heat flow experiment involved drilling two holes into the regolith to depths of 1.6–2.3 m. The temperature was measured at several depths within the hole. The rate at which temperature increases with depth is a measure of the heat flowing from the Moon's interior. The drilling caused some heating within the hole, although the effects of this heating decayed with time. Also, temperatures in the upper part of the regolith vary as the amount of incident sunlight changes throughout the lunar day and night. By monitoring temperatures in the drill holes over a long period of time, these effects can be accounted for, allowing a determination of the average heat flow rate at the landing site.

The results of these measurements indicate a heat flow of $21 \, \mathrm{mW/m^2}$ at the Apollo 15 landing site and of $16 \, \mathrm{mW/m^2}$ at the Apollo 17 landing site. To place these numbers in context, to power an ordinary 60-W light bulb would require the collection of all the thermal energy flowing out of a lunar region 60 m by 60 m across. The Moon's heat flux is 18–24% of the Earth's average heat flux of $87 \, \mathrm{mW/m^2}$. The small value of the lunar heat flow was expected, given the Moon's small size and the observation that it has been nearly dead volcanically for the last 3 billion years. As the heat flow was measured at only two locations, it is not known how representative these values are for the Moon as a whole. However, because both measurements were obtained near boundaries between mare and highland regions, it is thought that the measured heat flows are probably 10–20% higher than the average value for the entire Moon (Langseth and Keihm, 1972).

It should be mentioned that the Moon is the only planet for which this fundamental measure has finally been performed. Hopefully similar experiment will be done in other to characterize Martian subsurface.

Another essential geophysical measure was performed thanks to the magnetometer. The magnetic field of the Moon was measured at four Apollo landing sites. On Apollo

12, 15 and 16, the lunar surface magnetometer was included in the ALSEP experiment package. The measured magnetic field has contributions both from the Moon's intrinsic magnetic field as well as from external sources, primarily the Earth and Sun. These external sources vary with time as the Moon orbits Earth and moves through its magneto-sphere. By making measurements over several months, these time-varying fields can be separated from the Moon's steady, intrinsic magnetic field. On Apollo 14 and 16, the lunar portable magnetometer was carried to measure how the Moon's magnetic field varies in local regions. For these measurements, the contributions of external field sources were estimated using previously deployed ALSEP magnetometers and measurements by the lunar orbiter Explorer 35. Similar measurements were done also on the Luna 16 and 20 samples collected during the soviet exploration (Guskova *et al.*, 1974).

The results of these experiments indicated significant variations in the strength of the Moon's magnetic field on both local and regional scales. The field strength varied from a low of 6 gammas at the Apollo 15 site to a high of 313 gammas at the Apollo 16 site. For comparison, the Earth's magnetic field strength is about 100 times higher than the highest value measured on the Moon. At the Apollo 14 site, the field strength varied between 43 and 103 gammas at two points separated by just 1.1 km. At the Apollo 16 site, the field strength varied between 121 and 313 gammas at points separated by a maximum of 7.1 km. These variations indicate the presence of strong localized sources in the crust for the Moon's magnetic field, a conclusion that is also consistent with observations from lunar orbit. In contrast, the Earth's magnetic field is generated by flow of fluids in the core and has a global dipole geometry. There is no evidence for such a dipole field on the Moon.

The magnetometer observations can also help constrain the interior structure of the Moon. From measurements of changes in the magnetic field at the lunar surface and in orbit as the Moon enters or exits the Earth's magnetic field, it is possible to estimate how the Moon's electrical conductivity varies with depth. This technique is known as EM sounding. The electrical conductivity varies with both chemical composition and with temperature. It has been used to place limits on the size of the Moon's iron core and on temperatures within the mantle. Mantle temperature estimates depend somewhat on the relative amount of olivine and pyroxene in the mantle. The temperature is below 1000°C until a depth of at least 500 km. Temperature estimates from EM sounding approach the melting point at a depth between 800 and 1500 km. For comparison, the PSE indicated melting beginning near 1000-km depth. The magnetometer observations do not require the existence of an iron-rich core, but are consistent with a core that is up to about 450 km in radius. Other constraints on the size of the core come from the PSE and the laser ranging retroreflector experiment.

The geophysical data were integrated into a coherent picture. Following Apollo's four seismometers detected infrequent, weak tremors originating within the Moon, at depths of 800–1100 km (500 to 700 miles), much deeper than those on Earth. Forty-three moon-quake zones were identified, each showing periodic activity correlated with lunar tides. Artificial seismic events (impacts of LM ascent stages and spent S-IVB stages) – plus the fortuitous collisions of a large (1100 kg, 2400 lb) meteorite with the Moon (Latham *et al.*, 1972) produced signals that suggested a crust about 60-km thick (Toksoz *et al.*, 1974), overlying a different homogeneous layer. In the 20-km-thick upper layer, veloc-ity gradients are high and microcracks may play an important role. The 40-km-thick lower layer has a nearly constant 6.8-km/s velocity. There may be a thin high-velocity layer present beneath the crust. Different kind of heterogeneous layers are extending down to about 1000 km. Deeper still a partially molten core may exist, which, assuming it to be a silicate rock, would be at a temperature of about 1500°C. These data are consistent with the magnetometer achievements. Results from the laser altimeters carried on Apollo 15 and 16 confirmed that the Moon's centre of mass does not coincide with its geometrical centre. Some scientists suggested this was due to the presence of the low-lying maria on the near side. Others suggested that the crust on the far side was thicker than that on the near side, which would make the absence of maria on the far side easier to explain: a thicker crust could have prevented the extrusion of molten material except in the very deepest craters, such as Tsiolkovsky.

However, the most important discoveries made by the lunar exploration were obtained thanks to the collection of lunar samples. All together, there are about 385 kg of material originally from the Moon in the worldwide collection. Most of this material (about 382 kg) was collected by US astronauts during the Apollo program (1969–72), but the collection also includes material collected by the Soviet Union's Luna robotic landers (1970–6) and a small number of meteorites that were ejected from the Moon during impact events. The samples collected by the astronauts come from different geological regions, unfortunately however only from the near-side. The returned samples showed that the Moon was chemically different from the Earth, containing a smaller proportion of volatile elements and more radioactive elements than the cosmic average. Geochemical evidence indicated that three types of rocks predominate on the lunar surface (Warner, 1975):

• basalts rich in iron covering the maria;
• plagioclase or aluminum-rich anorthosites characteristic of the highlands;
• uranium- and thorium-rich basalts that also contain high proportions of potassium, rare-earth elements and phosphorus ("KREEP" basalts).

Mare rocks are basaltic lavas that are primarily composed of pyroxene, plagioclase, olivine and opaque metallic oxides of varying composition, all of which may have widely varying iron, titanium and aluminum abundances. Highlands rocks are dominated by plagioclase (anorthosites), pyroxene and olivine, and have typically been pulverized and vitrified by impacts into unsorted, welded aggregates of material called *breccias*. Other, rarer lunar phases such as quartz and potassium feldspar have been detected in returned samples, but for the most part the mineralogy of the Moon is made up primarily of plagioclase, pyroxene, olivine and ilmenite. The oldest rocks found on the Moon, belonging to the highlands, have an age of about 4.5 billion years. According to a general view, the Moon was never completely molten; only its outer layer, perhaps to a depth of 320 km was melted. This theory is the so-called magma ocean theory. The large amount of molten material was possibly generated by a heavy bombardment, that characterized the Moon in a short period of time. This body of molten rock later, when the heavy bombardment period ended, began to solidify (Warren, 1985). As it did so, different minerals crystallized at different temperatures. This crust is represented by the Ferroan Anorthosites (FANs), which are believed to be flotation cumulates from an incipient Lunar Magma Ocean (LMO). As cooling continued, crystals of different composition separated, giving rise to the chemical segregation observed in the lunar crust. Eventually, probably after some 200 million years, a rigid crust of considerable thickness formed, composed mostly of light-coloured minerals rich in calcium and aluminum (plagioclase). Several recent isotopic studies have shown the importance of FANs to our understanding of lunar petrogenesis. The chronology of FANs suggests a range of ages from 4.57 ± 0.08 to 4.44 ± 0.02 Ga. This age range indicates a need for additional isotopic and geochronology studies of FANs. Beneath the crust a mantle of iron- and magnesium-rich material settled, consisting predominantly of the minerals pyroxene and olivine. At the centre of the Moon a core of dense, partially melted material may have formed, rich in iron and sulphur. The presence of a non-solid core is indicated by the behaviour of seismic waves passing through the Moon and by the presence of residual magnetism in lunar rocks: at sometime in the past, it appears, the Moon had a magnetic field which has since almost entirely vanished. The original crust was further bombarded. Large impact basins were formed by impact of large meteorite. Some of these impacts broke up the crust. Later lavas formed by melting of rock within the Moon due to the decay of radioactive elements. The broken crust under the big impact craters allowed the lava to come to the surface. Over time the craters came to be filled with lava flows: the so-called "mare basalts". Since the most important process for altering the surface of the Moon, is impact, these impacts also break older rock, re-compacting them. This kind of lunar samples are named "breccias"; in other words, a breccia is made when meteorites break up the surface and the pieces are welded together by the heat and pressure of impact processes. The lunar mineralogy is a result of the previously described phenomena. Lunar sample mineralogy is relatively simple with only the following major minerals: plagioclase, pyroxene, olivine and ilmenite (Steele and Smith, 1976; Papike *et al.*, 1996). This simple mineralogy of lunar samples results because lunar rocks were formed in a completely dry and very reducing environment with no hydrous minerals were produced. Geological processes special or important to the formation of lunar samples include: shock metamorphism (Schaal, 1976), cratering mechanics (Roddy, 1976), basin formation (Howard *et al.*, 1974), breccia formation, regolith gardening and partial melting to form basalt (Basaltic Volcanism Team, 1981).

For perhaps 300 million years after the crust formed, fragments of primordial material, some of them 50–100 km in diameter, bombarded the Moon. The impacts caused local melting of the crust and chemical alteration of the original material, scattered debris over thousands of square miles of the Moon's surface, and fractured the crust to a considerable depth.

This theory, even if is not totally accepted and deserves a further verification, was derived mainly from the inspection of the microscopic and macroscopic structure of rocks, and by the knowledge of their absolute dating. Further information were obtained by looking at the texture of rocks, that is characterized by the presence also of impact generated minerals.

8.5.1.2. *The Mars exploration*

Viking 1 and 2. The pioneering exploration of Mars was concluded by the most spectacular mission of the 70s: the Viking mission. Viking was the last mission conceived with the generosity of the seventies, however the project was so expensive to require extra support. The completion of Viking mission required a special consideration directly from the government; congressional hearing was conducted to examine the NASA request for additional funds to be applied to the lunar and planetary exploration programme. The increase in the programme is necessary to provide for unanticipated growth in funding requirements for the Viking project caused by development problems in a number of systems and subsystems (Committee on Science and Astronautics, 1974). NASA's Viking mission to Mars was composed of two spacecraft, Viking 1 and Viking 2, each consisting of an orbiter and a lander. The primary mission objectives were to obtain high-resolution images of the Martian surface, characterize the structure and composition of the atmosphere and surface, and search for evidence of life. Viking 1 was launched on 20 August 1975 and arrived at Mars on 19 June 1976. The first month of orbit was devoted to imaging the surface to find appropriate landing sites for the Viking Landers. On 20 July 1976 the Viking Lander 1 separated from the orbiter and touched down at Chryse Planitia (22.4° North, 49.97° West planetographical, 1.5 km below the datum (6.1 mbar) elevation). Viking 2 was launched 9 September 1975 and entered Mars orbit on 7 August 1976. The Viking Lander 2 touched down at Utopia Planitia (47.97° North, 225.74° West, 3 km below the datum elevation) on 3 September 1976. The orbiters imaged the entire surface of Mars at a resolution of 150–300 m, and selected areas at 8 m. The landers were powered by two radioisotope thermal generator (RTG) units attached to opposite sides of the lander base, each containing plutonium 238, providing 70 W continuous power. Four nickel–cadmium 8-amp-h rechargeable batteries were also on-board to handle peak power loads. These allowed the landers to survive for a long period of time, sending back to Earth a huge amount of data (a synthesis of Viking project can be found in Ezell and Ezell, 1984).

The Viking measurement still represent a great patrimony of knowledge. The two lander survived on the surface of Mars thanks to their nuclear power generators. The lander consisted of a six-sided aluminum base with alternate 1.09- and 0.56-m long sides, supported on three extended legs attached to the shorter sides. The leg footpads formed the vertices of an equilateral triangle with 2.21 m sides when viewed from above, with the long sides of the base forming a straight line with the two adjoining footpads. Instrumentation was attached to the top of the base, elevated above the surface by the extended legs. Power was provided by two RTG units containing plutonium 238 affixed to opposite sides of the lander base and covered by wind screens. Each generator had a mass of 13.6 kg and provided 30 W continuous power at 4.4 V. Four wet-cell sealed nickel–cadmium 8-amp-h, 28 V rechargeable batteries were also on-board to handle peak power loads. The Viking mission was planned to continue for 90 days after landing. Each orbiter and lander operated far beyond its design lifetime. Viking Orbiter 1 functioned until 25 July 1978, while Viking Orbiter 2 continued for 4 years and 1489 orbits of Mars, concluding its mission 7 August 1980. Due to the variations in available sunlight, both landers were powered by radioisotope thermoelectric generators – devices that create electricity from heat given off by the natural decay of plutonium. That power source allowed long-term science investigations that otherwise would not have been possible. The last data from Viking Lander 2 arrived at Earth on 11 April 1980. Viking Lander 1 made its final transmission to Earth 11 November 1982. With the exception seismometer the science instruments of Viking were fully operative. The seismometer on Viking Lander 2 detected only one event that may been seismic. The three biology experiments discovered unexpected and enigmatic chemical activity in the Martian but provided no clear evidence for the presence living micro-organisms in soil near the landing sites. According to mission biologists, Mars is self-sterilizing. They believe the combination of solar UV radiation that saturates the surface, the extreme dryness of the soil and the oxidizing nature the soil chemistry prevent the formation of living organisms in the Martian soil. The interpretation of that data has been recently challenged by former Viking scientist that have reinterpreted the Viking results.

The X-ray fluorescence spectrometers measured elemental composition of the Martian soil. Viking measured physical and magnetic properties the soil. As the landers descended towards the surface they also measured composition and physical properties of the Martian upper atmosphere. It should be mentioned that at present we are unable to provide new lander with similar generators. New generation generators are now under study for the exploration of the outer solar system, particularly Europa.

The results from the Viking experiments can be summarized as follows: the geological structure of Mars at regional level, was established: volcanoes, lava plains, immense canyons, cratered areas, wind-formed features and evidence of surface water were apparent in the orbiter images. Two main regions were identified: northern low plains and southern cratered highlands. Superimposed on these regions were volcanic areas and an extensive system of canyons near the equator. The surface material at both landing sites can best be characterized as iron-rich materials. Measured temperatures at the landing

FIGURE 8.25. Historical picture of Pathfinder on Mars, taken the 4 July 1997. Mars Pathfinder air bags have been successfully retracted, allowing safe deployment of the rover ramps. The air bags visible prominently in the top image are noticeably retracted at the bottom of the second image.

sites ranged from $-123°C$ to $-23°C$ ($-189°F$ to $-9°F$), with a variation over a given day of $35–50°C$. Seasonal dust storms, pressure changes and transport of atmospheric gases between the polar caps were observed. The biology experiment produced no evidence of life at either landing site.

Back to Mars: Pathfinder. In July of 1997, a small lander was deployed on the surface of the planet. After many years it was the first safe landing on Mars, not of the same level of the Viking mission, but characterized, for the first time, by mobility. This aspect attracted the attention of the non-Martian scientist and of the media, creating the premises for a further Martian exploration. It was a small roving laboratory, protected in the descent, by air bags (Fig. 8.25). After the descent, the safe deployment of the rover ramps and scout called the Sojourner began to analyse some of the nearby rocks. At the height of the mission, some of the names given to the rocks by the mission scientists, such as Barnacle Bill, and Yogi, became household words. The principle instrument to probe the mineralogy was called an Alpha Proton X-ray Spectrometer (APXS). It is a kind of ion probe, but in this case, the ions were alpha particles or helium nuclei. The particles came from radioactive nuclei that emit alphas, such as plutonium. This was a part of the instrumental package. When these alphas hit the rock, they interact with the atoms and nuclei of the rock to produce X-rays, protons and simply backscattered alphas. From the energy spectra of these particles and photons, it is possible to determine relative atomic abundances in the sample. Unfortunately, an analysis for the relative proportions of the chemical elements leaves the mineralogical composition of the rocks open. Through modelling it is possible of getting from the atomic percentages to plausible mineralogical compositions.

Among the more significant discoveries of the Mars Pathfinder mission was the identification of possible conglomerate rocks, which suggests the presence of running water to smooth and round the pebbles and cobbles, and deposit them in a sand or clay matrix. This scenario supports the theory that Mars was once warmer and wetter, as stressed by Golombek (2004). Pathfinder renewed the interest of the scientific community, as well as of the general public on the Martian evolution and was an important precursor of the MER.

8.5.1.3. *The MER*

The most exciting discoveries on the Martian surface evolution were recently obtained by the two MER, that showed clearly that the surface has been affected by several local processes involving volcanic activity, aeolian processes such as deposition and removal of dust generating dunes but mainly the water. This last has been the most important discovery of the two MERs. The two rovers weigh 174 kg and are long less then 2 m. The rovers have six wheels. Despite the fact that the two rovers have an on-board computer, they are not automatically guided, but they have to be mainly directed from ground. This implies the need of a good communication system, using a low-gain antenna and a high-gain antenna they can communicate with Earth directly or relay through the MGS or Mars Odyssey spacecraft orbiting the planet. The rovers can travel up to 100 m a day, however, most daily trips will be less so that the rover can study the soil, rocks, atmosphere and so on. To keep the batteries and electronics warm there are eight heaters. The temperatures may be as low as $-105°C$ ($-157°F$) during the night. Solar panels will use the daytime Sun to recharge the batteries and operate the vehicle. The rovers are designed to last about 90 days on Mars.

Each rover has several cameras to view the landing area so that scientists on Earth can direct the science investigation: there are four Engineering Hazcams (hazard avoidance cameras) mounted on the lower portion of the front and rear of the rover. These black- and-white cameras use visible light to capture three-dimensional (3D) imagery. Two Engineering Navcams (navigation cameras) are also present, mounted on the mast. Also these are in black-and-white: they gather panoramic, 3D imagery. The Navcam is a stereo pair of cameras, each with a 45° FOV to support ground navigation planning by scientists and engineers. They work in co-operation with the Hazcams by providing a complementary view of the terrain. The scientific cameras are also stereo, but they can provide colour information in 11 filters plus two-colour, solar-imaging filters to take multispectral image. The small laboratory is also equipped with a small "rock abrasion tool" (RAT), that uses three electric motors to drive rotating grinding teeth into the surface of a rock. The RAT, uses three electric motors to drive rotating grinding teeth into the surface of a rock. Once a fresh surface is exposed, the abraded area can be examined in detail using mainly the small microscope. The microscopic imager is a combination of a microscope and a CCD camera that will provide information on the small-scale features of Martian rocks and soils. The MER are also fitted out with three spectrometers, a spectrometers the miniature TES (Mini-TES); the Mossbauer (MB) spectrometer and the APXS. This miniaturized set of spectrometers has been able to determine the chemical and mineralogical composition of the different rock encountered along the MERs path. MER-1, nicknamed "Spirit" was launched towards Mars on 10 June 2003. MER-2, called "Opportunity", was launched on 7 July 2003. Spirit arrived at the Red Planet on time and sent its rover to the surface with a successful landing at 8:35 p.m. (PST) on 3 January 2004.

The landing of Spirit in the Gusev Crater was successful is settled. Opportunity landed to Meridiani Planum in a rocky area, characterized by in hematite-rich materials. This choice was made on the basis of the results of Tess on Odyssey Satellite. The Mini-Thermal Emission Spectrometer (MTES) data collected by the Opportunity module has shown the presence of silicates and CO_2, obviously present in the Martian atmosphere. Spirit's analysis of nearby soil showed the presence of olivine, a magnesium–iron silicate that is a constituent of basalt. The mineral probably is present as loose grains blown to the site by winds. The Mossbauer experiment showed that the rock has basaltic composition. Following (Short, 2004) who made an excellent synthesis of the MER results, the main data obtained at Gusev by Spirit site are that the main rock type present is basalt; hematite is of the red variety; although several rocks may contain hydrous mineral(s), no other signs of lake water or water-related sediments, including strata of any kind, have as yet been detected; however, hydrothermal or groundwater activity from subsurface sources has been verified. Coatings containing Cl, Br and SO_3 remain enigmatic

FIGURE 8.26. This rock is splitting into thin flaky layers or chips. It resembles a fissile shale on Earth but is richer in iron than most shales. Such layering is usually a consequence of continuing deposition in a fluid medium, usually water (NASA picture).

(see Fig. 8.26). The first view of the Opportunity landscape showed the unusual rock outcropping nearby, as shown by the rover's panoramic camera. These rocks could be interpreted layered rocks are either volcanic ash deposits or sediments laid down by wind or water. It was given the name Opportunity ledge the rocks at Opportunity's Meridiani site visually are strongly suggestive of layered rocks, which may have been water-laid sediments, although altered basalt or even volcanic ash deposits were plausible alternatives (Joliff and the Athena team, 2005). At the moment each hypothesis has merit and thus both are still valid, but the activity of water is now firmly confirmed (Fig. 8.26).

The MER activity, as well as the Mars Express activity is successfully continuing. It should be mentioned that an effort has been made by European and American scientists to co-ordinate their activities. As an example the OMEGA hyperspectral imager (0.35–5.1 μm) covered the hematite-bearing plains and underlying etched terrains of Meridiani Planum during several orbits with spatial resolutions ranging from several hundred metres to approximately a kilometre, when the Opportunity rover acquired Pancam and Mini-TES observations of the surface and atmosphere, exactly during the times when OMEGA acquired its data (Arvidson *et al.*, 2004). Therefore, we shall expect new results from the Mars exploration in the near future, also thanks to the Mars Advanced Radar for Subsurface and Ionosphere Sounding (MARSIS) radar of Mars Express whose antenna will be finally deployed by May 2005.

8.5.2. *The outer solar system*

8.5.2.1. *Galileo and Huygens probes*

The exploration of outer solar system, after the pioneering work made by Voyager, the detailed investigation from orbit of Galileo, asks for dedicated probes. In the case of giant planets, the word "*in situ*" apply mainly to the atmosphere. After a 6-year journey through the solar system and after being inexorably accelerated to a speed of 170,700 km/h (106,000 mph) by Jupiter's gravitational pull, the Galileo probe successfully entered Jupiter's atmosphere at 22:04 UT (2:04 p.m. PST) on 7 December 1995. During the first 2 min, near-probe air temperatures twice as hot as the Sun's surface

FIGURE 8.27. Using the MB spectrometer, these spherules have now been shown to consist mainly of hematite.

and deceleration forces as great as 230 g (230 times the acceleration of gravity at Earth's surface) were produced as the probe was slowed down by Jupiter's atmosphere. During the parachute-descent phase of the mission, the Galileo probe successfully studied the atmosphere of Jupiter with seven scientific experiments and radioed its findings to the Galileo Orbiter, which was 215,000 km overhead. The Galileo probe and orbiter separated on 13 July 1995 and both arrived at Jupiter on 7 December on slightly different trajectories. The probe, which did not include a camera, transmitted data for 57.6 min. It penetrate the Jovian atmosphere until about 200 km below the visible cloud tops, where the communication system failed due to the high temperatures. Measurements of temperature, pressure and vertical winds were performed. During the descent in the Jupiter's atmosphere the Atmosphere Structure Instrument (ASI) measured the temperature, pressure and density structure of atmosphere from the uppermost regions down through an atmospheric pressure of about 24 bars.

The results of Galileo probe can be summarized as follows: the upper atmospheric densities and temperatures are significantly higher than expected therefore an additional source of heating beyond sunlight appears to be necessary to account for this result. At deeper levels the temperatures and pressures are close to expectations. The vertical variation of temperature in the 6–15 bar pressure range (about 100–150 km below visible clouds) indicates the deep atmosphere is dryer than expected and is convective. The ASI's initial results have various important implications. The ideas about the abundance and distribution of water on Jupiter was strongly modified. The ASI measurements will increase our understanding of the escape of Jupiter's internal heat – a power source for its dynamic atmosphere. In addition, because of the convective nature of the lower levels of the atmosphere, the deep atmosphere must be well mixed, and composition measurements obtained by other instruments must be representative of the deeper levels of Jupiter's atmosphere as well. The Neutral Mass Spectrometer (NMS) experiment's objective was to accurately determine the composition of the atmosphere. Initial results indicate the atmosphere has much less oxygen – mainly found as water vapour in Jupiter's atmosphere – than the Sun's atmosphere, implying a surprisingly dry atmosphere. On the other hand, the amount of carbon – mainly found as methane gas – is highly enriched with respect to the Sun, while sulphur – in the form of hydrogen sulphide gas – occurs at greater than solar values. The abundance of nitrogen, in the form of ammonia gas, is still pending. The abundance of neon, a Noble or "inert" gas, is highly depleted. Little evidence of organic molecules was found. From the previous considerations it is clear that the *in situ* measurements are extremely valuable. For this reason in the Cassini mission was also included the Huygens probe.

When the Cassini mission was firstly proposed, a discussion in the planetary science community was present about the opportunity to have a twin probe of the Galileo one, possibly released into Saturn. However the new discoveries on Titan, started with the Voyager measurements and the peculiarity of the Titan convinced the community, and the two main agencies involved in Cassini to support the idea of a Titan probe. The Huygens probe was then designed by ESA, becoming the first *"in situ"* experiment made in Europe.

The Huygens probe is an atmospheric probe designed to make *in situ* observations of the Saturnian satellite Titan. ESA's contribution to the Cassini mission, Huygens' objectives are to: (1) determine the physical characteristics (density, pressure, temperature, etc.) of Titan's atmosphere as a function of height; (2) measure the abundance of atmospheric constituents; (3) investigate the atmosphere's chemistry and photochemistry, especially with regard to organic molecules and the formation and composition of aerosols; (4) characterize the meteorology of Titan, particularly with respect to cloud physics, lightning discharges and general circulation and (5) examine the physical state, topography and composition of the surface. Similar in design to the Galileo probe, Huygens is a 1.3 m diameter descent module with a spherical nose and a conical aft section. A thermal protection aeroshell surrounds the descent module, decelerating it from 6 km/s at arrival to 400 m/s in about 2 min and protecting it from the intense heat of entry. A parachute will then be deployed and the aeroshell jettisoned. The probe will float down through the atmosphere making measurements. Instrumentation for the probe will include: an aerosol collector and pyrolyzer, a descent imager and spectral radiometer, a Doppler wind experiment, a gas chromatograph/mass spectrometer, an ASI and a surface science package (Table 8.8).

The Huygens results were much better then expected. The combination of VIMS data, the recent radar data have revealed that the Saturn satellite Titan does not have a liquid ocean covering it. This new discovery has been confirmed by the Huygens data, showing clearly a rocky surface (Fig. 8.28).

8.5.2.2. *The Rosetta lander*

The next *in situ* payload is the Rosetta lander. The landing on the comet Churyumov–Gerasimenko is foreseen May 2014. The Rosetta lander is aimed to obtain a first *in situ* composition analysis of primitive material from the early solar system, to study the composition and structure of a cometary nucleus, that possibly still preserves records of processes in the early solar system. As any *in situ* lander the most important goal is to provide ground truth data for the Rosetta orbiter experiments and to investigate dynamic processes leading to changes in cometary activity.

The primary objective of the Rosetta lander mission is the *in situ* investigation of the elemental, isotopic, molecular and mineralogical composition, and the morphology of early solar system material as it is preserved in the cometary nucleus. Interpretation of the chemical composition of surface material benefits strongly from the knowledge of fractionation and aging processes which occur in the upper surface layers since it allows to draw conclusions on the original cometary material. Physical and thermal properties

TABLE 8.8. Huygens instruments and Principle Investigators (PIs).

Instrument	PI
Descent Imager and Spectral Radiometer (DISR)	Dr Martin G. Tomasko
Doppler Wind Experiment (DWE)	Dr Michael K. Bird
Gas Chromatograph and Mass Spectrometer (GCMS)	Dr Hasso B. Niemann
Surface Science Package (SSP)	Prof. John C. Zarnecki
The Aeronomy of Titan (IDS)	Dr Daniel Gautier
Titan Atmosphere–Surface Interactions (IDS)	Dr Jonathan I. Lunine
The Chemistry and Exobiology of Titan (IDS)	Dr Francois Raulin
Aerosol Collector and Pyrolyzer (ACP)	Dr Guy M. Israel
Huygens Atmospheric Structure Instrument (HASI)	Prof. Marcello Fulchignoni

FIGURE 8.28. Raw data of the Cassini-Huygens descent (ESA/NASA/JPL/University of Arizona).

of near-surface material affect these fractionation processes. Therefore, the investigation of the composition together with the physical and thermal properties of surface and subsurface material, and the study of the thermal behaviour over many daynight and during possibly one seasonal cycles (along almost one cometary year).

The Rosetta lander is an extremely clever experiment since, for the first time a group of researchers has attempted to develop in a "laboratory style" a complex small spacecraft made by several state of the art experiments. Moreover, the Rosetta lander contains as well the so-called DS2 that is a small drill able to collect sample and distribute it to different experiments. The drill, that could be the prototype of a similar experiment to be flown on Mars, has been developed under the guidance of the Politecnico di Milano.

To conclude, only the combination of orbital and *in situ* science can give a definitive contribution to the knowledge of a complex reality, such as a planet or a satellite. For the future more challenging experiment are foreseen, that will couple the mobility, as the MER, with the possibility of autonomous movement. Possibly in the future also the subsurface of planets will be explored using new tools, as drills.

REFERENCES

ADAMS, J.B. & MCCORD, T.B. 1969 Mars: interpretation of spectral reflectivity of light and dark regions. *J. Geophys. Res.*, **74**, 4851–4856.

ADLER, I. & TROMBKA, J.I. 1980 Geochemical Exploration of the Moon and Planets. *Phys. Chem. Space*, **3**.

ANDERSON, S., HAMILTON, V.E. & CHRISTENSEN, P.R. 2003 Mineralogy of the valles mariners from TES and THEMIS. *Sixth International Conference on Mars*, LPI, 3280.

ARVIDSON, R.E., BIBRING, J., POULET, F., SQUYRES, S.W., WOLFF, M. & MORRIS, R. 2004 Coordinated Mars exploration rover and Mars Express OMEGA observations over Meridiani Planum. *Am. Geophys. Union*, abstract no. P24A-06.

BAINES, K.H., BELLUCCI, G., BIBRING, J.P., BROWN, R.H., BURATTI, B.J., BUSSOLETTI, E., CAPACCIONI, F., CERRONI, P., CLARK, R.N., CORADINI, A., CRUIKSHANK, D.P., DROSSART, P., FORMISANO, V., JAUMANN, R., LANGEVIN, Y., MATSON, D.L., MCCORD, T.B., MENNELLA, V., NELSON, R.M., NICHOLSON, P.D., SICARDY, B., SOTIN, C., HANSEN, G.B., AIELLO, J. & AMICI, S. 2000 Detection of sub-micro radiation from the surface of Venus by Cassini/VIMS. *Icarus*, **148**(1), 307–311.

BANDFIELD, J., JOSHUA, L., HAMILTON, VICTORIA, E. & CHRISTENSEN, P.R. 2000 A global view of Martian surface compositions from MGS–TES. *Science*, **287**, 1626–1630.

BASALTIC VOLCANISM TEAM 1981 *Basaltic Volcanism on the Terrestrial Planets*. Pergamon Press, Inc, New York.

BELLUCCI, G., BROWN, R.H., FORMISANO, V., BAINES, K.H., BIBRING, J.P., BURATTI, B.J., CAPACCIONI, F., CERRONI, P., CLARK, R.N., CORADINI, A., CRUIKSHANK, D.P., DROSSART, P., JAUMANN, R., LANGEVIN, Y., MATSON, D.L., MCCORD, T.B., MENNELLA, V., MILLER E., NELSON, R.M., NICHOLSON, P.D., SICARDY, B. & SOTIN, C. 2002 Cassini/VIMS observations of the Moon. *Adv. Space Res.*, **30**(8), 1889–1894.

BIBRING, J.P., ERARD, S., GONDET, B., LANGEVIN, Y., SOUFFLOT, A., COMBES, M., CARA, C., DROSSART, P., ENCRENAZ, T., LELLOUCH, E., ROSENQVIST, J., MOROZ, V.I., DYACHKOV, A.V., GRYGORIEV, A.V., HAVINSON, N.G., KHATUNTSEV, I.V., KISELEV, A.V., KSANFOMALITY, L.V., NIKOLSKY, YU.V., MASSON, P., FORNI, O. & SOTIN, C. 1991 Topography of the Martian tropical regions with ISM. *Planet. Space Sci.*, **39**, 225–236.

BIBRING, J.P., LANGEVIN, Y., POULET, F., GENDRIN, A., GONDET, B., BERTHÉ, M., SOUFFLOT, A., DROSSART, P., COMBES, M., BELLUCCI, G., MOROZ, V., MANGOLD, N., SCHMITT, B. & OMEGA TEAM 2004 Perennial water ice identified in the south polar cap of Mars. *Nature*, **428**(6983), 627–630.

BODE, M.F. 1995 Robotic observatories. *Wiley-Praxis Series in Astronomy and Astrophysics*. Chichester, New York.

BROOKS, C.G., GRIMWOOD, J.M. & SWENSON L. 1979 *Chariots for Apollo: A History of Manned Lunar Spacecraft SP NASA Special Publication-4205 in the NASA History Series*.

BROWN, R.H., BAINES, K., BELLUCCI, G., BIBRING, J.-P., BURATTI, B., CAPACCIONI, F., CERRONI, P., CLARK, R., CORADINI, A., CRUIKSHANK, D., DROSSART, P., FORMISANO, V., JAUMANN, R., MATSON D., MCCORD, T., MENNELLA, V., NELSON, R., NICHOLSON, P., SICARDY, B., SORTIN, C. & CHAMBERLAIN, M. 2002 Near-infrared spectroscopy of Himalya. *Planet. Sci. Con.*, abstract no. 200.

BROWN, R.H., BAINES, K.H., BELLUCCI, G., BIBRING, J.P., BURATTI, B.J., CAPACCIONI, F., CERRONI, P., CLARK, R.N., CORADINI, A., CRUIKSHANK, D.P., DROSSART, P., FORMISANO, V., JAUMANN, R., LANGEVIN, Y., MATSON, D.L., McCORD, T.B., MENNELLA V., NELSON, R.M., NICHOLSON, P.D., SICARDY, B., SOTIN, C., AMICI, S., CHAMBERLAIN, M.A., FILACCHIONE, G., HANSEN, G., HIBBITTS, K. & SHOWALTER, M. 2003 Observations with the visual an infrared mapping spectrometer (VIMS) during Cassini's fly-by of Jupiter. *Icarus*, **164**(2), 461–470.

BURNETT, D.S. 1975 Lunar science – The Apollo legacy. *Rev. Geophys. Space Phys.*, **13**, 13–34.

BURNS, R. 1993 *Mineralogical Applications of Crystal Field Theory*, 2nd Edn. Cambridge University Press, Cambridge, 551.

CASTRONUOVO, M.M. & ULIVIERI, C. 1995 Removal of Mars atmospheric gas absorption from Phobos-2/ISM spectra. *Adv. Space Res.*, **16**, 75.

CERRONI, P. & CORADINI, A. 1995 Multivariate classification of multispectral images: an application to ISM Martian spectra. *Astron. Astrophys. Suppl.*, **109**, 585–591.

CHAPMAN, C.R., POLLACK, J.B. & SAGAN, C. 1969 An analysis of the Mariner-4 cratering statistics. *Astron. J.*, **74**, 103.

CHRISTENSEN, P.R. 1986 The spatial distribution of rocks on Mars. *Icarus*, 68, 217–223.

CLARK R.N. 1981 Water frost and ice: the near infrared spectra reflectance 0.65–2.5 μm *J. Geophys. Res.*, **86**, 3087.

CLARK, R.N. 1999 Spectroscopy of rocks and mineral and principles of spectroscopy. In: *Manual of Remote Sensing*, A.N. Rencz (ed.), 3–58 John Wiley and Sons, New York.

CLARK, R.N., BROWN, R.H., JAUMANN, R., CRUIKSHANK, D.P., NELSON, R.M., BURATTI, B.J., McCORD, T.B., HOEFEN, T.M., CURCHIN, J.M., HANSEN, G., HIBBITS, K., MATZ, K.D., BAINES, K.H., BELLUCCI, G., BIBRING, J.P., BUSSOLETTI, E., CAPACCIONI, F., CERRONI, P., CORADINI, A., FORMISANO, V., LANGEVIN, Y., MATSON, D.L., MENNELLA, V., NICHOLSON, P.D., SICARDY, B. & SOTIN, C. 2004 VIMS The surface composition of Saturn's Moon Phoebe as seen by the Cassini visual and infrared mapping spectrometer. *America Astronomical Society*, DPS Meeting, **36**.

COMBES, M., CARA, C., DROSSART, P., ENCRENAZ, T., LELLOUCH, E., ROSENQVIST, J., BIBRING, J.-P., ERARD, S., GONDET, B., LANGEVIN, Y., SOUFFLOT, A., MOROZ, V.I., GRYGORIEV, A.V., KSANFOMALITY LU. V., NIKOLSKY, Y.V., SANKO, N.F., TITOV, D.V., FORNI, O., MASSON, P., SOTIN, C. 1991 Martian atmosphere studies from the ISM experiment. *Planet. Space Sci.*, **39**, 189–197.

COMMITTEE ON SCIENCE AND ASTRONAUTICS (U.S. HOUSE) 1974 Viking Project Hearings before Subcomm. on Space Sci. and Appl. of Comm. on Sci. and Astronaut. 93rd Congr., 2nd Sess., **51**, 21–22.

CORADINI, A., CAPACCIONI, F., DROSSART, P., SEMERY, A., ARNOLD, G., SCHADE, U., ANGRILLI, F., BARUCCI, M.A., BELLUCCI, G., BIANCHINI, G., BIBRING, J.P., BLANCO, A., BLECKA, M., BOCKELEE-MORVAN, D., BONSIGNORI, R., BOUYE, M., BUSSOLETTI, E., CAPRIA, M.T., CARLSON, R., CARSENTY, U., CERRONI, P., COLANGELI, L., COMBES, M., COMBI, M.; CROVISIER, J., DAMI, M., DeSANCTIS, M.C., DiLELLIS, A. M., DOTTO, E., ENCRENAZ, T., EPIFANI, E., ERARD, S., ESPINASSE, S., FAVE, A., FEDERICO, C., FINK, U., FONTI, S., FORMISANO, V., HELLO, Y., HIRSCH, H., HUNTZINGER, G., KNOLL, R., KOUACH, D., IP, W.H., IRWIN, P., KACHLICKI, J., LANGEVIN, Y., MAGNI, G., McCORD, T., MENNELLA, V., MICHAELIS, H., MONDELLO, G., MOTTOLA, S., NEUKUM, G., OROFINO, V., OROSEI, R., PALUMBO, P., PETER, G., PFORTE, B., PICCIONI, G., REESS, J.M., RESS, E., SAGGIN, B., SCHMITT, B., STEFANOVITCH, A., TAYLOR, F., TIPHENE, D. & TOZZI, G. 1998 Virtis: an imaging spectrometer for the Rosetta mission. *Planet. Space Sci.*, **46**, 9–10, 1291–1304.

CORADINI, A., CAPACCIONI, F., DROSSART, P., SEMERY, A., ARNOLD, G. & BENKHOFF, J. 1999 VIRTIS: an imaging spectrometer for the Rosetta mission. *Am. Astron. Soc.*, **31**.

CORADINI, A., FILACCHIONE, G., CAPACCIONI, F., CERRONI, P., ADRIANI, A., BROWN, R.H., LANGEVIN, Y. & GONDET, B. 2004 CASSINI/VIMS-V at Jupiter: radiometric calibration test and data results. *Planet. Space Sci.*, **52**(7), 661–670.

COSTES, N.C., CARRIER, W.D. & SCOTT, R.F. 1970 Apollo 11 soil mechanics investigation. *Science*, **167**(3918).

EL-BAZ, F. 1974 Surface geology of the Moon. *Ann. Rev. Astron. Astrophys.*, **12**, 135–165.

EL-BAZ, F. & WILHELMS, D.E. 1975 Photogeological, geophysical, and geochemical data on the east side of the Moon. In: *Lunar Science Conference*, **3**, 2721–2738.

ENCRENAZ, T., LELLOUCH, E., ROSENQVIST, J., DROSSART, P., COMBES, M., BILLEBAUD, F., DE PATER, I., GULKIS, S., MAILLARD, J.P. & PAUBERT, G. 1991 The atmospheric composition of Mars-ISM an ground-based observational data. *Ann. Geophys.*, **9**, 797–803.

ERARD, S. & CALVIN, W. 1997 New composite spectra of Mars 0.4–5.7 μm. *Icarus*, **130**(2), 449–460.

EVANS, J.V. 1969 Radar studies of planetary surfaces. *Ann. Rev. Astron. Astrophys.*, **7**, 201–248.

EVANS, L.G., REEDY, R.C. & TROMKA, J.I. 1993 Introduction to planetary remote sensing X and gamma ray spectroscopy. *Topic Remote Sens.*, 167–212.

EVANS, L.G., STARR, R.D., BRÜCKNER, J., REEDY, R.C., BOYNTON, W.V., TROMBKA, J.I., GOLDSTEIN, J.O., MASARIK, J., NITTLER, L.R. & MCCOY, T.J. 2001 Elemental composition from gamma RAY spectroscopy of the NEAR-Shoemaker landing site. *Meteorit. Planet. Sci.*, **36**(12), 1639–1660.

EZELL, E.C. & NEUMAN EZELL, L. 1984 Exploration of the Red Planet 1958–1978 The NASA History Series Scientific and Technical Information Branch.

FILIPPENKO, A.V. 1992 The scientific potential of automatic CCD imaging telescopes in *ASP Conf. Ser. 34: Robotic Telescopes in the 1990s*, 55–66.

FORMISANO, V. & THE PFS TEAM 2004 First preliminary result of PFS–MEX. *Mars Geophys. Res. Abstr.*, **6**, 7336.

FRANKS, F. (Ed.) 1972 *Water – A Comprehensive Treatise*. Plenum Press, New York.

GAFFEY, S.J., MCFADDEN, L.A., NASH, D. & PIETERS, C.M. 1993 Ultraviolet, visible, and near-infrared reflectance spectroscopy laboratory spectra of geologic materials. In: *Remote Geochemical Analysis: Elemental and Mineralogical Composition* (C.M. Pieters, P.A.J. Englert (Eds.), 43–78. Cambridge University Press.

GARY, V., LATHAM, M.E., FRANK, P., GEORGE, S., JAMES, D., YOSIO, N., NAFI, T., DAVID, L. & FRED, D. 1972 Passive Seismic Experiment, *Apollo 16 Preliminary Science Report*, NASA SP-315, 9–1.

GEISS, J., EBERHARDT, P., BCHLER, F., MEISTER, J. & SIGNER, P. 1970 Apollo 11 and 12 solar wind composition experiments: fluxes of He and He isotopes. *J. Geophys. Res.*, **75**, 5972–5979.

GOLOMBEK, M. 2004 Climate Change on Mars: From wet in the Noachian at Meridiani to dry and desiccating in the Hesperian/Amazonian plains of Gusev. *Am. Geophys. Union*, abstract no. P13B-08.

GUINNESS, E.A., ARVIDSON, R.E., JOLLIFF, B.L., DEAL, K.S., SEELOS, F.P., MORRIS, R.V., MING, D.W. & GRAFF, T.G. 2003 Mapping Hydrothermal Alteration Zones on Mauna Kea Using AVIRIS Data: A Analog for Mars *AGU Fall Meeting Abstracts*, B1043.

GUSKOVA, E.G., GORSHKOV, E.S., IVANOV, A.V., POCHTAREV, V.I. & FLORENSKII, K.P. 1974 Magnetic properties of lunar rocks obtained from ALS "Luna-16" and "Luna-20". *Kosmicheskie Issledovaniia*, **12**, 748–757 (in Russian).

HANEL, R.B., CONRATH, W., HOVIS, V., KUNDE, P.D., LOWMAN, W., MAGUIRE, J.P., PIRRAGLIA, J., PRABHAKARA, C., SCHLACHMAN, B., LEVIN, G., STRAAT, P. & BURKE, T. 1972 Investigation of the Martian environment by infrared spectroscopy on Mariner-9. *Icarus*, **7**, 423–442.

HANEL, R.B., CONRATH, B.J. & JENNINGS, D.E. 1992 *Exploration of the Solar System by Infrared Remote Sensing*. Cambridge University Press, Cambridge.

HAPKE, B. 1981 Bidirectional reflectance spectroscopy I. Theory. *J. Geophys. Res.*, **86**, 3039–3054.

HAPKE, B. 2001 *Introduction to the Theory of Reflectance and Emittance Spectroscopy.* Cambridge University Press.

HAPKE, B. & WELLS, E. 1981 Bidirectional reflectance spectroscopy II. Experiments an observations. *J. Geophys. Res.*, **86**, 3055–3060.

HOLLAS, J.M. 2004 *Modern Spectroscopy.* John Wiley & Sons Inc., Chichester.

HOWARD, K.A., WILHELMS, D.E. & SCOTT, D.H. 1974 Lunar basin formation and highland stratigraphy. *Rev. Geophys. Space Phys.*, **12**, 309–327.

HUNT, G.R. & SALISBURY, J.W. 1970 Visible and near infrared spectra of minerals and rocks. 1. Silicate minerals. *Mod Geol.*, **1**, 283–300.

KAHLE, A.B. 1994 Preface of the volume. *Adv. Space Res.*, **14**(3), 3.

KIRKLAND, K.C. HERR & McAFEE, J.M. 1999 Utilizing night spectra of mars for mineralogy, *30th Annual Lunar and Planetary Science Conference*, abstract no. 168.

KIRKLAND, L.E., FORNEY, P.B. & HERR, K.C. 2000 Comparison of Thermal Infrared Spectral Data Sets of Mars: 1969 IRS, 1971 IRIS and 1996 TES, *1st Annual Lunar and Planetary Science Conference*, abstract no. 192.

KITCHIN, C.R. 1998 Astrophysical Techniques *Institute of Physics Publishing.*

KRUSE, F.A. & TARANIK, D.L. 1988 Mapping hydrothermally altered rocks with the airborne imaging spectrometer (AIS) and the airborne visible/infrared imaging spectrometer (AVIRIS), *Proceedings of IGARSS '89 and Canadian Symposium on Remote Sensing*, 952–956.

LANGSETH, M.B. & KEIHM, S.J. 1977 *In-situ* measurements of lunar heat flow. In NASA The Soviet-Am. Conf. on Cosmochem. of the Moon and Planets, pt. 1, 283–293.

LATHAM, G., EWING, M., DORMAN, J., LAMMLEIN, D., PRESS, F., TOKSOZ, M.N., SUTTOU, G., DUENNEHIER, F., NAKAMURA, Y. 1972 Earth, Moon and Planets, *4*, 373.

LEIGHTON, R.B., MURRAY, B., SHARP, R.P., ALLEN, J.D. & SLOAN, R.K. 1967 Mariner-4 Pictures of Mars *Technica Report TR-32-884* Jet Propulsion Laboratory.

LENA, P., LEBRUN, F. & MIGNARD, F. 1998 *Observational Astrophysic,* Springer Verlag, Berlin.

MACKAY, C.D. 1986 Charge-coupled devices in astronomy. *Ann. Rev. Astron. Astrophys.*, **24**, 255–283.

MATURILLI, A., FORMISANO, V. & GRASSI, D. 2002 Albite a component of martian dust: evidences from Iris and TES data COSPAR, 2002.

McCORD, T.B., BROWN, R.H., BAINES, K., BELLUCCI, G., BIBRING, J.P., BURATTI, B., CAPPACCIONI, F., CERRONI, P., CLARK, R.N., CORADINI, A., CRUIKSHANK, D., DROSSART, P., FORMISANO, V., JAUMANN, R., LANGEVIN, Y., MUTSON, D.M., MEVELLA, V., NELSON, R., NICHOLSON, P., SICARDY, B., SOTIN, C., HAUSEN, G.B., HIBBITTS, C.A. 2001 Galilean Satellite Surface Non-Ice Constituents: New Results from the Cassini/Huygens VIMS Jupiter Flyby in the context of the Galileo NIMS Results, 32nd Ann. Lunav and Planetary Science Conf., abstract 1247.

McCORD, T.B., CORADINI, A., HIBBITTS, C.A., CAPACCIONI, F., HANSEN, G.B., FILACCHIONE, G., CLARK, R.N., CORRONI, P., BROWN, R.H., BAINES, K.H., BELLUCCI, G., BIBRING, J.P., BURATTI, B.J., BUSSOLETTI, E., CAUBES, M., CRUIKSHANK, D.P., DROSSART, P., FORMISANO, V., JAUMANN, R., LANGEVIN, Y., MATSON, D.L., NELSON, R.M., NICHOLSON, P.D., SICARDY, B., SOTIN, C. 2004 CASSINI VIMS Observations of the Galilean Satellites including the VIMS calibration procedure, Icarus, *172*, 104–126.

McCORD, T.B., GRIFFITH, C.A., HANSEN, G.B., LUNINE, J.I., BAINES, K.H., BROWN, R.H., BURATTI, B., CLARK, R.N., CRUIKSHANK, D.P., FILACCHIONE, G., JAUMANN, R., HIBBITTS, C.A., SOFINE, C. 2004 Titan's Surface Composition from the Cassini Visual and Infrared Mapping Spectrometer (VIMS) Investigation, AAS, DSP meeting 36, 06.02.

McLEAN, I. 1997 Electronic Imaging in Astronomy Detectors and Instrumentation. In: *Wiley-Praxis Series in Astronomy and Astrophysics.* Chichester.

MORRIS, R.V., LAWSON, C.A., GIBSON, E.K., LAUER, H.V., NACE, G.A. & STEWART, C. 1985 Spectral and other physicochemical properties of submicron powders of hematite (alpha-Fe$_2$O$_3$) maghemite (gamma-Fe$_2$O$_3$), magnetite (Fe$_3$O$_4$), goethite (alpha-FeOOH) and lepidocrocite (gamma-FeOOH). *JGR*, **90**, 3126–3144.

MUSTARD, J.F. & BELL, J.F. 1994 New composite reflectance spectra of Mars from 0.4 to 3.14 μm. *Geophys. Res. Lett.*, **21**(5), 353–356.

MUSTARD, J.F., MURCHIE, S., ERAD, S., SUNSHINE, J. 1997 In situ composition of Martian volcanics: Implications for the Mantle, J&R, *102*, 25605.

NAKAMURA, Y., LATHAM, G.V. & DORMAN, H.J. 1982 Apollo lunar seismic experiment – final summary. *J. Geophys. Res. Suppl.*, **87**, A117–23.

NEUKUM, G. 1982 Multispectral imaging of Mars ESA-SP 185. *The Planet Mars*, 117–120.

NORWOOD, V.T. & LANSING, J.C. 1983 Electro-optical imaging system in Colwell, R.N., eds, Manual of remote sensing, ch. 8, 335–367, Falls Church, VA.

PAPIKE, J.J., HODGES, F.N., BENCE, A.E., CAMERON, M. & RHODES, J.M. 1976 Mare basalts – crystal chemistry mineralogy, and petrology. *Rev. Geophys. Space Phys.*, **14**, 475–540.

RAVA, B. & HAPKE, B. 1987 An analysis of the Mariner-10 color ratio map of Mercury. *Icarus*, **71**, 397–429.

RODDY, D.J. 1976 *Lunar Science Conference Proceedings*, **3**, 3027–3056.

SALISBURY, J.W. & D'ARIA, D.M. 1992 Emissivity of terrestrial materials in the 8–14-μm atmospheric window. *Remote Sens. Environ.*, **42**, 83–106.

SALISBURY, J.W., D'ARIA, D.M. & JAROSEWICH, E. 1991 Midinfrare (2.5–13.5 microns) reflectance spectra of powdered stone meteorites. *Icarus*, **92**, 280–297.

SAUNDERS, R.S. & MEYER, M.A. 2001 Mars Odyssey Mission Geologic questions for global geochemical and mineralogica mapping. *LPI*, **32**, 1945.

SCHAAL, R.B., HOERZ, F. & GIBBONS, R.V. 1976 Shock metamorphic effects in lunar micro-craters. *Lunar Science Conference Proceedings*, **1**, 1039–1054.

SCOTT, R.F. 1975 Apollo program soil mechanics experiment. Final Report. California Institute of Technology, Pasadena.

SNYDER, L. 1979 Television optics for the Voyager mission to Jupiter and Saturn in *Space optics*, *SPIE*, 1979, 274–282.

SODERBLOM, L.A., EDWARDS, K., ELIASON, E.M., SANCHEZ, E.M. & CHARETTE, P. 1978 Global color variations of the Martian surface. *Icarus*, **34**, 446–464.

STEELE, I.M. & SMITH, J.V. 1976 Mineralogy and petrology of complex breccia. *Lunar Science Conference Proceedings*, **2**, 1949–1964.

STEUTEL, D., LUCEY, P.G. & HAMILTON, V.E. 2003 Reinterpretation of ISM data: quantitative analysis of pyroxen compositions. *Lunar Planet. Sci.*, **XXXIV**, 2.

SUNSHINE, J.M. & PIETERS, C.M. 1998 Determining the composition of olivine from reflectance spectroscopy. *J. Geophys. Res.*, **103**, 13675–13688.

TAYLOR, S.R. 1975 *Lunar Science: A Post-Apollo View*. Pergamon Press, Inc. New York.

TAYLOR, J. 2001 Resources, mars (planet), water, soils, Mars surface in *Workshop on Science and the Human Exploration of Mars*, 184.

TITOV, D.V., MOROZ, V.I., GRIGORIEV, A.V., ROSENQVIST, J., COMBES, M., BIBRING, J.-P. & ARNOLD, G. 1994 Observations of water vapour anomaly above Tharsis volcanoes on Mars in the IS (Phobos-2) experiment. *Planet. Space Sci.*, **42**(11), 1001–1010.

TOKSOZ, M.N., DAINTY, A.M., SOLOMON, S.C. & ANDERSON, K.R. 1974 Structure of the Moon. *Rev. Geophys. Space Phys.*, **12**, 539–567.

TROMBKA, J.I., NITTLER, L.R., STARR, R.D., EVANS, L.G., McCOY, T.J., BOYNTON, W.V., BURBINE, T.H., BRÜCKNER, J., GORENSTEIN, P., SQUYRES, S.W. & 12 CO-AUTHORS 2001 The NEAR-Shoemaker X-ray/gamma-ray spectrometer experiment: overview and lessons learned. *Meteorit. Planet. Sci.*, **36**(12), 1605–1616.

ULABY, F.T., MOORE, R.K. & FUNG, K. 1983 *Microwaves Remote Sensing*. Vol. II: *Scattering and Emission Theory, Advanced Systems and Applications*, Addison-Wesley Pub. Co, MA.

URBAN, M.G. 1987 Voyager image data compression an block encoding. *Proceedings of the International Telemetering Conference*, 137–162.

VANE, G., CHRIEN, T.G., MILLER, E.A. & REIMER, J. 1988 Spectral and radiometric calibration of the airborne visible/infrared imaging spectrometer. *Imaging Spectroscopy II, Proceedings of the Meeting*, San Diego, 91–105.

WARNER, J. 1975 Mineralogy, petrology, and geochemistry of the lunar samples. *Rev. Geophys. Space Phys.*, **13**, 107–13.

WARREN, P.H. 1985 The magma ocean concept and lunar evolution. Ann. review of earth and planetary sciences, *13*, 201–240.

WEISSTEIN, E.W. 2004 Fourier cosine transform. *http://mathworld.wolfram.com/FourierCosine Transform.htm*

WHITE, W.B. 1974 The Carbonate Minerals in The Infrared Spectra of Minerals, V.C. Faruer, ed., 227–284, Mineralogical Society, London.

ZASOVA, L., FORMISANO, V., GRASSI, D., MATURILLI, A. 1999 Martian atmosphere in the Region of Great Volcanoes: Mariner 9 IRIS Data Revisited, AAS, DPS meeting 31, 68.01.